Polymer Waste Recycling and Management II

Polymer Waste Recycling and Management II

Guest Editors

Cristina Cazan
Mihaela Cosnita

MDPI

Basel • Beijing • Wuhan • Barcelona • Belgrade • Novi Sad • Cluj • Manchester

Guest Editors
Cristina Cazan
Department of Product Design,
Mechatronics and Environment
Transilvania University of Brasov
Brasov
Romania

Mihaela Cosnita
Department of Product Design,
Mechatronics and Environment
Transilvania University of Brasov
Brasov
Romania

Editorial Office
MDPI AG
Grosspeteranlage 5
4052 Basel, Switzerland

This is a reprint of the Special Issue, published open access by the journal *Polymers* (ISSN 2073-4360), freely accessible at: www.mdpi.com/journal/polymers/special_issues/polym_waste_recycl_manag_ii.

For citation purposes, cite each article independently as indicated on the article page online and using the guide below:

Lastname, A.A.; Lastname, B.B. Article Title. *Journal Name* **Year**, *Volume Number*, Page Range.

ISBN 978-3-7258-2900-2 (Hbk)
ISBN 978-3-7258-2899-9 (PDF)
https://doi.org/10.3390/books978-3-7258-2899-9

Contents

Preface

Responsible management of natural resources and reducing negative environmental impacts are essential challenges of the 21st century. Innovative and sustainable solutions for recycling and reusing materials are more necessary in a global context marked by increasing consumption and continuous accumulation of waste. This collection of articles reflects the significant efforts of researchers in transforming environmental challenges into opportunities, proposing advanced methods for the reuse of residual materials, and contributing to the circular economy.

The included studies cover various topics, illustrating the interdisciplinary potential of materials science, engineering, and applied chemistry. From processing complex plastics, such as polyethene terephthalate (PET), to developing environmentally friendly composites from renewable materials, each paper contributes to expanding the frontiers of knowledge in the recycling field. Thus, methods such as electrospinning are being explored in order to transform recycled PET into high-performance fibers, and solar-assisted cutting processes are being used to valorize waste from plastic injection molding.

A remarkable aspect of this research is its focus on innovation and applicability. For example, studies dedicated to regenerating polyurethane by alcoholysis highlight the possibility of transforming waste into rigid foams with varied applications, demonstrating a pragmatic approach to chemical recycling. The article on transforming polylactic acid (PLA) into value-added chemicals, such as 1,2-propanediol, highlights advanced solutions for integrating a widely used polymer into the circular economy.

Another promising direction is valorizing agricultural waste, such as protein hydrolysates used for food preservation. By integrating these wastes into the food chain, the authors promote the efficient use of natural resources, reduce waste, and offer natural alternatives for extending food shelf life. At the same time, digital traceability initiatives and physical markings proposed for ABS plastics demonstrate how technology can support the monitoring of recycling flows, ensuring better waste management throughout the product life cycle.

In addition to technological solutions, the studies also address aspects related to the mechanical and thermal properties of the obtained materials, emphasizing the importance of maintaining performance standards. For example, the analysis of composites made from recycled metal powders or materials made from recovered textiles reveals the direct impact of recycling processes on the quality of the final products, highlighting the benefits brought to different industries.

This collection is addressed not only to researchers and practitioners but also to industry and policymakers, offering a broad perspective on the opportunities and challenges related to recycling.

Cristina Cazan and Mihaela Cosnita
Guest Editors

polymers MDPI

Article

Performance Evaluation of Out-of-Spec Carbon Prepregs for Upcycling Purposes

Sanjeev Rao * and **Dereck Bastienne**

Department of Aerospace Engineering, Khalifa University of Science and Technology, Abu Dhabi P.O. Box 127788, United Arab Emirates; 100035546@kustar.ac.ae

* Correspondence: sanjeev.rao@ku.ac.ae; Tel.: +971-2-312-4143

Abstract: In view of exploring the possibility of upcycling aerospace scrap, cure characteristics of out-of-spec carbon fiber prepregs are investigated in this study. The cure behavior of the prepreg is examined in the form of the mechanical cure conversion state of the material using a Dynamic Mechanical Analyzer (DMA). Cure kinetics is modeled by comparing the storage modulus at the start of the reaction (E'_0) and instantaneously (E'_t) during isothermal experiments with those of the fully cured material (E'_∞) obtained from dynamic scans. The glass transition temperature T_g and the extent of reaction before gelation are modeled using the DiBenedetto model, where the T_g of each laminate is determined in a DMA, per standard ASTM D7028. The mechanical properties, the extent of cure, and the glass transition temperature of the cured laminates were determined according to industry and international standards. The maximum conversion at temperatures between 100 °C and 140 °C is approximately 80% (±5%). The modeled rate of conversion shows a reasonable match with the experimental data, exhibiting a maximum reaction rate at about 30–40% of the cure conversion. The predicted evolution of the T_g as a function of cure conversion using the DiBenedetto model provides a 94% match with the experimental data. The multi-stage cure cycle based on the models offers shorter cycle times and high-quality laminates. The mechanical test results indicate approximately a 13% and 15% decrease in tensile strength and modulus, respectively, compared to pristine ones. The experimental extent of cure of the cured laminates (95.4%) is in close agreement with that predicted by the model (97%). The porosity in the laminates is estimated to be approximately 2.4%, which is acceptable in several industries.

Keywords: automated fiber placement; upcycling; out-of-spec prepregs; cure kinetics; DMA

Citation: Rao, S.; Bastienne, D. Performance Evaluation of Out-of-Spec Carbon Prepregs for Upcycling Purposes. *Polymers* **2024**, *16*, 1625. https://doi.org/10.3390/polym16121625

Academic Editors: Cristina Cazan and Mihaela Cosnita

Received: 29 February 2024
Revised: 3 April 2024
Accepted: 8 April 2024
Published: 7 June 2024

1. Introduction

The Automated Fiber Placement (AFP) technology evolving from the Automated Tape Laying (ATL) process has significantly advanced the aerospace manufacturing sector. Since its introduction in the 1980s, it has been widely implemented across several aerospace industries. The process uses narrow slit tapes in the range of $\frac{1}{4}$ to $\frac{1}{2}$ inch width compared to the 6-to-11-inch tapes used in the conventional tape laying process. The narrow width of the tape provides the user control over steering angles during placement, reduces fiber buckling associated with conventional ATL methods, and reduces material wastage. The AFP provides on-the-fly material debulking due to in situ compaction from the rollers and application-based cutting alternatives, which are a few of the advantages of the process over the traditional ATL method. This progression in technology also led to the development of Out-of-Autoclave (OoA) slit tapes, presenting a cost-efficient alternative to conventional autoclave processing, and expanding the possibilities for material application. The oven curing process is a two-step process based on a primary low-temperature vacuum bag curing and a high-temperature free-standing post-cure to obtain a high glass transition temperature (T_g) of the laminate. Given that autoclaves are not involved, the technology drastically reduces the initial investment of the process and can be easily adapted to a variety of components' shapes and sizes, resulting in a much more versatile methodology. However, irrespective of the advantages the process has to offer, there is some form of material wastage arising from offcuts from ply-cutting operations, skeletons, trim waste, end-of-roll waste, and out-of-spec material (within the aerospace industry, defined as the materials surpassing their expiration date as provided by the manufacturer). It is cost-prohibitive in the aerospace industry to re-certify such out-of-spec material, and it is, therefore, disposed of in landfills, donated to research institutes, or consumed in-house for research purposes.

In recent times, legislative changes, particularly the End-of-life Vehicle Directive (2000/53/EC) [1], have significantly propelled the focus toward the recycling and upcycling of materials, emphasizing the need for sustainable practices. Recent advancements in the recycling of carbon fiber composites, especially through pyrolysis and solvolysis, have shown promise in reclaiming fibers with minimal loss of integrity, aligning with the goals of environmental sustainability [2,3]. Recycling carbon fiber composites provides great potential in reducing both cost and environmental effects and can generally be categorized into three main groups: mechanical, thermal, and chemical. In mechanical recycling, composite material waste is shredded between blades in mechanical shredders, and the resulting pieces or powder form are used as fillers in other applications, such as roads, concrete bricks, etc. This process is mostly effective for fiber glass parts that have been manufactured using bulk molding or sheet molding compounds, and reclaiming fibers using expansive and expensive processes may not be commercially or environmentally beneficial [4]. A popular and effective process where expensive carbon fibers can be reclaimed is pyrolysis, a thermal recycling process where the resin portion of the composite is eliminated via decomposition in a high-temperature furnace with an inert atmosphere. In recent times, this process has received substantial attention because of the possibility of reclaiming fibers that may possess properties near their virgin counterparts. For example, in a recent work by Abdou et al. [5], the authors have reported that it is possible to obtain fiber surface morphologies close to that of the virgin fibers when the pyrolysis is conducted at 550 °C for a period of 1 h. Yang et al. [6] have shown that it is possible to retain approximately 80% of fiber tensile strength if the composite material was processed at 650 °C in an atmosphere consisting of 5% oxygen for a period of 45 min. Kim and co-workers [7] have been able to retain 90% of the tensile strength by exposing carbon fiber-reinforced composites to superheated water steam at 550 °C for 30 min, followed by oxidation in midair at 550 °C for 30–75 min. Jiang et al. [8] conducted pyrolysis in a nitrogen-purged microwave furnace at various temperatures of 400, 500, and 600 °C for 30 min and revealed that it was not possible to obtain completely clean fibers, even at high temperatures. Furthermore, Wei and S.A. Hadigheh [9] described the low-temperature pyrolysis combined with solvolysis pre-treatment as an effective recycling method for carbon fiber-reinforced plastic (CFRP) waste. Stefania Termine et al. [10] reported that the pyrolysis process can reclaim carbon fiber fabrics for reuse, and post-pyrolysis treatment improves fiber wettability but decreases mechanical properties. Gabriel Nicolai [11] illustrated that thermochemical recycling is an effective way to reuse carbon fiber, and recycled materials preserved physical structure as well as tensile strength. In another work, Wong and co-workers [12] reported that conventional pyrolysis techniques are sufficient to yield fibers that can be comparable to their virgin counterparts, and other elaborate methods using mechanical, combustion, gasification, or slow pyrolysis techniques are costly and may indeed yield poor-quality fibers. Although thermal recycling methods may yield fibers with properties closer to virgin ones, the recycling process, especially pyrolysis, results in fibers with considerable char formation on the surface. Another efficient way is chemical recycling, where the matrix material is separated from the fibers using solvolysis, glycolysis, and hydrolysis techniques [13–16]. The process is able to extract fibers with no surface char and negligible loss in their mechanical properties. Researchers [17–19] have worked on the chemical recycling process and have reported the properties of the recovered fibers to be at least 80% compared to virgin ones. Whilst several researchers have focused on recycling and methods to recover and reclaim reinforcing fibers for reuse, few have paid attention to upcycling carbon scrap that is generated during production. The process involves reusing scrap prepreg materials resulting from ply cutter waste, end-of-roll, or expired material to manufacture new composite parts. Nilakantan and Nutt [20] have presented a comprehensive study on a practical and scalable way to process carbon fiber/epoxy prepreg scrap. The resulting material can then be readily used to manufacture new products using hot pressing and compression molding techniques. Kiss and co-workers [21] have used tattered layers between fiber films to produce compression-molded plates that exhibit impact properties as high as 50% of those manufactured using virgin materials. Several research [20,22,23] regarding upcycling focused on the mechanical properties of the byproduct, and few have focused on the characterization of the scrap in order to use them efficiently. Therefore, in view of exploring the viability of upcycling out-of-spec prepreg materials to manufacture sports equipment, investigations have been carried out in this work to understand the change in cure characteristics of the out-of-spec material. This work differs from those existing in the literature by presenting the bottom-up approach to productively upcycle prepreg scrap. The cure kinetics of the out-of-spec prepreg material is modeled based on the mechanical cure conversion state of the material via a Dynamic Mechanical Analyzer (DMA), in essence allowing the evaluation of the mechanical properties of the entire fiber-resin system.

2. Materials and Methods

An Out-of-Autoclave (OoA) towpreg, CYCOM® 5320-1, supplied by Cytec/Solvay Engineered Materials Inc., Woodland Park, NJ, USA was used in this study. The OoA slit tape is specifically designed for the Automated Fiber Placement (AFP) process, consisting of IM7 carbon fibers impregnated in toughened CYCOM® 5320 epoxy resin. The material had surpassed its shelf life by a year but was stored in refrigerated conditions during that period of time.

The mechanical cure characteristics of the out-of-spec towpreg were investigated via experiments in a DMA 8000 equipment from Perkin Elmer®, Waltham, MA, USA. Isothermal and dynamic scan experiments were carried out at a frequency of 1 Hz under strain control mode, using a single cantilever setup [24,25]. Isothermal experiments were conducted at 10 °C intervals between 100 °C and 140 °C, and dynamic scans were carried out at 2 °C/min between −30 °C and 300 °C. The dry glass transition temperature 'T_g dry' was determined as per standard AITM-0003 [26] on cured samples that were prepared in accordance with standard EN 2565 [27]. Five samples were tested in total, and the average T_g dry was determined using tangents across loss moduli traces and tan δ peaks.

Simple balanced symmetric laminates [90/+45/0/−45/90] were manufactured using an in-house fiber placement system that consisted of a roller and a 20 kN load cell (Omegadyne LCM202) housed between the roller and the tool arbor of a Bridgeport® milling machine. The layup temperature was monitored using an infrared camera (FLIR A655sc), and a hot air gun from Bosch®, Abu Dhabi, United Arab Emirates, was used as the heat source [24]. The obtained thin laminates (<2 mm) were cured using only a vacuum bag in a convection oven. The two-step curing cycle with two ramps and two holds at 121 °C and 177 °C was designed based on mechanical curing experiments using DMA techniques. The multi-stage cure cycle consisted of a temperature ramp from ambient until approximately 120 (±5) °C at the rate of 2 °C/min, followed by a temperature isotherm dwell stage for approximately 1 h, during which edge devolatilization and temperature uniformity between the tooling and laminate is attained. An aluminum flat plate tooling of 5 mm was used to cure all the plies, and it was determined via experimentation in a convection oven that 2 °C/min provided uniform heating throughout the plate. The layup and consumables used in this study, shown in Figure 1, consisted of an edge dam with 80 g/m^2 plain weave glass for edge breathing/bleeding and a perforated release film over which a 140 g/m^2 polyester bleeder/breather (N4 from Airtech®, Huntington Beach, CA, USA) was used for face bleeding. High-temperature vacuum tape AT-200Y from Airtech® was used to seal the nylon vacuum bags (Airtech®, Wrightlon® 7400). Further details of the hardware set-up and the processing conditions for the automated layup can be found in [28].

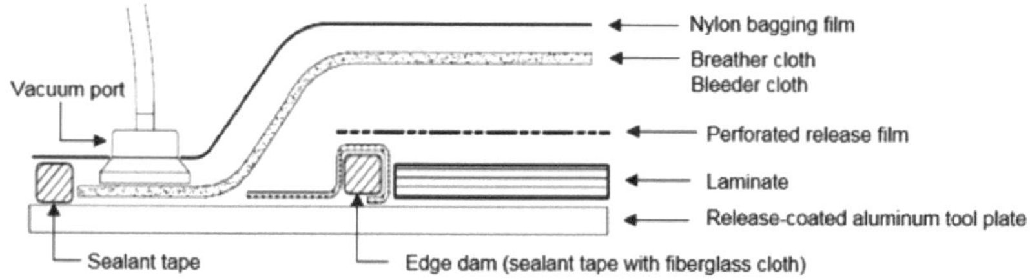

Figure 1. Laminate bagging setup sequence.

Tensile tests were conducted on manufactured coupons as per standard ASTM C 3039 using an Instron universal testing machine with a 50 kN load cell. The calibration results of prior experimentations showed a maximum error at 125 N to be 0.05% (0.0625 N) with a repeatability of 0.05%. An extensometer with a gauge length of 50 mm was used to record the strain. The voids and porosities were determined using a Matlab® (R2023a) program from 2D optical images of the sample cross-sections using a Zeiss AX10 stereo microscope.

The measurement of the degree of cure α' was established in accordance with the prescribed AITM 3-0008 standard [26], employing the Perkin Elmer DSC 8500 instrument subsequent to the conditioning of the specimens, following the guidelines outlined in EN2743 [29]. Determination of the degree of cure was carried out via the utilization of Equation (1), with an accompanying adjustment for 100% resin content achieved through the implementation of EN 2559 [30], thereby facilitating the determination of the authentic resin content present within the system.

$$\alpha' = \frac{\Delta H_A - \Delta H_B}{\Delta H_A} 100[\%] \quad (1)$$

where ΔH_A is the reaction of enthalpy in Joules of the A curve, which is heated at a rate of 10 °C/min, and ΔH_B is the reaction of enthalpy in Joules of the B curve, which is subjected to the curing cycle that is being investigated.

Thermal scans were conducted for the uncured prepreg from either ambient or sub-ambient (−40 °C) up to 300 °C for a series of ramp rates: 1 °C/min, 1.5 °C/min, and 2 °C/min. It is a quick method used to study thermal transitions and also provides estimates for the glass transition temperature of the uncured and fully cured material. Isothermal curing experiments were conducted within a temperature range of 100 °C to 140 °C, with a 10 °C increment, in order to investigate the cure progression as a function of time and temperature.

Due to thermal overshoots resulting from the swift temperature rise, the temperature was allowed to stabilize within ±2 °C of the target isotherm, which took about 14 to 20 min after the starting time, depending on the temperature. For each isothermal temperature, the test was made to run up until the reaction was effectively quenched, that is, when the rate of increase in storage modulus tends to zero, which is analogous to the heat flow approaching zero in DSC experiments.

2.1. Cure Kinetics Modeling

The cure conversion of the towpregs was calculated using Equation (2) [24]. For each curing temperature, the storage modulus at the beginning of the reaction (E'_0) and the instantaneous storage modulus at different times (E'_t) were obtained from the isothermal tests, while the maximum storage modulus value of the fully cured material (E'_∞) was acquired from the dynamic scan.

$$\alpha = \frac{E'_t - E'_0}{E'_\infty - E'_0} \quad (2)$$

The rate of reaction ($d\alpha/dt$) was obtained by differentiating the cure conversion curves with respect to time. Within the isothermal processing conditions considered, a good correlation between the model and the experimental data was obtained using the modified autocatalytic kinetic model developed by Kamal and Sourour [31], which accounts for the diffusion reaction control due to vitrification, as outlined below.

$$\frac{d\alpha}{dt} = (k_1 + k_2\alpha^m)(1-\alpha)^n \quad (3)$$

where α is the cure conversion and m and n are the reaction orders. k_1 and k_2 are the reaction constants and follow the Arrhenius temperature-dependent relationship given by

$$k_i = k_{0i}\exp\left(\frac{-E_{ai}}{RT}\right) \text{ for } i = 1,2 \quad (4)$$

where k_0 is the pre-exponential factor, E_a is the apparent activation energy, R is the gas constant, and T is the absolute temperature.

2.2. Gel Time and Vitrification Time Evaluation and Modeling

Gelation was estimated to occur at the inflection point of the storage modulus curve during the isothermal curing. The theoretical gel time as a function of cure kinetics is expressed in Equation (5) [32]

$$t_{gel} = \frac{1}{K(T)}\int_0^{\alpha_{gel}} \frac{1}{f(\alpha)}d\alpha \quad (5)$$

where $K(T)$ is the kinetic parameter and $f(\alpha)$ is the function of the reaction mechanism and cure conversion. Flory's gelation theory [33] can then be used to linearize it (Equation (6)). The apparent activation energy (E_a) is obtained from the slope of Equation (6).

$$\ln\left(t_{gel}\right) = \frac{E_a}{RT} + \text{constant} \quad (6)$$

Vitrification during curing appears when the glass transition temperature of the crosslinked polymer reaches the curing temperature. In this experiment, the start of vitrification on the storage modulus curve is representative of E', attaining a maximum value asymptotically.

2.3. DiBenedetto Model

Several laminates were cured at 120 °C for various periods of time in order to reach a pre-defined fractional conversion (x) estimated from the cure kinetics model. Subsequently, the T_g of each laminate was determined by a dynamic DMA scan, in accordance with ASTM D7028 [34]. Since T_g is sensitive to the overall dimension of the sample, each test sample was one layer thick in order to minimize random errors when it comes to the computation of the T_g values. The onset of the sharp drop in the storage modulus was used as the criterion for determining the glass transition temperature. Three samples were tested for each degree of cure, and an average was taken. The variation of the T_g with respect to the cure conversion was modeled with the DiBenedetto equation [35] as follows:

$$T_g = T_{g0} + \frac{\lambda x(T_{g\infty} - T_{g0})}{1-(1-\lambda)x} \quad (7)$$

where T_{g0} and $T_{g\infty}$ are the glass transition temperatures of the unreacted and the fully cured system, respectively. λ is obtained by fitting Equation (7) to experimental data of the degree of cure and glass transition temperature, which is a limiting case; when $\alpha = 0$, then $T_g = T_{g0}$, and when $\alpha = 1$, $T_g = T_{g\infty}$.

3. Results and Discussion

3.1. Cure Kinetics

The cure kinetics of the Out-of-Autoclave (OoA) towpreg investigated via DMA analysis in this work enables the evaluation of the mechanical properties of the entire fiber-resin system during the curing process. The instantaneous cure conversion (α) as a function of time was determined as per Equation (2) using a series of isothermal curing traces. In Figure 2, the conversion is characterized by three stages. The first one is a flat region, indicating negligible cure that monotonically decreases with the increase in testing temperatures; the second one is a rapid increase in conversion due to the change in the rate of the reaction until it asymptotically attains a maximum value, which is followed by the third region, where the rate of reaction sufficiently slows down to exhibit a plateau. This is caused because of the crosslinking-induced vitrification, whereby the molecular mobility is hindered, and the reaction assumes diffusion-controlled kinetics. The time evolution of the reaction rate for the series of isothermal cure processes (Figure 2) was obtained by differentiating the cure conversion curves with respect to time. All patterns are well described by Gaussian functions, and a monotonic increase in the reaction rate with the curing temperature is verified.

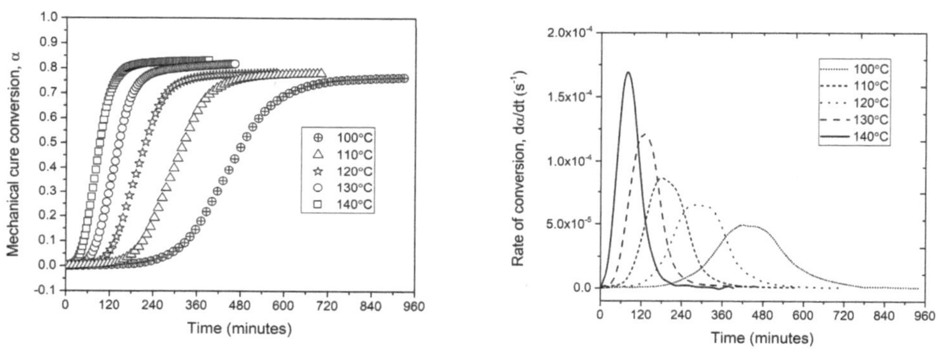

Figure 2. Time evolution of the mechanical cure at various isothermal temperatures (**left**) and time-evolution of the conversion rate at various isothermal temperatures (**right**).

The key observation in Figure 2 is that the rate of the reaction progresses with the increase in temperature, confirming the ability of the system to achieve a cure. The cure conversion percentage is an estimate of how much resin has already been cured at that temperature, and the information is required to apply autoclave pressure such that the voids and porosities are squeezed out of the laminate. A higher conversion rate means higher resin viscosity, and the propensity of voids and trapped porosities in the laminate is high, and at a lower conversion, the resin viscosity is low, and the application of pressure may lead to resin bleeding and dry spots.

The maximum conversion at temperatures between 100 °C and 140 °C is approximately 80% (±5%). The modeled rate of conversion plotted as a function of the mechanical cure conversion in Figure 3 shows a reasonable match with the experimental data, exhibiting the maximum reaction rate at about 30–40% of the cure conversion. The maximum fractional conversion (α_{max}) was found to be dependent on the cure temperature and is expressed as a linear relationship based on the experimental data (Figure 3). Furthermore, the obtained values for the maximum achievable conversion at each temperature are in agreement with those previously reported in the literature [36–38].

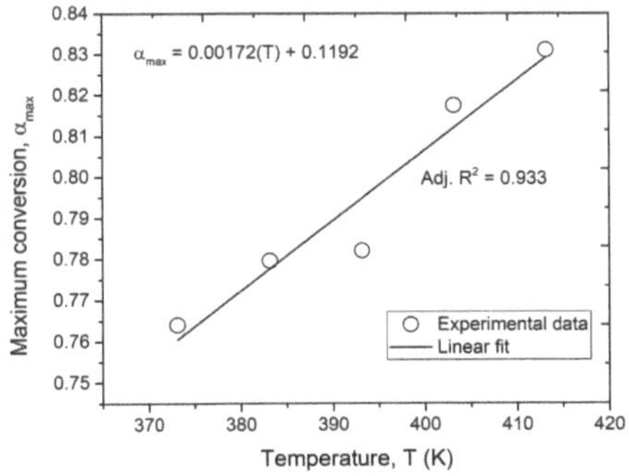

Figure 3. Maximum conversion as a function of the isothermal curing temperature.

The constants and exponents of the autocatalytic model (k_1, k_2, m, and n) were determined using a non-linear least-square fit, based on the Levenberg Marquardt algorithm, for each isothermal 'rate of reaction' curve. The total reaction order (m + n) was found to be relatively insensitive to the cure temperature, resulting in an average value of approximately 2. This is in close agreement with several thermoset polymerization reactions found in the literature [39,40], further validating the assumption of the cure mechanism to remain unchanged in the temperature band considered in this study. The reaction constants and the activation energies were calculated from the linear region of ln k vs. the reciprocal of the cure temperature (1/T), as listed in Table 1; these findings are in coherence with Francucci et al. [41].

Table 1. Calculated kinetic parameters of the autocatalytic model.

Pre-Exponential Factors		Apparent Activation Energy		Reaction Order		Maximum Conversion	
k_{01} (s^{-1})	16.45	E_{a1} (kJ/mol)	54.25	m	0.86	α_{max}	0.00172(T) + 0.119
k_{02} (s^{-1})	3.43	E_{a2} (kJ/mol)	28.44	n	1.11		

3.2. Gel Time and Vitrification Time Modeling

The gel times for the series of isothermal cure temperatures are summarized in Figure 4. The combination of the viscoelastic results with the cure conversion of each curing temperature allowed us to determine the degree of cure at the gel point (α_{gel}). In the review works by Bilyeu et al. [42], the authors have summarized the various criteria that have been proposed in the literature to determine the gel point, such as the crossover point between the storage and loss moduli, the peak of the tangent delta, the onset of the increase in storage modulus, the peak in loss modulus, the inflection point of the storage modulus curve, and the onset of the plateau of the maximum value of the storage modulus. Therefore, in this study, the gelation time is estimated to occur at the onset of the increase in the storage modulus and complex viscosity curves.

A statistical average value of α_{gel} =0.486 was obtained from the isothermal curing experiments, which appears to be comparably lower than that reported in the literature [43,44]. This may indeed be due to the aging of the prepreg. However, as the standard deviation of the data set is within the acceptable range (5.3%) to apply Flory's gelation theory [33], Equation (6) was used to obtain a linear relationship between the gel time as a function of the cure temperature, as seen in Figure 4. The gradient of the straight line in Figure 5 is then used to obtain the apparent activation energy for the gel time model (48.92 kJ/mol).

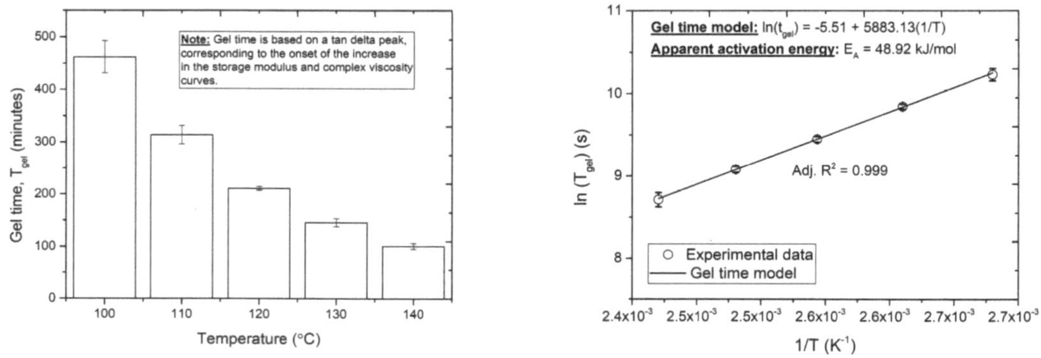

(a) (b)

Figure 4. Gel time (**a**) as a function of isothermal cure temperature and (**b**) a comparison of the measured and predicted gel time with respect to the reciprocal of the isothermal cure temperature.

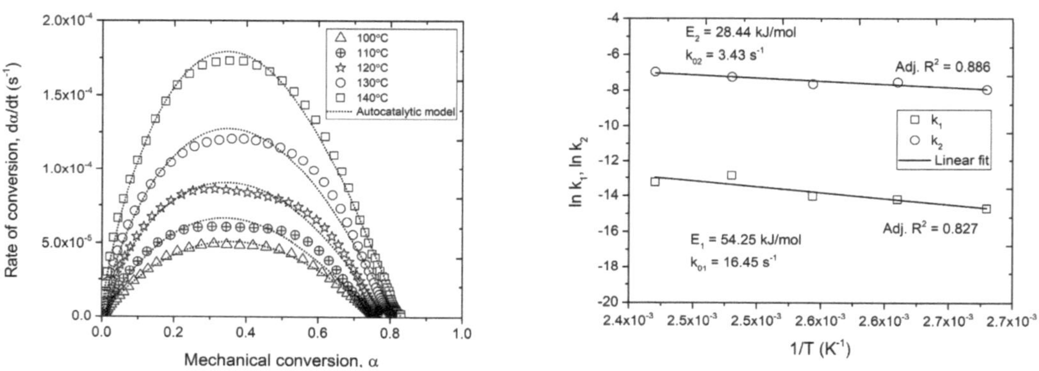

Figure 5. Rate of conversion as a function of the mechanical conversion for various isothermal curing (**left**), Arrhenius linear fit for the calculated kinetics constants (**right**).

3.3. T_g-α *Conversion*

The DiBenedetto model (Equation (7)) [35] is used to relate the glass transition temperature and the extent of reaction before gelation. The glass transition temperatures of the unreacted (T_{g0}) and the fully cured system ($T_{g\infty}$) were determined from a dynamic scan from −30 °C to 300 °C at a rate of 2 °C/min, while applying a strain amplitude of 50 μ at a frequency of 1 Hz. The experimental data to obtain the pre-defined degree of cure at T_g was estimated by letting the towpregs cure at 120 °C for various periods of time such that they reached the required pre-defined degree of cure (estimations from the cure kinetics model), as seen in Figure 6b. Following this, the T_g of all pre-cured samples were determined in a DMA as per ASTM D7028 [45] (Figure 6a). The wet glass transition temperature of the polymer, T_{g0}, and the fully cured $T_{g\infty}$ are determined in Figure 6, and the DiBenedetto model was used to predict the evolution of the T_g as a function of cure conversion. In Figure 6d, the theoretical model provides a reasonable fit (94% match) with the experimental data of the T_g evolution as a function α [46].

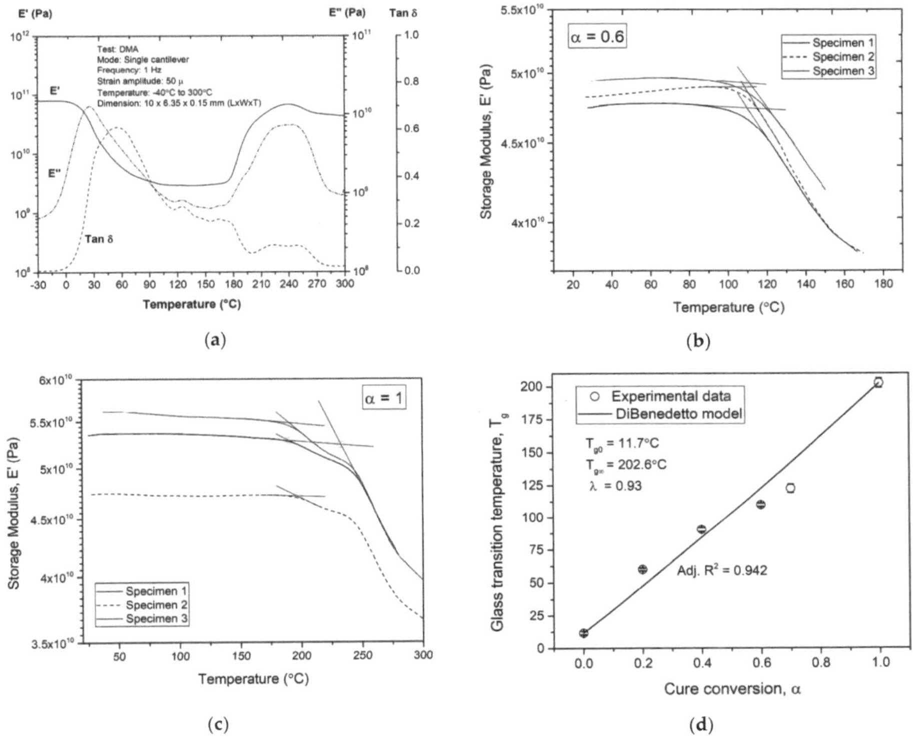

Figure 6. Glass transition temperature as a function of fractional conversion and relative DiBeneteddo modeling: (**a**) dynamic scan, (**b**) fractional degree of cure (0.6), (**c**) fully cured, (**d**) DiBeneteddo model.

3.4. Manufacturing Cure Cycle

The manufacturer's recommended cure cycle for the vacuum bag only cure consists of several ramps and holds [47], which would invariably drive the cost of manufacturing. As sporting industry products are price sensitive compared to those in aerospace, it is important to keep the manufacturing costs to a minimum for product viability [48]. Therefore, in this work, a multi-step cure with two ramps is designed to cure the relatively thin laminates (<2 mm) using only a vacuum bag in a convection oven [49]. The two-step curing cycle with two ramps and two holds at 121 °C and 177 °C was designed based on the mechanical curing experiments using DMA techniques (Figure 7a). The multi-stage cure cycle (Figure 7b) consisted of a temperature ramp from ambient until approximately 120 (±5) °C at a rate of 2 °C/min, followed by a temperature isotherm dwell stage for approximately 1 h, during which edge devolatilization and temperature uniformity between the tooling and laminate is attained. The vacuum bag pressure was maintained at approximately 0.5 bar to prevent premature curing due to high vacuums. At the chosen temperature ramp rate, the resin attains its minimum viscosity at approximately 151 min, during which the vacuum pressure was increased to −1 bar to facilitate face bleeding (normal to the top laminate face) to remove any entrapped air and volatiles from the central areas of the laminate. The second stage of the temperature ramp was applied at the end of the temperature dwell until a cure temperature of 177 °C, where the actual curing of the resin takes place. At the end of the cure temperature dwell, the oven temperature is reduced gradually at approximately 2 °C/min until the laminate reaches at least 50 °C, following which the vacuum is bled [49].

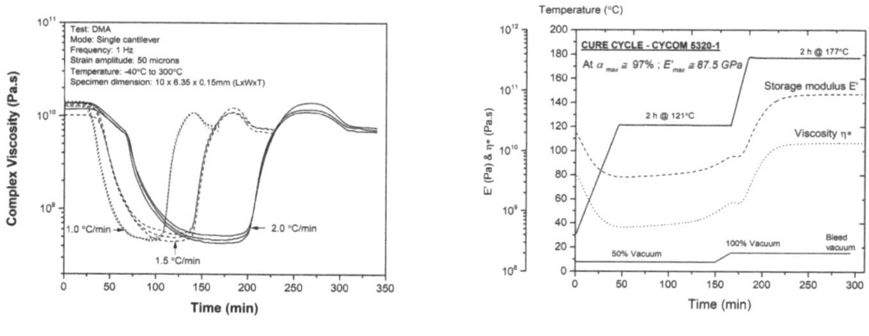

(**a**) (**b**)

Figure 7. Cure cycle development for out-of-spec Cycom 5320-1: (**a**) complex viscosity curves used to determine the time taken to reach minimum viscosity, (**b**) temperature and pressure stages along with the achievable maximum cure.

3.5. Mechanical Tests, Extent of Cure, and Glass Transition T_g

The mechanical test results show an approximately 13% and 15% decrease in tensile strength and modulus values, respectively (Table 2), compared to pristine material [47]. The reduction in tensile properties is analogous to the decrease in the cure properties and can be attributed to the maximum achievable cure of the out-of-spec material. However, the marginal change in properties may indeed not hinder their applicability in sports and recreational goods, as the structural requirements in those sectors are relatively lower compared to those in aerospace applications.

Table 2. Mechanical properties from the tensile tests conducted per standard ASTM C 3039.

	Tensile Strength (MPa)	Tensile Modulus (GPa)
Manufacturer's data	917	59
Test coupons	800 (±20)	50 (±5)

The extent of cure determined using the DSC methods on three cured samples exhibits an average cure of 95.4% (±1.2%) (Table 3), correlating well with the results of cure kinetics that were obtained using DMA methods (97% cure at 177 °C). While out-of-spec materials may not be suitable for the aerospace industry due to stringent rules and regulations around aged materials, it is still possible to use them in other practical applications, as they result in relatively good-quality laminates.

Table 3. The extent of cure via DSC analysis conducted per standard AITM 3-0008.

Sample	ΔH_{100} (J/g)	T_{onset} (°C)	T_{peak} (°C)	ΔH (J/g)	α' (%)
1	443.1	190.7	250.5	25.84	94.2
2	443.1	185.9	249.0	20.60	95.3
3	443.1	183.0	250.3	13.66	96.9

The void fraction was determined using cross-sectional micrographs of three laminates (Figure 8). Three regions of interest (ROIs) of each sample were evaluated at 10× and 20× in Matlab® (R2023a) environment, and the results are listed in Table 4. An example of one region that was considered for evaluation is shown in Figure 8a. The micrographs show an average of 2.4% in all the regions considered as seen in Figure 8b. This indicates that the out-of-spec material is still able to produce laminates with minimum voids compared to the non-expired material (0.15%) [50].

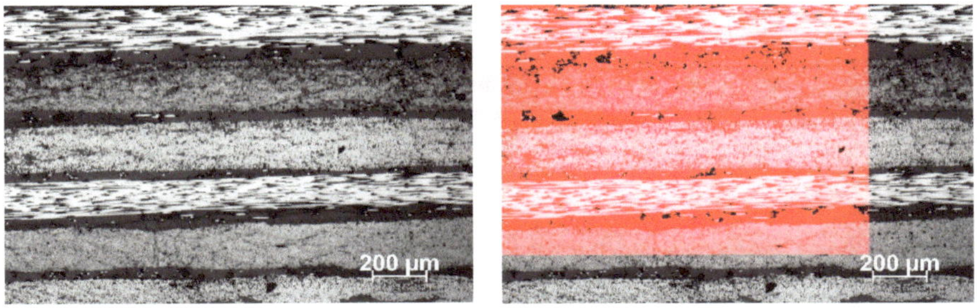

(a) (b)

Figure 8. Optical micrograph of the cross-section of sample 1, with the red shaded region indicating ROI 1 that is used for void calculations.

Table 4. Void content of manufactured samples at two magnifications.

		Resolution	
		10×	20×
Sample 1	ROI 1 (%)	2.5	1.9
	ROI 2 (%)	2.6	2.5
	ROI 3 (%)	2.3	2.8
Sample 2	ROI 1 (%)	2.8	2.4
	ROI 2 (%)	2.1	2.2
	ROI 3 (%)	2.2	1.8
Sample 3	ROI 1 (%)	2.5	2.6
	ROI 2 (%)	2.8	2.6
	ROI 3 (%)	1.8	2.8
Average		2.4	2.4
Standard deviation		0.3	0.3

The glass transition temperature (T_g) of the cured laminate was determined in accordance with standard AITM-0003 [26] (Figure 9), considering the onset of the drop of the storage modulus (E′). The experimental T_g value (193.1 °C) is slightly lower compared to the one indicated by the manufacturer (196.7 °C). This expected result, on the one hand, proves that the maximum degree of cure and mechanical properties practically achievable for this prepreg cannot be obtained with an out-of-spec (expired) tape, and on the other hand, the quality of the laminate is not much lower. Therefore, while the material cannot be used for aerospace applications, it can be used for other purposes, such as sports equipment, where the necessary specifications are less strict.

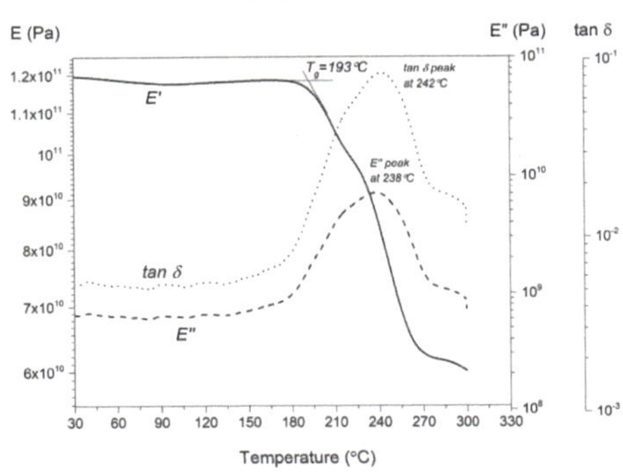

Figure 9. Mechanical properties' time evolution during the DMA two-step cure cycle and evaluation of the T_g from the onset of the drop of the storage modulus.

4. Summary

In this study, cure kinetics of aged carbon prepregs were modeled using the mechanical cure conversion state of the material that was determined using a DMA machine. This method simulates the cure behavior of the entire fiber-resin system closely to the actual manufacturing process as opposed to the conventional DSC methods. The autocatalytic model, which was developed by comparing the storage modulus at the start of the reaction (E'_0) and instantaneously (E'_t) during the isotherm experiments with those of the fully cured material (E'_∞) obtained from the dynamic scans, shows a reasonable match with experimental data. The maximum conversion at temperatures between 100 °C and 140 °C does not exceed 80% (±5%) and is found to be linearly related to the cure temperature. The aging of the prepreg has evidently lowered the degree of cure at the gel point to 0.486 compared to pristine ones. However, the evolution of the T_g as a function of cure conversion that was modeled using the DiBenedetto equation provides a reasonable fit (94% match) with the experimental data, undermining the effect of viscoelastic results with the cure conversion α_{gel}. The shorter multi-stage cure cycle that was developed results in shorter cycle times with negligible loss in the mechanical properties that can be attributed to the aging of the prepregs. The maximum cure that can be obtained in the laminates manufactured from the out-of-spec prepregs is 95.4%, which may indeed not be adequate for aerospace applications but is sufficient for other non-aerospace applications.

Following the analysis, the out-of-spec material was used to manufacture tubing that was used to produce framesets for professional touring bicycles. The Mandel wrapping technique was used to manufacture the tubing that was cured in a convection oven using the designed cure cycle (Figure 7b). The bicycle was designed to withstand a 150 kg rider and handle severe crosswinds, and the weight of the bicycle frame is well under 1 kg. The product demonstrator developed showed that the aged prepregs can indeed be upcycled rather than being discarded as landfill.

Author Contributions: D.B. is a BSc student and has contributed to the investigation, testing, validation, and preparation of the original draft; S.R. is the principal investigator of this project and has contributed to the conceptualization of methodology, procuring funding, project administration, original draft preparation, and supervision of this work. All authors have read and agreed to the published version of the manuscript.

Funding: This research was performed as part of the Khalifa University of Science and Technology internal grants FSU-2020-35, and this study also received financial support from the Abu Dhabi Sports Council (ADSC).

Institutional Review Board Statement: Not applicable.

Informed Consent Statement: Not applicable.

Data Availability Statement: Data are contained within the article.

Acknowledgments: The authors express sincere gratitude to all lab staff at Khalifa University for their assistance in manufacturing and testing the test panels, Passoni Titanio s.r.l. for their assistance in building and testing bicycle frames, Luca Desidery and Harsha Medicharla for their assistance in typesetting this manuscript.

Conflicts of Interest: The authors declare no conflicts of interest.

References

1. European Parliament and Council of the European Union. *Directive 2000/53/EC of The European Parliament and of The Council of 18 September 2000 on End-of Life Vehicles*; European Parliament and Council of the European Union: Brussels, Belgium, 2000; p. 34.
2. Genna, S.; Papa, I.; Lopresto, V.; Tagliaferri, V. Mechanical Characterisation of CFRP Laminates with Recycled Carbon Fiber Obtained by Resin Infusion under Flexible Tooling (RIFT) Technology. *Compos. Sci. Technol.* **2020**, *199*, 108328. [CrossRef]
3. Liu, Z.; Turner, T.A.; Wong, K.H.; Pickering, S.J. Development of High Performance Recycled Carbon Fibre Composites with an Advanced Hydrodynamic Fibre Alignment Process. *J. Clean. Prod.* **2021**, *278*, 123785. [CrossRef]

4. Naqvi, S.R.; Prabhakara, H.M.; Bramer, E.A.; Dierkes, W.; Akkerman, R.; Brem, G. A Critical Review on Recycling of End-of-Life Carbon Fibre/Glass Fibre Reinforced Composites Waste Using Pyrolysis towards a Circular Economy. *Resour. Conserv. Recycl.* **2018**, *136*, 118–129. [CrossRef]
5. Abdou, T.R.; Botelho Junior, A.B.; Espinosa, D.C.R.; Tenório, J.A.S. Recycling of Polymeric Composites from Industrial Waste by Pyrolysis: Deep Evaluation for Carbon Fibers Reuse. *Waste Manag.* **2021**, *120*, 1–9. [CrossRef] [PubMed]
6. Yang, J.; Liu, J.; Liu, W.; Wang, J.; Tang, T. Recycling of Carbon Fibre Reinforced Epoxy Resin Composites under Various Oxygen Concentrations in Nitrogen-Oxygen Atmosphere. *J. Anal. Appl. Pyrolysis* **2015**, *112*, 253–261. [CrossRef]
7. Kim, K.W.; Lee, H.M.; An, J.H.; Chung, D.C.; An, K.H.; Kim, B.J. Recycling and Characterization of Carbon Fibers from Carbon Fiber Reinforced Epoxy Matrix Composites by a Novel Super-Heated-Steam Method. *J. Environ. Manag.* **2017**, *203*, 872–879. [CrossRef]
8. Jiang, L.; Ulven, C.A.; Gutschmidt, D.; Anderson, M.; Balo, S.; Lee, M.; Vigness, J. Recycling Carbon Fiber Composites Using Microwave Irradiation: Reinforcement Study of the Recycled Fiber in New Composites. *J. Appl. Polym. Sci.* **2015**, *132*, 42658. [CrossRef]
9. Wei, Y.; Hadigheh, S.A. Development of an Innovative Hybrid Thermo-Chemical Recycling Method for CFRP Waste Recovery. *Compos. Part B Eng.* **2023**, *260*, 110786. [CrossRef]
10. Termine, S.; Naxaki, V.; Semitekolos, D.; Trompeta, A.-F.; Rovere, M.; Tagliaferro, A.; Charitidis, C. Investigation of Carbon Fibres Reclamation by Pyrolysis Process for Their Reuse Potential. *Polymers* **2023**, *15*, 768. [CrossRef]
11. Nicolai, G.; Bôas, A.P.; Méndez, M.O.; Taranto, O.P. Thermochemical Recycling of Epoxy Resin Composites Reinforced with Carbon Fiber. *J. Compos. Mater.* **2023**, *57*, 3201–3212. [CrossRef]
12. Wong, K.; Rudd, C.; Pickering, S.; Liu, X.L. Composites Recycling Solutions for the Aviation Industry. *Sci. China Technol. Sci.* **2017**, *60*, 1291–1300. [CrossRef]
13. El Gersifi, K.; Durand, G.; Tersac, G. Solvolysis of Bisphenol A Diglycidyl Ether/Anhydride Model Networks. *Polym. Degrad. Stab.* **2006**, *91*, 690–702. [CrossRef]
14. Liu, Y.; Meng, L.; Huang, Y.; Du, J. Recycling of Carbon/Epoxy Composites. *J. Appl. Polym. Sci.* **2004**, *94*, 1912–1916. [CrossRef]
15. Braun, D.; Von Gentzkow, W.; Rudolf, A.P. Hydrogenolytic Degradation of Thermosets. *Polym. Degrad. Stab.* **2001**, *74*, 25–32. [CrossRef]
16. Oliveux, G.; Dandy, L.O.; Leeke, G.A. Degradation of a Model Epoxy Resin by Solvolysis Routes. *Polym. Degrad. Stab.* **2015**, *118*, 96–103. [CrossRef]
17. Pegoretti, A. Towards Sustainable Structural Composites: Recycling of Continuous-Fiber-Reinforced Thermoplastics. *Adv. Ind. Eng. Polym. Res.* **2021**, *4*, 105–115. [CrossRef]
18. Oliveux, G.; Dandy, L.O.; Leeke, G.A. Current Status of Recycling of Fibre Reinforced Polymers: Review of Technologies, Reuse and Resulting Properties. *Prog. Mater. Sci.* **2015**, *72*, 61–99. [CrossRef]
19. Karuppannan Gopalraj, S.; Kärki, T. A Review on the Recycling of Waste Carbon Fibre/Glass Fibre-Reinforced Composites: Fibre Recovery, Properties and Life-Cycle Analysis. *SN Appl. Sci.* **2020**, *2*, 433. [CrossRef]
20. Nilakantan, G.; Nutt, S. Reuse and Upcycling of Thermoset Prepreg Scrap: Case Study with out-of-Autoclave Carbon Fiber/Epoxy Prepreg. *J. Compos. Mater.* **2018**, *52*, 341–360. [CrossRef]
21. Kiss, P.; Stadlbauer, W.; Burgstaller, C.; Stadler, H.; Fehringer, S.; Haeuserer, F.; Archodoulaki, V.M. In-House Recycling of Carbon- and Glass Fibre-Reinforced Thermoplastic Composite Laminate Waste into High-Performance Sheet Materials. *Compos. Part A Appl. Sci. Manuf.* **2020**, *139*, 106110. [CrossRef]
22. Cestari, S.P.; Mendes, L.C.; Altstädt, V.; Lopes, L.M.A. Upcycling Polymers and Natural Fibers Waste—Properties of a Potential Building Material. *Recycling* **2016**, *1*, 205–218. [CrossRef]
23. Nilakantan, G.; Nutt, S. Reuse and Upcycling of Aerospace Prepreg Scrap and Waste. *Reinf. Plast.* **2015**, *59*, 44–51. [CrossRef]
24. Xie, M.; Zhang, Z.; Gu, Y.; Li, M.; Su, Y. A New Method to Characterize the Cure State of Epoxy Prepreg by Dynamic Mechanical Analysis. *Thermochim. Acta* **2009**, *487*, 8–17. [CrossRef]
25. Stark, W. Investigation of the Curing Behaviour of Carbon Fibre Epoxy Prepreg by Dynamic Mechanical Analysis DMA. *Polym. Test.* **2013**, *32*, 231–239. [CrossRef]
26. *AITM 1-0003*; Airbus Test Method, Determination of the Glass Transition Temperatures. Airbus: Applus Illescas, Spain, 2010.
27. *DIN EN 2565*; Aerospace Series—Preparation of Carbon Fibre Reinforced Resin Panels for Test Purposes. Deutsches Institut für Normung: Berlin, Germany, 2013.
28. Rao, S.; Umer, R.; Thomas, J.; Cantwell, W.J. Investigation of Peel Resistance during the Fibre Placement Process. *J. Reinf. Plast. Compos.* **2016**, *35*, 275–286. [CrossRef]
29. *BS EN 2743:2002*; Aerospace Series. Fibre Reinforced Plastics. Standard Procedures for Conditioning Prior to Testing Unaged Materials. British-Adopted European Standard. BSI Group: London, UK, 2002.
30. *BS EN 2559:2022*; Aerospace Series. Carbon, Glass and Aramid Fibre Preimpregnates. Determination of the Resin and Fibre Content and the Mass of Fibre per Unit Area. British-Adopted European Standard. BSI Group: London, UK, January 2023.
31. Kamal, M.; Sourour, S. Kinetics and Thermal Characterization of Epoxy-Amine Systems. *Polym. Eng. Sci.* **1973**, *13*, 59–64. [CrossRef]
32. Sun, L. Thermal Rheological Analysis of Cure Process of Epoxy Prepreg. Ph.D. Thesis, Louisiana State University, Baton Rouge, LA, USA, 2002.

33. Flory, P.J. *Principles of Polymer Chemistry*; Cornell University Press: New York, NY, USA, 1953.
34. *ASTM D7028-07(2015)*; Standard Test Method for Glass Transition Temperature (DMA Tg) of Polymer Matrix Composites by Dynamic Mechanical Analysis (DMA). ASTM: West Conshohocken, PA, USA, 2008.
35. Dibenedetto, A.T. Prediction of the Glass Transition Temperature of Polymers: A Model Based on the Principle of Corresponding States. *J. Polym. Sci. Part B Polym. Phys.* **1987**, *25*, 1949–1969. [CrossRef]
36. Rahmani, N.; Willard, B.; Lease, K.; Legesse, E.T.; Soltani, S.A.; Keshavanarayana, S. The Effect of Post Cure Temperature on Fiber/Matrix Adhesion of T650/Cycom 5320-1 Using the Micro-Droplet Technique. *Polym. Test.* **2015**, *46*, 14–20. [CrossRef]
37. Kratz, J.; Hsiao, K.; Fernlund, G.; Hubert, P. Thermal Models for MTM45-1 and Cycom 5320 out-of-Autoclave Prepreg Resins. *J. Compos. Mater.* **2013**, *47*, 341–352. [CrossRef]
38. Moroni, A.; Mijovic, J.; Pearce, E.M.; Foun, C.C. Cure Kinetics of Epoxy Resins and Aromatic Diamines. *J. Appl. Polym. Sci.* **1986**, *32*, 3761–3773. [CrossRef]
39. Du, W.; Tan, L.; Zhang, Y.; Yang, H.; Chen, H. Rheological and Kinetic Investigation into Isothermal Curing of a Thermoset Polythiourethane System. *Polym.-Plast. Technol. Mater.* **2020**, *59*, 63–71. [CrossRef]
40. Hosseinpour, A.; Nazockdast, H.; Behzad, T.; Salimijazi, H.R. Investigation of the Cure Kinetics of an Epoxy Resin by Advanced Isoconversional and Model-Fitting Methods. *AIP Conf. Proc.* **2016**, *1713*, 110004. [CrossRef]
41. Francucci, G.; Cardona, F.; Manthey, N.W. Cure Kinetics of an Acrylated Epoxidized Hemp Oil-Based Bioresin System. *J. Appl. Polym. Sci.* **2013**, *128*, 2030–2037. [CrossRef]
42. Bilyeu, B.; Brostow, W.; Menard, K.P. Epoxy Thermosets and Their Applications. II. Thermal Analysis. *J. Mater. Educ.* **2000**, *22*, 107–109.
43. Soltani, S.; Keshavanarayana, S. Development of Time-Temperature Viscosity Diagram for Effective Cure Monitoring of Thermosetting Composite Materials. In Proceedings of the International SAMPE Technical Conference, Paris, France, 10–11 March 2014.
44. Reddy, B.R.; Eoff, L.; Dalrymple, E.D.; Black, K.; Brown, D.; Rietjens, M. A Natural Polymer-Based Cross-Linker System for Conformance Gel Systems. *SPE J.* **2003**, *8*, 99–106. [CrossRef]
45. *ASTM D7028-07e1*; Glass Transition Temperature (DMA Tg) of Polymer Matrix Composites by Dynamic Mechanical Analysis (DMA). ASTM: West Conshohocken, PA, USA, 2007.
46. Hardis, R.; Jessop, J.L.P.; Peters, F.E.; Kessler, M.R. Cure Kinetics Characterization and Monitoring of an Epoxy Resin Using DSC, Raman Spectroscopy, and DEA. *Compos. Part A Appl. Sci. Manuf.* **2013**, *49*, 100–108. [CrossRef]
47. Cytec Industries Inc. *CYCOM® 5320-1 Epoxy Resin System*; Cytec Industries Inc.: Smyrna, GA, USA, 2015.
48. Centea, T.; Nutt, S.R. Manufacturing Cost Relationships for Vacuum Bag-Only Prepreg Processing. *J. Compos. Mater.* **2016**, *50*, 2305–2321. [CrossRef]
49. Ma, Y.; Centea, T.; Nutt, S.R. Defect Reduction Strategies for the Manufacture of Contoured Laminates Using Vacuum BAG-Only Prepregs. *Polym. Compos.* **2017**, *38*, 2016–2025. [CrossRef]
50. Hwang, S.S.; Park, S.Y.; Kwon, G.C.; Choi, W.J. Cure Kinetics and Viscosity Modeling for the Optimization of Cure Cycles in a Vacuum-Bag-Only Prepreg Process. *Int. J. Adv. Manuf. Technol.* **2018**, *99*, 2743–2753. [CrossRef]

polymers MDPI

Review

Review and Assessment of Existing and Future Techniques for Traceability with Particular Focus on Applicability to ABS Plastics

Ignacy Jakubowicz and Nazdaneh Yarahmadi *

RISE Research Institutes of Sweden, 40022 Gothenburg, Sweden; ignacy.jakubowicz@ri.se
* Correspondence: nazdaneh.yarahmadi@ri.se; Tel.: +46-705495921

Abstract: It is generally recognized that the use of physical and digital information-based solutions for tracking plastic materials along a value chain can favour the transition to a circular economy and help to overcome obstacles. In the near future, traceability and information exchange between all actors in the value chain of the plastics industry will be crucial to establishing more effective recycling systems. Recycling plastics is a complex process that is particularly complicated in the case of acrylonitrile butadiene styrene (ABS) plastic because of its versatility and use in many applications. This literature study is part of a larger EU-funded project with the acronym ABSolEU (Paving the way for an ABS recycling revolution in the EU). One of its goals is to propose a suitable traceability system for ABS products through physical marking with a digital connection to a suitable data-management system to facilitate the circular use of ABS. The aim of this paper is therefore to review and assess the current and future techniques for traceability with a particular focus on their use for ABS plastics as a basis for this proposal. The scientific literature and initiatives are discussed within three technological areas, viz., labelling and traceability systems currently in use, digital data sharing systems and physical marking. The first section includes some examples of systems used commonly today. For data sharing, three digital technologies are discussed, viz., Digital Product Passports, blockchain solutions and certification systems, which identify a product through information that is attached to it and store, share and analyse data throughout the product's life cycle. Finally, several different methods for physical marking are described and evaluated, including different labels on a product's surface and the addition of a specific material to a polymer matrix that can be identified at any point in time with the use of a special light source or device. The conclusion from this study is that the most promising data management technology for the near future is blockchain technology, which could be shared by all ABS products. Regarding physical marking, producers must evaluate different options for individual products, using the most appropriate and economical technology for each specific product. It is also important to evaluate what information should be attached to a specific product to meet the needs of all actors in the value chain.

Keywords: ABS plastics; traceability; circular economy; physical marking; recycling

Citation: Jakubowicz, I.; Yarahmadi, N. Review and Assessment of Existing and Future Techniques for Traceability with Particular Focus on Applicability to ABS Plastics. *Polymers* **2024**, *16*, 1343. https://doi.org/10.3390/polym16101343

Academic Editors: Cristina Cazan and Mihaela Cosnita

Received: 6 March 2024
Revised: 26 April 2024
Accepted: 7 May 2024
Published: 9 May 2024

1. Introduction

During the last decade, EU regulations and the work of the European Telecommunications Standards Institute have encouraged more traceability across the value chain [1], especially in the food and pharma sectors, to guarantee to end users and companies that counterfeiting is being prevented.

Because of the more extensive regulations, more severe security threats to the goods and materials transported along supply chains and the demand for the fast, flexible and secure exchange of information, traceability has become the responsibility of all organizations in the value chain, from raw material providers to retailers who sell products to end customers. The aspects of traceability are unique identification, data capture and recording, link management and data communication. For the implementation of these principles, different technologies have been used, such as automated identification, electronic data processing and electronic data interchange [2].

One of the objectives of the EU-funded project with the acronym ABSolEU (Paving the way for an ABS recycling revolution in the EU) is to propose a suitable traceability system for ABS products through physical marking with a digital connection to a suitable data management system to facilitate automated identification, electronic data processing and electronic data interchange for the circular use of ABS. Using this system is expected to facilitate the selective sorting and recycling of ABS materials as well as the circular use of ABS in general. In addition to this, the system is also expected to provide information by collecting and managing data from different processes throughout the value chain about the origin of feedstock, including the proportion of recycled material, production/manufacturing instructions, service life conditions and disassembly instructions to allow for communication and the interchange of data with actors in the whole value chain, such as consumers, waste managers, recyclers, second-life users and producers, and provide a basis for their decision making. The biggest difficulty in creating a traceability system for ABS materials is the lack of information and data sharing between actors along the value chain and the complexity of ABS plastics.

1.1. Characteristics of ABS

ABS resin is a terpolymer consisting of three monomers: acrylonitrile, butadiene and styrene. In general, ABS consists of a continuous phase of copolymers of styrene and acrylonitrile (SAN), whereas butadiene usually forms the dispersed phase. ABS is produced mainly using two different polymerization processes, viz., mass and emulsion processes. In the first of these, styrene and acrylonitrile react in the presence of a polybutadiene substrate, whereas in the second of these, ABS is produced in two steps. In step one, butadiene is produced in an aqueous emulsion using radical initiators and emulsifiers, followed by a grafting step in which styrene and acrylonitrile are emulsion polymerized onto the polybutadiene substrate [3]. The most important mechanical properties of ABS are its impact resistance and toughness. Various modifications of ABS are made to alter its impact resistance, toughness and heat resistance. For example, its impact resistance is heightened by increasing the proportions of polybutadiene in relation to styrene and acrylonitrile, although this causes changes in other properties. In addition to the proportion of polybutadiene, morphology has a significant additional effect on its material properties. For example, increasing the size of rubber particles can increase its toughness, whereas increasing the polymer chain length produces a stronger material [4]. Different manufacturing methods also affect the properties of ABS plastics [5]. In particular, injection moulding conditions should be controlled carefully to ensure that shear is kept to a minimum, to avoid excessive pressures during the packing phase in the mould and to ensure the linear formation of the butadiene chains.

The ageing characteristics of ABS are largely influenced by the polybutadiene content. The thermo-oxidative degradation of ABS leads to the formation of oxidation products and a decrease in unsaturated bonds in the butadiene phase. In addition, cross-linking in the butadiene phase has been observed through studies of molar mass changes during oxidation [6]. Cross-linking and chain scission in the polybutadiene phase are also observed as a result of the photodegradation of ABS but this degradation is mainly limited to the exposed surface of the material [7]. Consequently, it is common to use antioxidants in the composition as well as additives or a surface coating to protect ABS materials against ultraviolet (UV) radiation.

As can be seen clearly from the above description, by changing the proportions of ABS components and additives, ABS materials can be prepared in many different grades for two major manufacturing categories, viz., for extrusion and for injection moulding. Consequently, ABS due to its versatility is used in many applications, such as car parts, toys, electronic housings, consumer products, pipe fittings and many more, and creates the second most amount of waste after polyolefins [8]. The great variety of its uses and the variability in the product composition as well as the presence of polymer mixtures constitute the main obstacles concerning ABS recycling.

1.2. Recycling of ABS

Global ABS demand in 2023 was 20.6 million tons, growing at a compound annual growth rate of 4.5% from 2018 to 2023 [9]. This can be compared to the global consumption volume of recycled ABS of 2.6 million tons in 2022 [10]. This means that to realize the European vision of more than half of the plastic waste generated in Europe to be recycled by 2030 (EUR-Lex—52018DC0028), we need to accelerate plastic circularity significantly.

The main problems concerning ABS recycling are the variability in its applications, material compositions, the presence of polymer mixtures and the use of additives. Despite these, there are already European and national requirements on how much material must be recycled within various industrial sectors. For example, one of the biggest areas of use for ABS is the automotive industry. This sector is regulated in the EU by the end-of-life vehicles (ELV directive) [11], which requires 85% of products to be reused and recycled and 95% to be reused and recovered. The new proposal from the EU Commission in July 2023 that aims to enhance the circular design of vehicles requires a minimum of 25% of recycled plastic to be used in vehicle manufacturing, with 25% of that coming from recycled ELVs, which will be achieved soon. The proportion of ABS in plastics that are used in electric and electronic equipment is about 30%. This industry sector is also regulated in the EU by the Waste of Electric and Electronic Equipment (WEEE) directive [12], which requires 70%–80% of waste to be recovered in the form of energy and/or materials.

Mechanical recycling is the most used recycling method for ABS and refers to processing waste into secondary raw materials without significantly changing the materials' chemical structure. The limitation of the mechanical recycling process is that it may cause chain scission and thermo-oxidative degradation, which degrades the material properties, leading to downcycling [13]. Another common problem is contamination with other polymers, which usually leads to a significant deterioration of the material properties [14]. In addition, even small amounts of some other contaminants can reduce the quality of recycled materials, limiting their end-use options. However, there are also methods to repair recycled materials. The successful renovation of molecular chains and phase interface of recycled ABS was achieved by an in situ chain extension reaction between ABS and pyromellitic dianhydride by modifications via melt blending [15]. Many other additives can be used during the mechanical recycling process to modify/upgrade recycled plastics and make them usable for higher-grade applications.

Another emerging recycling option for ABS is physical recycling, often referred to as solvent extraction. It is a process in which ABS material is subjected to dissolution and purification steps to separate ABS from other polymers and additives, which normally results in pure ABS or SAN. Lu and Chen developed a physical recycling process for toys that produces a recovered ABS suitable for plastic reprocessing if proper additives are added to improve its thermal stability and mechanical performance [16]. Achilias et al. reported that ABS can be recycled from WEEE using the dissolution method. They reported that ABS was dissolved in acetone at room temperature, thus eliminating the risk of thermal degradation [17].

1.3. Circular Use of ABS

The circular economy goes beyond the simple concept of material recycling; it approaches a systematic level that integrates economic, environmental and social sustainability [18]. The circular use of ABS materials implies that products and materials are maintained at their highest value for the longest time possible and are designed for the reduction in waste from the outset, avoiding and eliminating hazardous and toxic substances. It also means that manufacturers are willing to introduce recycled materials into their manufacturing processes to limit the number of raw materials used. It is also important to reuse parts from products in the same application as well as convert materials or parts from products at the end of their lives into new applications. This requires the transparent flow of product information and other data between suppliers, manufacturers, logistics, recyclers and customers. Lack of traceability and transparency in value chain transactions have been identified as barriers to the circular use of plastics [19].

The circular approach implies that we need to change the whole life cycle of products and materials, including their design, technological and production systems, distribution, consumption, collection, recycling and final disposal options [20].

1.4. Aim and Methods

This article is based on the assumption that traceability has great potential to accelerate the transition to the circular use of plastics. The purpose of this survey is therefore to review and evaluate existing and future traceability technologies that could be suitable for this purpose, in particular, focusing on products made of ABS. Based on this information, we will prepare at a later stage guidelines on how a traceability system for ABS products should be designed to facilitate the transition towards the circular use of ABS products. The innovative value of this paper relies on the fact that it elaborates on the vision of the European Strategy for Plastics in a Circular Economy, which sets ambitious targets for plastic recycling and circularity, within which we are focusing on a very important group of ABS plastics. To our knowledge, it is the first paper that focuses on traceability technologies for products made from ABS.

Our assessment refers to articles, books and works across scholarly and non-scholarly platforms such as Scopus, Web of Science and Google Scholar, European and international standards as well as interviews with stakeholders. We also examined some news articles about the latest developments in the field of traceability. Searches in the Scopus database using a combination of 'traceability AND ABS' resulted in 0 publications.

In general, we found a large number of papers that consider different aspects of both materials and information flows as well as technological developments and their evaluation. Our ambition with this article was not a comprehensive literature review of the entire area of traceability. The purpose of this article was to find and evaluate existing and future traceability technologies that could be suitable for use in products made of ABS to enable their circular use. We tried to use the most simplified terms, such as traceability, information sharing, physical marking, plastics and ABS, which resulted in many unrelated results, and we missed several articles from some specific areas of traceability. At the same time, we tried to compensate by using a citation analysis, which gave us additional articles. Due to the limitations of our study, it does not reflect research in the entire field. However, we believe that our article provides valuable information to those who have an ambition to design a product-based information sharing system for a plastic product or product group in general and for ABS products in particular.

2. Traceability—A Key to Successful Transition to Circular Economy

2.1. Traceability for Circular Use of Plastics

The large variety of plastics used, the varying lifespan of different products and technical challenges in the identification and separation of plastics create difficulties in tracking plastic flows, from the manufacturing of a product to waste management and preparation for the second life of a material. It is generally recognized that traceability has the potential to contribute decisively to improving environmental sustainability and supporting the global objectives of the circular economy. However, the first obstacle many encounter is the definition of this term. There are many definitions of the concept of 'traceability' both in international standards and in various dictionaries, and in addition, the respective interpretations of what traceability is are neither precise nor consistent [21]. A reasonably suitable definition can be found in Webster's Online Dictionary under 'Environment' as follows: 'The ability to trace the history, application, or location of an item, data, or sample using recorded documentation'. This definition can be expanded as follows: 'Traceability refers to the completeness of the information about every step in a process chain'. In the literature, two different types of traceability are often mentioned, viz., internal traceability and chain traceability. In this paper, traceability refers to chain traceability, which ranges from procuring raw materials through manufacturing, distribution and sale, usage, and handling as waste.

Despite the general belief in the positive effects of traceability especially for the improved recycling of plastics, there is still a lack of reliable and consistent traceability systems. The recycling of plastics is a complex process, which requires a customized infrastructure and cooperation between different actors in a value chain to be able to unleash its full potential. If fully implemented, the benefits for the environment, consumers and society are obvious. The system must contain complex information about the plastic material so that different actors in the value chain can extract the piece of information that is important to them. The information should be attached to the product by labelling it, with possibilities for digital reading, automated identification and electronic data processing. The need for improved material recycling is also mentioned in several publications in which improved sorting processes for mixed plastic flows, increased traceability and knowledge about chemical content are often highlighted as important factors (e.g., The New Plastic Economy, Ellen MacArthur Foundation, 2016).

A crucial role of the traceability system in plastic recycling is ensuring that the new material produced has the appropriate quality for a second life by suitable reusing/recycling processes. For this reason, industry and public institutions implement physical labelling, which verifies all the stages that each batch of product has gone through and establishes its date and place of origin. In addition, the important information that is needed to evaluate the quality of recycled material is the composition, the state of degradation and the level of contaminants, which are not included in any labelling system yet. Moreover, some consumer products made of plastics are subjected in the EU to chemical restrictions, bans, labelling and lab testing requirements. A product label that enables digital information management can also provide advantages during the product's usage phase. An example of this is a quite recently introduced digital QR marking system (SCANNECT®) for tyre manufacturers, in which QR codes are engraved on vehicle tyres for improved internal tyre logistics.

2.2. *Example of the Scientific Literature in the Field of Traceability*

The scientific literature in the field of traceability can be divided into two main disciplines, viz., information sharing (IS) to enable a circular economy and technologies for the physical marking of products able to carry the necessary information, which are particularly important to consumers and recycling organizations. The main research disciplines that contribute to IS are operations research, which studies the economic impact of IS, computer science, which evaluates solutions for reliable and secure IS, engineering, which studies how to record and manage product information throughout a product's entire life cycle, and environmental management, which assesses the environmental impact of IS [22]. A good example of operations research is a recently published article reporting on a tool that has been developed for economic analyses and financial risk assessments to assist actors in inter-organizational circular systems in evaluating benefits throughout the different phases [23]. Hsieh et al. refined the existing framework that is applied by circular economy supply chains by screening out key factors that are applied to remanufacturing products. They concluded that "optimizing the production process", "effectively tracking and recycling products", "redesigning remanufactured rubber products" and "improving resource efficiency" are the criteria that need to be considered first in the manufacturing of remanufactured products [24]. The impacts of materials on human health and the environment are of great significance. The complex interactions between materials and the environment require special consideration in the design of databases. Amos et al. have presented two data processing methods that can be integrated into the experimental process and can be applied across multiple subfields of material science [25]. Bindel et al. have described the development of a system capable of supporting the collection and visualisation of product data that are used in life cycle monitoring systems for electronic products. They also determined what opportunities and barriers exist for embedded information in this domain [26]. However, the general conclusion from our review of the scientific literature on IS is that it is fragmented, and research and practise are disconnected. Thus, to fully explore the value of IS to enable a circular economy, the scientific literature in this multidisciplinary field must be synthesised, and a common understanding must be established [27].

3. Examples of Labelling and Traceability Systems Currently in Use

3.1. Society of the Plastics Industry (SPI) Codes

In 1988, the SPI developed the international resin identification coding system, which is a set of symbols placed on plastics to identify the polymer type. The primary purpose of the codes was to allow for the efficient separation of different polymer types for recycling as well as to help consumers better understand some differences between materials and products as well as how they affect one's health and the environment. The recycling symbols are designed as a triangle formed by three circling arrows. The number in the triangle indicates the type of plastic. SPI codes 1 to 6 denote specific commodity resins, whereas number 7 is used to designate miscellaneous types of plastic, including ABS, that are not defined by the other six codes (see Figure 1). The labelling system is described in the newly revised ASTM international standard ASTM D7611 [28].

Figure 1. SPI codes.

A major disadvantage of the system concerns the fact that the symbols are not trademarked and therefore not subjected to official regulation for their use. As a result, they can be applied to any item and do not necessarily indicate recyclability. Another disadvantage is the lack of specific symbols for engineering plastics and new materials such as copolymers (e.g., ABS), as well as bio-based, biodegradable and compostable plastics. More recently, the logo of ABS, or the number 9 surrounded by arrows, has been used to denote ABS plastic, as shown in Figure 1.

3.2. EU Medical Device Regulations (MDRs)

One of the mandatory traceability systems is applied on medical device products, for which, according to EU Regulation 2017/745 (EU MDR) [29] medical devices must demonstrate that they meet legal requirements to ensure they are safe and are used as intended. It is mandatory to affix the medical device symbol, which shows that the product is a medical device. All the labelling requirements are detailed in MDRs, including the symbols to be used in the labelling of medical devices and their packaging, the trade name and the device's original name. All the symbols covering the required information in the labelling of the device and the documents such as booklets and manuals that accompany it must be included. The main purpose of labelling is to provide safety information to users such as healthcare professionals, consumers or other relevant persons. If a device carries a CE mark which confirms that it meets the basic requirements, it must be allowed in all EU countries.

Most medical devices must also be marked with a unique device identifier (UDI), which is a unique code that identifies and tracks medical devices throughout their distribution and use. The UDI system was established by the US Food and Drug Administration to improve patient safety and provide a consistent way to identify medical devices. A significant role of a UDI is to simplify traceability, facilitate corrective measures on the market and make product counterfeiting more difficult. The use of the UDI system can also improve purchasing and waste disposal policies and stock management by health institutions, recyclers and other operators, as well as ensure compatibility with other authentication systems. Figure 2 below illustrates a typical label on a medical device.

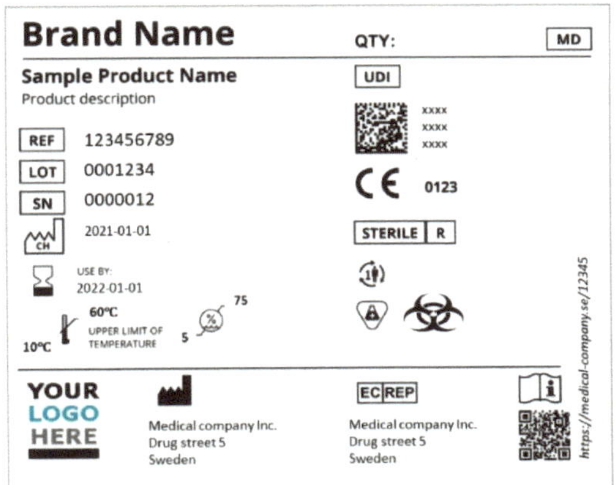

Figure 2. Labelling in accordance with MDR.

3.3. EU Toy Safety Directive (TSD) (Directive 2009/48/EC)

A traceability system is also applied on toys for the EU, where all toys must comply with the TSD and must carry a CE mark. This mark is an indication that the toy meets the safety requirements. The directive sets out the labelling requirements for general toy products, including plastic toys. In addition to the CE mark, the required labelling information includes product traceability information (e.g., the manufacturer's location and contact information, batch ID, model number and date of production) as well as all relevant details related to the sourcing of the product, or the raw material used in making the product. Figure 3 below illustrates a typical label on a toy.

Figure 3. Labelling in accordance with TSD.

3.4. EU Electronic Product Regulations

A traceability system for electronic and electrical products is mandatory in the EU, and products intended for the EU market must meet requirements in regards to safety standards, labelling, documentation and testing. Products covered by one or more EU directives (LVD, RED, EMC, RoHS, etc.) must carry the CE mark, which indicates conformity with the requirements. The CE mark must be permanently affixed to the product and its packaging. Importers also need to attach a permanent traceability label to the products and packaging, showing information about the company and the product. Both importers and manufacturers also need to provide documents such as the Declaration of Conformity, user instructions and technical files showing that the product is compliant with certain product directives and standards. Figure 4 below shows an example of electronic product traceability labelling. The information is limited, non-specific, and effective for waste management or the recycling sector, though it could be supplemented with appropriate information to help some actors in the electronic supply chain.

Figure 4. Product traceability labelling for electronic products.

4. Information-Based Traceability of Plastics

Today's traceability systems for plastic products are not designed to provide appropriate information to all actors in a value chain; the reasons for this are many and depend on the product. Some of the reasons are related to a lack of collaboration between partners in a value chain, the poor utilization of new technical possibilities and an insufficient understanding of the benefits of traceability. The most challenging parts are understanding and acceptance of the benefits of traceability for the environment, society, a circular economy and individual actors. To achieve the set goals for plastic recycling, a suitable traceability system must be introduced with broad acceptance from all players along the value chain and that is clearly supported by society. In the near future, traceability and information exchange between all actors of a value chain in the plastics industry will be crucial to establishing more effective recycling systems. A suitable traceability system should combine physical marking with data sharing platforms that provide access to information about products and materials. A recent scientific review paper analyses four different technological areas for information-based tracking solutions for plastic materials, viz., physical markers and tracers, blockchain, Digital Product Passports and a certification system [30]. The authors summarized the strengths, weaknesses, opportunities and challenges for all four tracking technologies in an SWOT analysis, as summarized in Table 1.

Table 1. Summary of the SWOT analysis (based on [30]).

	Strengths	Weaknesses	Opportunities	Challenges
Physical marker and tracer	Solution to improve sorting efficiency immediately	Tracer materials remain in plastic	Few technical hurdles for implementation	Single initiatives without a common standard
Blockchain	High transparency and security	Energy-intense verification system Risk of false information	The ability to build reliable trust can create incentive opportunities	Many technical hurdles
Digital Product Passport	Building on established tools and machinery in line with EU regulations	Non-transparent data handling	Offers link for combining physical and digital solutions	Data ownership and security are unclear
Certification system	Can build on established certification systems	Slow implementation with global alignment. No traceability on object level	Can build on global established knowledge	It is unclear as for how to solve the traceability issue and does not cover all aspects or participants of VC

However, in our opinion, data sharing systems and physical marking should be discussed separately. For data sharing, there are various digital technologies that enable information that is attached to a product to identify it as well as store, share and analyse data throughout its life cycle. This is described more in Section 4.1. A physical mark must be applied directly onto the plastic material, using a label on the product's surface or by the addition of a specific material to a polymer matrix which can be identified at any point in time with the use of a specific light source or device. Physical marks can be accomplished with several different methods as described in Section 4.2.

4.1. Data Sharing Systems

In a circular economy, actors must be able to access data, information and knowledge from across the entire value chain, meaning they must have access to an advanced data sharing system that can allow for the exchange of relevant data without trespassing on another actor's intellectual property rights. Currently, there are still barriers because of historically weak networks and communication. Data sharing systems are digital platforms for information exchange such as Digital Product Passports, blockchain or standard and certification systems. A well-functioning system makes it possible to make reliable information about a product available throughout the value chain. The shared information about regulatory compliance, value chain efficiency, material composition, environments of use and consumer engagement is crucial to understanding the full life cycle of a product and to facilitate separation, recycling and valorisation techniques.

4.1.1. Digital Product Passport

A Digital Product Passport (DPP) is a tool for collecting and sharing product data throughout its entire life cycle. It is used to provide information about a product's sustainability as well as its environmental impact and recyclability. A DPP creates 'a digital twin of a physical product and securely records event, transactional and sustainability-based data from across the product's life cycle. The digital twin is commonly associated with the physical product via a QR code, barcode, or other technology—with the DPP being accessible via a smart device application or similar' [31]. The expected benefits of a DPP are access to data, flexibility, transparency and the accountability of customers.

However, many challenges remain such as different interests of stakeholders, missing underlying standards, missing infrastructure, data ownership and intellectual property rights. Other challenges include balancing transparency with the safeguarding of sensitive information and the use of centralized databases that store copies of data, which can easily become outdated, leading to poor data quality.

4.1.2. Blockchain

By definition, a blockchain is "a distributed database that maintains a continuously growing list of ordered records, called blocks". These blocks are linked using cryptography. Each block contains a cryptographic hash (a digital fingerprint or unique identifier) of the previous block, a timestamp and transaction data. A blockchain is "a decentralized, distributed, and public digital ledger that is used to record transactions across many computers so that the record cannot be altered retroactively without the alteration of all subsequent blocks and the consensus of the network" [32].

Blockchain solutions enable businesses, regulatory bodies, governments and consumers to exchange information and cooperate to improve and facilitate waste management. It offers the highest level of transparency, data integrity and security, and for these reasons, it is regarded as one of the key emerging technologies for Europe by the European Commission. The commission has therefore released a blockchain strategy to describe Europe's digital future [33]. There are several already existing initiatives in the area of plastics that utilize blockchain solutions. One of these is reciChain, a solution created by BASF in Sao Paolo, Brazil, in which circularity was successfully demonstrated in 2020 by tracking a product's life cycle "from pellet to pellet". ReciChain comprises two technology components, viz., a physical tracer that enables the connection of plastic to a digital twin and a blockchain marketplace which creates and translates the digital twin, providing a secure, auditable transfer of ownership and assigning incentives [34]. Gopalakrishnan et al. [35] propose a blockchain-based solid waste management model that can help to enhance the efficiency of waste management. They also estimated the cost aspects associated with blockchain implementation based on several use cases obtained from companies providing blockchain solutions. Arora et al. presented a framework that focusses on the use of blockchain technology from the entry to the exit of any process in a system as well as the identification of important areas to which blockchain technology can be applied and implemented [36]. El-Rayes et al. performed a thorough evaluation of the status of plastic management in research. They found a shortage of publications when the term "blockchain" was combined with six selected areas, viz., "supply chain", "sustainability", "environment", "finance", "fintech" and "plastic". Furthermore, among the few documents found in the area of "plastic and blockchain", only 3% mentioned the circular economy [33,37].

An important prerequisite for a successful implementation of blockchain technology is that it must mirror physical reality to be credible. This means that in production, each batch of materials needs to be audited by an independent third party that can verify that the material's properties, its origin, composition, process and sustainability claims are consistent with its real conditions.

4.1.3. Standards and Certification Systems

Certification systems ensure compliance with standards, rules and regulations. The most reliable certification systems and standards for a circular economy rely on annual auditing to ensure that a manufacturing company has quality procedures that guarantee that its products meet the set requirements. An older example of a standard for recycling is The International Resin Identification Coding System described in Section 3.1. A new European certification system for recycled post-consumer plastic waste is the EuCertPlast Certification. The certification system is based on EN 15343:2007 [38], which specifies the procedures needed for the traceability of recycled plastics [38]. As a complement to EuCertPlast, there is the Recycled Plastics Traceability Certification, which is a certification scheme that provides both evidence of the traceability of recycled plastics from the source and the specific recycled content in products [39].

There is an ISO standard for the computer-interpretable representation and exchange of product manufacturing information with the potential to be used for the purpose of promoting materials recycling, but so far it has not been used for this sole purpose. The ISO 10303 [40] standard 'Industrial automation systems and integration' is a comprehensive and modular standard that defines product data, such as geometry, topology, materials and properties, and also specifies the methods and protocols for exchanging product data between different systems, such as CAD and CAM. However, there are major obstacles to this standard being able to be generally accepted and implemented by the plastics industry's value chains. In particular, SMEs, which are often part of the value chain, and consumers will likely not be able to use the standard due to its complexity and diversity, which makes it difficult to understand, implement and maintain. In addition, it is not always compatible or compliant with existing software applications or platforms used by the plastics manufacturing and recycling industry.

In current standards and certification systems, the traceability of individual products is only possible in rare cases, and the information provided by the system does not cover all aspects needed to facilitate recycling. To increase the usage of standards and certification for traceability, it is important to increase their broad acceptance along the value chain and to speed up the implementation process.

4.2. Methods for Physical Marking of Plastic Products

The physical marking of plastic products is an important part of the design for the environment. Plastics must be easily identified and then separated at the end of their life according to their material type, chemical structure and quality to maximize their value in their next usage. Permanent marking is the most desired option for all actors who want to track information about a product. Permanent marking or a physical tracer can be accomplished with several different methods, as described below. However, it is necessary to install relevant equipment within sorting facilities and to establish databases to retrieve information from physical markers for all marking technologies. Moreover, it is important that the readability of the tracer during a product's entire service life is robust even when the material becomes waste.

4.2.1. Photoluminescent Labelling

Photoluminescence is a well-known phenomenon that involves the emission of light from a material after the absorption of electromagnetic radiation. The process is initiated by photoexcitation with subsequent relaxation processes in which other photons are emitted. Because specific chemical substances emit light of unique wavelengths, this technique can be used as a robust method for the identification and separation of waste plastics, without interfering with other sorting techniques. There are three main types of labels depending on the excitation source, viz., UV—vis, NIR and X-ray radiation.

UV–vis markers can be inorganic chemical compounds such as metal oxides or organic compounds containing aromatic components. In some cases, already-existing NIR light sources can be used as the excitation source. This is because some chemical compounds can undergo a so-called photoluminescent up-conversion. This is accomplished by a two-photon absorption process through an intermediate excited state, leading to the emission of radiation of a higher energy level than that absorbed [41]. The most studied NIR markers are inorganic crystals. However, black-coloured plastics present a major issue and remain an obstacle for waste sorting using this technique. The third type of label works through X-ray fluorescence which can be used to identify specific elements in a sample. Because commodity plastics are hydrocarbon based, it is necessary to insert a chemical compound into a plastic with unique identifiable elements for the method to work. In addition to emission wavelengths, photoluminescence can also be used to identify a marker by measuring the emission lifetime. The emission lifetime is determined by the exponential decay of the intensity of the emission when an excitation source is removed from a material.

A particularly interesting photoluminescent-based sorting system uses carbon quantum dots (CQDs) in a plastic matrix through the formation of a polymer nanocomposite. CQDs are carbon nanoparticles which are less than 10 nm in size and have some form of surface passivation. These nanocomposites can respond to stimuli by emitting different fluorescence wavelengths in the range from the UV to the NIR region and thus can monitor many material parameters due to their wide sensing range. However, more studies are needed to show if these nanomaterials can be incorporated into polymers using scalable manufacturing methods, such as extrusion, instead of the currently used solvent-based methods. A recent review summarizes progress regarding the 'green' synthesis of CQDs [42].

Labelling plastics using photoluminescence has great potential to facilitate waste sorting and to improve both the quantity and quality of recycled plastic materials. However, it is quite a new approach and still needs to be standardized and implemented across the whole plastic value chain with the active participation of all its actors. Initially, this technique could be used in the recycling of especially valuable plastics or for products which are used in high volumes, such as in packaging.

4.2.2. Digital Watermarks

Digital watermarks are invisible optical codes embedded in a material that can carry a wide range of product information and can be identified in a sorting plant via optical sensors.

Between 2016 and 2019, various stakeholders from the packaging value chain gathered in a project named HolyGrail to investigate possibilities for improving recycling using chemical tracers and digital watermarks. The conclusion of the project drawn by the large majority of stakeholders was that digital watermarks are the most promising technology. In the follow-up project, HolyGrail 2.0, a semi-industrial testing was performed that mimicked real-life conditions. Characteristics such as detection efficiency, ejection efficiency, purity, prototype stability and routine function, ease of programming the sorting operation and counting capabilities of the prototype were evaluated [43]. The project was a success and led to other potential opportunities, such as the creation of smart packaging and its use in other areas. These include enhanced consumer engagement, linking to a data sharing platform, supply chain visibility, retail operations and more.

4.2.3. Laser Marking Systems

A laser is a device that emits coherent light that allows a beam to be focused to a tight spot. The beam is produced through a process of optical amplification based on the stimulated emission of electromagnetic radiation. Laser marking is a broad category of methods to produce marks on an object using beam exposure. A laser creates various surface changes which can include colour changes due to chemical/molecular alterations, charring, foaming, melting, ablation, photoreduction and more. A variety of laser devices and media are used. The differences depend on how the laser is generated and what beams are produced. A critical factor is of course the legibility of the mark which can be affected by the mark contrast width of the surrounding area. Because plastics have a wide range of colours, surface structures, densities and material designs, specific lasers work better with different materials. It is thus important to choose the right laser (in regards to energy density, pulse width and wavelength) relative to the physical and chemical properties of the material, e.g., the absorptivity. For example, a green laser (532 nm) has a very high absorption ratio that can produce suitable marks on PVC, ABS and PS. An ultraviolet laser (375 nm) is a good replacement for ink marks on HDPE material. Fibre and vanadate lasers (1064 nm) work well on many high-density plastics, such as POM, ABS and many engineering plastics. Infrared and visible lasers can induce black and white changes by the carbonization or foaming of samples, respectively, and other colour changes can be laser-induced on pigmented polymers [44].

The laser marking of plastic products is durable and flexible and does not have any negative effects on the material or environment. The marking process can be applied to a wide range of materials by selecting an appropriate laser beam. Furthermore, the numerically controlled beam motion can create a wide variety of marks directly using a computer. Most of the plastic materials can be laser marked, but if a plastic does not have that ability, it can usually be modified with a suitable additive. Thus, the results of laser marking on plastics depend not only on the type of plastic but also on the specific additive used [45]. Suitable laser marking additives create better opportunities to produce high-contrast marks. The lasers that are most commonly used for marking plastics are the following:

A **CO_2 laser** produces a beam of infrared light with wavelength bands at 9.6 and 10.6 μm. It is suitable for marking most plastics, such as PE, POM, PC, PP, PET and ABS. The beam only removes a surface layer, leaving a permanent mark. CO_2 lasers have a very low beam divergence (the angle at which a laser beam spreads out from its source), which is a big advantage if a very fine, detailed mark is needed [46].

An **excimer laser (exciplex laser)** is a form of a UV laser (at 193 or 248 nm) that makes patterns on or etches a plastic product. The marking speed depends on the mark complexity and the depth of the mark, which is created as a series of dots. Excimer lasers are commonly used in the production of electronic devices. The UV energy from excimer lasers causes high levels of electron excitation in irradiated materials, leading to bond breaking and photoablation. Excimer lasers have high-energy photons that can induce photochemical reactions, which may result in colour changes with negligible thermal side effects [47].

Nd:YAG (neodymium-doped yttrium aluminium garnet; Nd:$Y_3Al_5O_{12}$) is a crystal that is used as a lasing medium for solid-state lasers. Nd:YAG lasers typically emit light with a wavelength of 1064 nm in the infrared region and are one of the most common types of laser, being used for many different applications, e.g., in manufacturing for engraving, etching or marking a variety of metals and plastics. Nd:YAG lasers are also used to make subsurface markings on transparent materials, such as glass or acrylic glass, and on white and transparent polycarbonate for identity documents. Clemente et al. used a frequency-tripled Nd:YAG laser emitting at 355 nm to produce aesthetical marks on white ABS. They found that the marks had a good level of scratch resistance similar to that of pad-painted marks and a good level of resistance to chemical agents and climate tests [48].

A **fibre laser** is a laser in which the active gain medium is an optical fibre doped with rare-earth elements such as erbium, ytterbium and neodymium. They are related to doped fibre amplifiers, which provide light amplification without lasing [49].

4.2.4. Printing

There are several specialized techniques for printing on plastics, such as inkjet, screen, flexo, UV litho, pad printing and hot stamping, and each one of these requires expertise to ensure the best results. One of the most important features of printing is the tendency of ink to remain attached to a substrate when acted on by different forces such as mechanical (rubbing or abrading) and environmental (sunlight, heat, moisture and chemicals). Adhesion is the result of the physical and chemical interaction between the ink and the substrate in which the contact surface area between the ink and the substrate is the major factor in adhesion. Chemical interactions are influenced by the substrate and ink's chemical compositions which are specially designed for different plastic types. Two of the most used printing methods for plastics include inkjet printing and pad printing.

Digital Inkjet Printing

Inkjet printing can be used to print traceability information, variable data, barcodes, multiline codes and more. It is a modern printing method that uses digital files containing information to be printed, which are created using computer software. An inkjet printer managed through digital files ejects droplets of ink onto the plastic surface to create a mark, image or text. An inkjet printer requires the nozzles to size drops precisely with a high degree of accuracy. High-quality inkjet printing systems must simultaneously integrate printheads, electronic controllers, a suitable pretreatment of the surface and curing. All items must work together to produce the intended results. Used correctly, inkjet printing is fast, flexible and can produce high-quality marks. However, there are some concerns that must be considered, such as the following:

- operating costs over time because of the expendable items involved;
- harmful chemicals that can add environmental challenges;
- ink is not permanent even if it can be made very durable;
- some chemicals can break down even durable inks, and the mark can be lost.

Pad Printing

Pad printing, or tampography, is a printing process that allows for the transfer of complex, detailed graphics onto flat or irregularly shaped surfaces. The process uses an image that has been created on a printing plate which then is coated with a layer of suitable printing ink. After the excess ink has been removed from the plate, a thin film of ink remains in the image. Then, a silicone pad is pressed onto the plate to pick up the ink. The pad is then pressed against the surface of a plastic product, transferring the ink. The process ends with the drying procedure until a full cure. Pad printing is often chosen because of its versatility, accuracy and cost-effectiveness and because it can be used on a variety of plastic materials, including ABS. However, there are some factors that must be considered. The first one is the necessity of removing any impurities from the surface that might affect ink adhesion. The second is the need to use a pretreatment of the plastic surface, such as primers, flame, corona or plasma. The third is the need to use, in some cases, a heat source such as an oven to speed up the ink drying process and chemical reaction [50].

4.2.5. Scribe Marking

Scribe (scratch) marking is a mechanical process whereby a mark is produced using a scribing pin with a carbide or diamond tip that penetrates and indents the surface and creates a continuous and homogeneous line. The amount of pressure the pin applies determines the depth of the mark in the material. Even very lightly scratched characters can be achieved with this method. A scribe marker can work on most plastics, including ABS, among many other types of plastic materials. Scribe marking machines are computer-controlled and are specifically designed to be used on a variety of part shapes and sizes. They work satisfactorily on flat surfaces as well as on round, curved, convex and concave parts. Scribe marking systems are not portable and are quite expensive; however, they are less expensive than laser marking systems but are significantly slower [51].

4.2.6. Dot Peen Marking

Dot peen marking uses rapid vibrations to produce marks. Pneumatically or electromagnetically powered carbide or a diamond stylus assembly is stroked against a surface creating depressions or indentations, which results in a succession of dots with negligible separation.

The markings are precise and thus complex shapes and data matrices can be produced on plastics. The equipment is inexpensive, and the process requires few consumables and little maintenance. Its disadvantages include the amount of noise being made by dot peen markers because of the vibrations, and the resolution of the markers needs to be considered if very clear markings are required [51].

4.2.7. In-Mould Marking

Date stamps have been used for many years in the injection moulding process to enable the traceability of plastic items. A date stamp is a cylinder-shaped metal which is composed of an outer ring and an inner arrow. The stamp looks like a clock with an outer ring containing letters, numbers and other characters. The inner arrow can turn to indicate different positions on the outer ring. In this way, a permanent mark is created, showing information such as the manufacture date, lot numbers, types of plastic material and other information depending on one's needs. Manual date stamps are operated by operators responsible for manually changing the date stamps. It is sometimes difficult to see to which position the stamp needs to be rotated. If they miss the mark and the stamp moulds show incorrect data, a part may be deemed unusable. These problems can nowadays be avoided by automating the process and using a reliable dating system, which allows the process to continue operating while providing accurate, real-time information. This is achieved through the use of a small electronic control unit that is mounted on the interface to enable the programming of the daters as well as to monitor cycles. In-mould marking also includes the permanent marking of a QR code on a plastic product during injection by inserting a QR stamp into the mould. This is an easy and cheap way to mark plastic products directly. The QR code can contain information about the installation manual, maintenance method, material composition and handling of waste.

4.2.8. Stickers

Stickers are basically pre-printed labels that are fixed by a pressure roller to the surface of a product. However, choosing the right label can be a big challenge because the label materials must be precisely matched to the respective application. First, it is important to take into consideration the surface properties of a plastic product as well as the areas of application. For instance, the surface tension of the plastic plays an important role. Consequently, labels with a rubber type of adhesive adhere better to low-surface-tension plastics than many acrylic adhesives, whereas labels with an acrylic type of adhesive are more suitable for high-surface-tension plastics. Adhesion requirements vary greatly depending on the product and customer requirements and range from very durable with an ultra-aggressive adhesive to removable stickers. The latter are made using pressure-sensitive adhesives which form a bond with the medium without a chemical reaction to facilitate its removal and without leaving any sticky residue behind. Of course, the subsequent areas of application must also be considered. For example, if the labels will be exposed to UV light or if they must be chemical-resistant or seawater-proof, it is important that the legibility of the labels remains satisfactory throughout the entire lifetime of the product.

5. Concluding Remarks

A life cycle approach to the use of plastics allows people to make informed decisions that can protect human health and reduce environmental impacts. This approach has the potential to reduce resource consumption, improve the performance of products and extend their lifespan in each life cycle stage. There is a general consensus that correctly designed information-based traceability systems can support the circular economy of plastics, remove obstacles, improve their quality and increase the number of recycled plastics. However, the detailed examination of the current situation regarding plastic waste destinations recently performed by Hsu et al. shows that plastics in the EU are still far from being circular [52]. In the scientific literature, there are many new proposals and developments in information-based technologies and technologies for the physical marking of plastics. Our conclusion is that it is possible to find a suitable combination of the smartest physical and digital tracking technologies to establish material tracking system sustainably for various applications in the market. All ABS products could use the same data management system, most probably the blockchain strategy, which has been pointed out by the European Commission as Europe's digital future [33]. The situation is different when it comes to the choice of physical marking. Which technology is the best option depends on the product's size, appearance, manufacturing method, area of use, price sensitivity and more. The situation is also complicated when considering what information a product should carry. Some products need to be tracked even at the individual level to enable the reuse of some parts (e.g., car parts and electronics), whereas other products need specific information to facilitate recycling.

Specially designed information to choose the right recycling method and a possible need for upgrading are also important because the recycling of ABS materials presents big technological challenges due to the variety of material compositions, applications, periods of use and usage environments. In our opinion, the most effective way to achieve the circularity of ABS products is by providing necessary, reliable and easily accessible information to all stakeholders in the value chain of an ABS product. To provide guidelines on how a suitable traceability system for ABS products containing all necessary information should be designed, it is important to have state-of-the-art knowledge of current techniques. This paper has therefore reviewed and evaluated various existing, emerging and future techniques for traceability with a special focus on ABS plastics. Our goal is to identify the most suitable techniques for the identification of ABS materials by marks connected to information-based platforms that enable data management and access to information for the actors along the value chain. Additionally, these techniques enable selective sorting, consumer communications, processing instructions, information about the origin of feedstock and more. Ultimately, the decisive step in the introduction of a traceability system will be the acceptance and involvement of all actors in the ABS plastic industry and society in general.

Funding: European Commission, grant no. 101058636.

Institutional Review Board Statement: Not applicable.

Data Availability Statement: Data are contained within the article.

Acknowledgments: The authors express their gratitude to the European Health and Digital Executive Agency (HADEA) under the power delegated by the European Commission for its support of the project ABSolEU—'Paving the way for an ABS recycling revolution in the EU'.

Conflicts of Interest: The authors declare no conflict of interest.

References

1. Probst, L.; Frideres, L.; Pedersen, B. *Traceability across the Value Chain—Advance Tracking Systems*; European Commission: Brussels, Belgium, 2015.
2. Verzijl, D.; Rouwmaat, E.; Dervojeda, K.; Probst, L.; Flíderes, L. *Traceability across the Value Chain—Standards, Processes and Traceability*; European Union: Brussels, Belgium, 2015.
3. Adams, M.E.; Buckley, D.J.; Colborn, R.E.; England, W.P.; Schissel, D.N. *Acrylonitrile-Butadiene-Styrene: Review Reports*; Rapra Technology: Shawbury, UK, 1993.
4. Han, Y.; Lach, R.; Grellmann, W. Effects of Rubber Content and Temperature on Unstable Fracture Behavior in ABS Materials with Different Particle Sizes. *J. Appl. Polym. Sci.* **2001**, *79*, 9–20. [CrossRef]
5. Shaik, Y.P.; Naidu, N.K.; Yadavalli, V.R.; Muthyala, M.R. The Comparison of the Mechanical Characteristics of ABS Using Three Different Plastic Production Techniques. *OAlib* **2023**, *10*, 1–18. [CrossRef]
6. Freymond, C.; Mackré-Delannoy, X.; Guinault, A.; Charbuillet, C.; Fayolle, B. Thermal Oxidation of Acrylonitrile-Butadiene-Styrene: Origin of the Ductile/Brittle Transition Thermal Oxidation of Acrylonitrile-Butadiene-Styrene: Origin of the Ductile/Brittle Transition. *Polym. Degrad. Stab.* **2022**, *206*, 110186. [CrossRef]
7. Davis, P.; Tiganis, B.E.; Burn, L.S. The Effect of Photo-Oxidative Degradation on Fracture in ABS Pipe Resins. *Polym. Degrad. Stab.* **2004**, *84*, 233–242. [CrossRef]
8. Plastics—The Fast Fact 2023, Plastics Europe. Available online: https://plasticseurope.org/knowledge-hub/plastics-the-fast-facts-2023/ (accessed on 22 February 2024).
9. ABS Polymer Growth Statistics 2023. Available online: https://polymer-compounders.com/ (accessed on 22 February 2024).
10. Tiseo, I. Global ABS Plastic Recycling Market Consumption Volume 2017–2028, by Region. Available online: https://www.statista.com/statistics/1378544/abs-recycling-consumption-volume-globally-by-region/ (accessed on 22 February 2024).
11. Directive 2000/53/EC of the European Parliament and of the Council of 18 September 2000 on End-of Life Vehicles. 2000. Available online: https://eur-lex.europa.eu/legal-content/EN/ALL/?uri=celex:32000L0053 (accessed on 22 February 2024).
12. Directive 2002/96/EC of the European Parliament and of the Council of 27 January 2003 on Waste Electrical and Electronic Equipment (WEEE). Available online: https://eur-lex.europa.eu/resource.html?uri=cellar:ac89e64f-a4a5-4c13-8d96-1fd1d6bcaa49.0004.02/DOC_1&format=PDF (accessed on 22 February 2024).
13. Boldizar, A.; Möller, K. Degradation of ABS during Repeated Processing and Accelerated Ageing. *Polym. Degrad. Stab.* **2003**, *81*, 359–366. [CrossRef]
14. Signoret, C.; Girard, P.; Guen, A.L.; Caro-Bretelle, A.-S.; Lopez-Cuesta, J.-M.; Ienny, P.; Perrin, D. Degradation of Styrenic Plastics during Recycling: Accommodation of PP within ABS after WEEE Plastics Imperfect Sorting. *Polymers* **2021**, *13*, 1439. [CrossRef]
15. Li, Y.; Wu, X.; Song, J.; Li, J.; Shao, Q.; Cao, N.; Lu, N.; Guo, Z. Reparation of Recycled Acrylonitrile-Butadiene-Styrene by Pyromellitic Dianhydride: Reparation Performance Evaluation and Property Analysis. *Polymer* **2017**, *124*, 41–47. [CrossRef]
16. Lu, T.; Chen, W.-T. Material Recycling of Acrylonitrile Butadiene Styrene (ABS) from Toy Waste Using Density Separation and Safer Solvents. *Resour. Conserv. Recycl.* **2023**, *197*, 107090. [CrossRef]
17. Achilias, D.S.; Antonakou, E.V.; Koutsokosta, E.; Lappas, A.A. Chemical Recycling of Polymers from Waste Electric and Electronic Equipment. *J. Appl. Polym. Sci.* **2009**, *114*, 212–221. [CrossRef]
18. Bucknall, D.G. Plastics as a Materials System in a Circular Economy. *Philos. Trans. R. Soc. A Math. Phys. Eng. Sci.* **2020**, *378*, 20190268. [CrossRef]
19. Milios, L.; Holm Christensen, L.; McKinnon, D.; Christensen, C.; Rasch, M.K.; Hallstrøm Eriksen, M. Plastic Recycling in the Nordics: A Value Chain Market Analysis. *Waste Manag.* **2018**, *76*, 180–189. [CrossRef] [PubMed]
20. Paletta, A.; Leal Filho, W.; Balogun, A.-L.; Foschi, E.; Bonoli, A. Barriers and Challenges to Plastics Valorisation in the Context of a Circular Economy: Case Studies from Italy. *J. Clean. Prod.* **2019**, *241*, 118149. [CrossRef]
21. Olsen, P.; Borit, M. How to Define Traceability. *Trends Food Sci. Technol.* **2013**, *29*, 142–150. [CrossRef]
22. Jäger-Roschko, M.; Petersen, M. Advancing the Circular Economy through Information Sharing: A Systematic Literature Review. *J. Clean. Prod.* **2022**, *369*, 133210. [CrossRef]
23. Cafruni Gularte, A.; Carísio de Paula, I.; Siqueira de Souza, J.; Flores Sum, F. Economic-Financial Analysis Procedure: Implementation of Inter-Organizational Circular Systems. *J. Clean. Prod.* **2024**, *452*, 142242. [CrossRef]
24. Hsieh, H.-H.; Yao, K.-C.; Wang, C.-H.; Chen, C.-H.; Huang, S.-H. Using a Circular Economy and Supply Chain as a Framework for Remanufactured Products in the Rubber Recycling Industry. *Sustainability* **2024**, *16*, 2824. [CrossRef]
25. Amos, J.D.; Zhang, Z.; Tian, Y.; Lowry, G.V.; Wiesner, M.R.; Hendren, C.O. Knowledge and Instance Mapping: Architecture for Premeditated Interoperability of Disparate Data for Materials. *Sci. Data* **2024**, *11*, 173. [CrossRef]
26. Bindel, A.; Conway, P.P.; West, A.A. Information Structure Required for Life-Cycle Monitoring of Electronic Products. *Proc. Inst. Mech. Eng. B J. Eng. Manuf.* **2012**, *226*, 1612–1627. [CrossRef]
27. Zhang, A.; Wang, J.X.; Farooque, M.; Wang, Y.; Choi, T.-M. Multi-Dimensional Circular Supply Chain Management: A Comparative Review of the State-of-the-Art Practices and Research. *Transp. Res. E Logist. Transp. Rev.* **2021**, *155*, 102509. [CrossRef]
28. *ASTM D7611/D7611M-10*; Standard Practice for Coding Plastic Manufactured Articles for Resin Identification. ASTM: West Conshohocken, PA, USA, 2013.

29. Regulation (EU) 2017/745 of the European Parliament and of the Council of 5 April 2017 on Medical Devices, Amending Directive 2001/83/EC, Regulation (EC) No 178/2002 and Regulation (EC) No 1223/2009 and Repealing Council Directives 90/385/EEC and 93/42/EEC. Available online: https://eur-lex.europa.eu/legal-content/EN/TXT/?uri=CELEX:32017R0745 (accessed on 22 February 2024).
30. Rumetshofer, T.; Fischer, J. Information-Based Plastic Material Tracking for Circular Economy—A Review. *Polymers* **2023**, *15*, 1623. [CrossRef]
31. Protokol Digital Product Passport (DPP): The Complete Guide. Available online: https://www.protokol.com/insights/digital-product-passport-complete-guide/ (accessed on 22 February 2024).
32. Synopsys What Is Blockchain and How Does It Work? Available online: https://www.synopsys.com/glossary/what-is-blockchain.html#:~:text=A%20blockchain%20is%20%E2%80%9Ca%20distributed,a%20timestamp,%20and%20transaction%20data (accessed on 22 February 2024).
33. European Union. *European Blockchain Strategy—Brochure*; European Union: Brussels, Belgium, 2023.
34. reciChain Pilot Consortium Members Positioning Canada as a Leader in Plastics Circularity through the ReciChainTM Program. Available online: https://www.basf.com/ca/en/who-we-are/sustainability/Sustainability-in-Canada/reciChain.html (accessed on 22 February 2024).
35. Gopalakrishnan, P.K.; Hall, J.; Behdad, S. Cost Analysis and Optimization of Blockchain-Based Solid Waste Management Traceability System. *Waste Manag.* **2021**, *120*, 594–607. [CrossRef] [PubMed]
36. Arora, A.; Sharma, M.; Bhaskaran, S. Use of Blockchain Technology in Enterprise Management. *Int. J. Recent Technol. Eng.* **2019**, *8*, 4938–4943. [CrossRef]
37. El-Rayes, N.; Chang, A.; Shi, J. Plastic Management and Sustainability: A Data-Driven Study. *Sustainability* **2023**, *15*, 7181. [CrossRef]
38. *EN 15343*; Plastics—Recycled Plastics—Plastics Recycling Traceability and Assessment of Conformity and Recycled Content. European Commission: Brussels, Belgium, 2007.
39. European Union. Recycled Plastics Traceability Certification. 2020. Available online: https://circulareconomy.europa.eu/platform/en/news-and-events/all-news/recycled-plastics-traceability-certification (accessed on 22 February 2024).
40. *ISO 10303:2024*; Industrial Automation Systems and Integration Product Data Representation and Exchange 2024. ISO: Geneva, Switzerland, 2024.
41. Larder, R.R.; Hatton, F.L. Enabling the Polymer Circular Economy: Innovations in Photoluminescent Labeling of Plastic Waste for Enhanced Sorting. *ACS Polym. Au* **2023**, *3*, 182–201. [CrossRef]
42. Feng, Z.; Adolfsson, K.H.; Xu, Y.; Fang, H.; Hakkarainen, M.; Wu, M. Carbon Dot/Polymer Nanocomposites: From Green Synthesis to Energy, Environmental and Biomedical Applications. *Sustain. Mater. Technol.* **2021**, *29*, e00304. [CrossRef]
43. Scott, J. Progress of the HolyGrail Initiative. VP Global Creative Services. Available online: https://www.berryglobal.com (accessed on 22 February 2024).
44. Devi, M.; Wang, H.; Moon, S.; Sharma, S.; Strauss, V. Laser-Carbonization—A Powerful Tool for Micro-Fabrication of Patterned Electronic Carbons. *Adv. Mater.* **2023**, *35*, 2211054. [CrossRef]
45. Wissemborski, R.; Klein, R. Welding and Marking of Plastics with Lasers. *Laser Tech. J.* **2010**, *7*, 19–22. [CrossRef]
46. Lamberton, H.M.; Roper, V.G. Beam Divergence of a Highly Multimode CO_2 Laser. *J. Phys. E* **1978**, *11*, 1102–1103. [CrossRef]
47. Santo, L.; Trovalusci, F.; Davim, J.P. Laser Applications in the Field of Plastics. In *Comprehensive Materials Processing*; Elsevier: Amsterdam, The Netherlands, 2014; pp. 243–260.
48. Clemente, M.J.; Lavieja, C.; Peña, J.I.; Oriol, L. UV-laser Marking of a TiO_2-containing ABS Material. *Polym. Eng. Sci.* **2018**, *58*, 1604–1609. [CrossRef]
49. Sabreen, S. Fiber Laser Enables Marking of Advanced Plastics. *Ind. Laser Solut.* **2016**, *2016*, 32–34.
50. Screen Printing, The Secrets of Successful Pad Printing. In *Screen Printing Magazine*; Screen Printing: Ottawa, ON, Canada, 2003.
51. Hang, R. The Scribe Marking Machine—Complete Guide. Available online: https://www.heatsign.com/the-only-best-article-for-scribe-marking/ (accessed on 22 February 2024).
52. Hsu, W.-T.; Domenech, T.; McDowall, W. How Circular Are Plastics in the EU?: MFA of Plastics in the EU and Pathways to Circularity. *Clean. Environ. Syst.* **2021**, *2*, 100004. [CrossRef]

polymers

MDPI

Article

Polymer Waste Recycling of Injection Molding Purges with Softening for Cutting with Fresnel Solar Collector—A Real Problem Linked to Sustainability and the Circular Economy

Ma. Guadalupe Plaza [1], Maria Luisa Mendoza López [1,*], José de Jesús Pérez Bueno [2,*], Joaquín Pérez Meneses [1] and Alejandra Xochitl Maldonado Pérez [2]

[1] Tecnológico Nacional de México, Instituto Tecnológico de Querétaro, Av. Tecnológico s/n Esq. M. Escobedo Col. Centro, Santiago de Querétaro C.P. 76000, Querétaro, Mexico; joaquin.pm@queretaro.tecnm.mx (J.P.M.)

[2] Centro de Investigación y Desarrollo Tecnológico en Electroquímica, S. C., Parque Tecnológico Querétaro-Sanfandila, Pedro Escobedo C.P. 76703, Querétaro, Mexico; amaldonado@cideteq.mx

* Correspondence: maria.ml@queretaro.tecnm.mx (M.L.M.L.); jperez@cideteq.mx (J.d.J.P.B.)

Abstract: A plastic injection waste known as "purge" cannot be reintegrated into the recycling chain due to its shape, size, and composition. Grinding these cannot be carried out with traditional mills due to significant variations in size and shape. This work proposes a process and the design of a device that operates with solar energy to cut the purges without exceeding the degradation temperature. The size reduction allows reprocessing, revalorization, and handling. The purges are mixtures of processed polymers, so their characterization information is unavailable. Some characterizations were conducted before the design of the process and after the cut of the purges. Some of the most representative purges in a recycling company were evaluated. The flame test determines that all material mixtures retain thermoplasticity. The hardness (Shore D) presented changes in four of the purges being assessed, with results in a range of 59–71 before softening and 60–68 after softening. Young's modulus was analyzed by the impulse excitation technique (IET), which was 2.38–3.95 GPa before softening and 1.7–4.28 after softening. The feasibility of cutting purges at their softening temperature was evaluated. This was achieved in all the purges evaluated at 250–280 °C. FTIR allowed for corroboration of no significant change in the purges after softening. The five types of purges evaluated were polypropylene-ABS, polycarbonate-ABS-polypropylene, yellow nylon 66, acetal, and black nylon 66 with fillers, and all were easily cut at their softening temperature, allowing their manipulation in subsequent process steps.

Keywords: recycling; purges; polymers; processing; waste

Citation: Plaza, M.G.; Mendoza López, M.L.; Pérez Bueno, J.d.J.; Pérez Meneses, J.; Maldonado Pérez, A.X. Polymer Waste Recycling of Injection Molding Purges with Softening for Cutting with Fresnel Solar Collector—A Real Problem Linked to Sustainability and the Circular Economy. *Polymers* **2024**, *16*, 1012. https://doi.org/10.3390/polym16071012

Academic Editors: Cristina Cazan and Mihaela Cosnita

Received: 27 February 2024
Revised: 4 April 2024
Accepted: 6 April 2024
Published: 8 April 2024

1. Introduction

Polymeric materials can be repeatedly recycled and reused. Gu et al. [1] studied life cycle environmental issues associated with recycling plastic waste in China. Recycled raw materials were shown to be ecologically preferable to unprocessed materials, as manufacturing products with the latter has more environmental impact than those made from reclaimed material. Recycling generally makes high-value products from waste and significantly reduces environmental impact [2–6]. The use of these recycled polymers in the reinforcement of construction materials is one of the alternatives that has been most evaluated in some studies [7–15], concluding that they present improvements in properties, such as resistance to high temperatures, mechanical durability, resistance to moisture damage, less porosity, higher resistance to cracking, higher durability, among others.

L. Gu and Ozbakkaloglu [16] critically reviewed studies on using plastics in concrete [17,18] and found that recycled plastic fibers can improve their properties. Their review indicates that its use can contribute to a more sustainable construction industry and recommends studies focused on environmental aspects, such as the long-term behavior and ecological consequences of recycling this type of concrete after use.

Vila-Cortavitarte et al. [19] explored the benefits of substituting polystyrene for tar in the asphalt mix. They used three types of recycled polystyrene waste in the concrete mix: general-purpose polystyrene, high-impact polystyrene, and reinforced polystyrene. This work was mainly focused on reducing the concentration of tar since its carbon emissions can pollute the environment. The mixture evaluated in this investigation was asphalt concrete. The two replacement rates of tar with polystyrene were 1% and 2% in a mix, representing about 23% and 46% of the total amount of tar. The 2% mixture was discarded due to a cohesion problem. Each sample was subjected to a series of tests to compare the modified mechanical properties. A blend with 1% polystyrene showed promising results with improved mechanical properties. The results showed that substituting polystyrene for tar reduced the environmental problem and contributed to the life cycle evaluation.

Recycling plastic waste for the construction industry is considered one of the best methods of disposing of plastic waste. Chaukura et al. [20] concentrate on some proposals for potential uses for these wastes in the manufacture of adhesives, paints, artifacts, garden furniture, and wastewater treatment products, including mosquito control. Currently, the methods for recycling plastics are grouped into primary, secondary, tertiary, and quaternary [21–26]. Since the process followed in primary and secondary recycling is the same, both are usually called "mechanical recycling", which is frequently used [3,27]. It consists of two steps: separating the plastic from any other material and cutting it (sometimes, it is required to introduce it into an extruder to make pellets). Tertiary or chemical recycling seeks to recover the monomers or basic chemicals from the polymer. The most widely used techniques are flame, gasification, and liquefaction, although new technologies are also being developed to convert plastic waste into fuel [28,29]. Although this type of recycling is widely used, there are still many opportunities. Some works present separation analyses based on solvents or catalysts [30,31]. In general, solvent extraction separation includes removing impurities and some plastic additives, dissolution, and reprecipitation or devolatilization, which occurs when the polymers dissolve in the solvents and selectively crystallize each polymer. Selective dissolution can be used when the dissolution of the polymer of interest or all polymers except the target is feasible. The significant factor for the dissolution process is finding a selective solvent. However, several factors influence the dissolution of the polymer, such as composition, molecular weight, polymer structure, solvent composition, type, etc.

Dissolution is the most widely used technology, particularly in the automotive industry. However, it is estimated that only 20% of plastics can be recycled this way [32]. Regarding transforming polymers into hydrocarbons, Ragaert et al. [33] compiled state-of-the-art techniques, such as pyrolysis, chemolysis, hydrogen techniques, fluid catalytic cracking, and gasification. In addition, they discussed the main challenges and potential methods for this type of recycling. Finally, in quaternary recycling (known as energy recovery), as its name suggests, plastic waste is used as an energy source through incineration. Thakur et al. [34] mention that power generation by recycling plastic waste is currently one of the most eco-efficient methods, due to the massive process, and follows almost all strict emission standards and energy requirements. However, they point out that many hazardous chemicals, such as gases and dioxins, can evolve from incomplete solid waste combustion, leading to severe environmental pollution by carbon dioxide, toxic chemicals, and harmful ashes. For this reason, some incineration methods are being studied to propose those that reduce the emission of toxic substances [35,36].

Plastics have caused severe environmental pollution since they involve extracting petroleum and are thrown away. Even collecting and recycling them implies a highly energy-consuming process. Nonetheless, compared with using new raw materials, recycling is accepted as less harmful. The process of recycling plastics, even considering that there is an implicit deterioration per cycle, is an industry that follows more regulations. One of many factors in polymer environmental pollution is the extreme dependency that modern society has on them. The cost/benefit ratio for each person does not consider the collateral consequences of the items' post-use destiny. Sustainability and the circular economy precisely consider the post-use of plastic products through purge reutilization. We are using plastics at an extreme level, considering that they degrade oppositely to glass, metal (even corroded metal can be melted again), and ceramics.

It is important to note that when any of the methods mentioned above do not recover plastic waste, it is sent to landfills. This is more than a solution; it is a severe problem due to its reduced degradability and the ample space it occupies, together with runoff and leakage of microplastics (MP) or nanoplastics (NP) [37,38]. Huang et al. [39] mentioned that landfills could have increased plastic-degrading microorganisms, causing biodegradation that, together with environmental oxidation, become drivers for plastic degradation. They show that polymers have differences in their generation rates of microplastics, with polyethylene above polypropylene (PP) or polystyrene (PS). Also, they proposed the carbonyl index to quantify the polymer degradation as a function of the disposal age. So, biodegradation through biological treatments is a widely evaluated alternative. It proposes using enzymes and organisms that can digest plastic, trap carbon, and return it to the environment as a disposable resource [4,40–42].

Vimala and Mathew [43] tested Bacillus subtilis for its potential to use polyethylene as its sole carbon source. They discovered that some microbial species produce surfactant compounds (biosurfactants) that enhance the degradation process. Furthermore, UV-treated polymer films facilitated their viability to feed microorganisms, which increases biodegradation.

Restrepo-Flórez et al. [44], in their study of the microbial degradation and deterioration of polyethylene, show a comprehensive summary of the microorganisms that, according to some studies, participate in the biodegradation of polyethylene. They also present their effects on the properties of polyethylene and summarize the process of its degradation. They stated that, although slow, the speed can be modulated by the intensity and physicochemical factors, such as ultraviolet light or other oxidizing agents. Therefore, the biodegradation of polymers could be a cooperative process in natural ecosystems.

Nowak et al. [45] and Mohan et al. [46] present some proposals for the biodegradation of plastics with different types of microorganisms, achieving a weight reduction of up to 2.3%. In primary, secondary, and tertiary recycling, it is important that plastic waste be as small as possible. For this, it is necessary to cut and grind it. There is a residue from the cleaning processes of plastic injection machines known as "purging" that cannot be reintegrated into the recycling chain. These are usually a mixture of polymers with some reinforcements. They have a paste consistency of size and hardness that is not possible to grind in conventional equipment due to the cost of wear and replacement of the blades, as well as the energy required for the motors. Currently, recycling companies only receive this type of material, classify it, store it temporarily, resell it, or export it. It should be noted that due to the management described, the profit is minimal, and there tend to be losses.

Derived from international agreements for the import of waste, companies have limitations for the export of purges. This generates an accumulation within the company's facilities, with all the problems that this entails, including the lack of space, the costs of movement of the material, the prices of containers, and the severe danger due to the risk of combustion.

Some landfills are authorized to receive this type of material. However, this implies an additional environmental and economic cost to the expense already made. Also, it can only be made in restricted quantities. Likewise, various national and international technical standards describe the safety, prevention, and protection conditions for storing the material due to fire risk in the workplace, which determines having a maximum volume. Regarding quaternary recycling, there is no information on the environmental aspects that the incineration of the purge material brings. In any case, it would be necessary to characterize the material and study its emissions. Any contribution added to the recovery of polymeric waste represents an economic and environmental gain.

The case presented in this study seeks to reintegrate purges into the recycling chain through their characterization, evaluation of the conditions to reduce their size, equipment design, and softening and cutting processes that use renewable energy sources, taking care that the change processes do not degrade the material and do not significantly impact the purges that they cannot be reintegrated.

2. Materials and Methods

Purges are discarded materials in an injection process due to the cleaning carried out when there is a change in the type of part produced. The equipment is cleaned to guarantee the production of plastic objects without contamination of the previously used material, causing the residue called purge. The traditional way is to gradually introduce the new material, dragging the previous one until it is eliminated. This form of purging results in a mixture of more than one polymer, fillers, or additives with irregular sizes and shapes, some reaching up to around 30 cm per side. Figure 1 shows some purges.

Figure 1. Purges of injection processes collected from industrial sources.

This material forms with a paste consistency that solidifies in such size and hardness that it is not possible to grind it in conventional equipment. It is nearly impossible to cut purges. It is costly in terms of labor hours, security, materials, tools for use, rotary cutter, different saws (band saw, hacksaw, jigsaw), CNC router, grinder, laser cutter, waterjet cutter, plasma cutter, etc. So, exploring alternatives for industrial applications, especially renewable energies, is desirable.

2.1. Selection and Characterization

Five of the most frequent purges in the company Reciclajes Victoria (Queretaro, Mexico) were selected. Although all samples were from the same waste management company, they originated from different plastic processing companies. Intrinsically, purges are made of a type of polymer with or without a dye that changes from piece to piece. That condition makes the properties change from one piece to another.

First, they were evaluated for how they behaved when trying to soften them and whether it was possible to cut them. Afterward, a series of characterization tests were carried out, such as flame tests, hardness, thermogravimetric analysis (TGA), Fourier transform infrared spectroscopy (FTIR), and resonant frequency damping analysis (RFDA) (used to calculate Young's modulus). Figure 2 shows the test equipment or instruments for (a) flame, (b) hardness, and (c) RFDA.

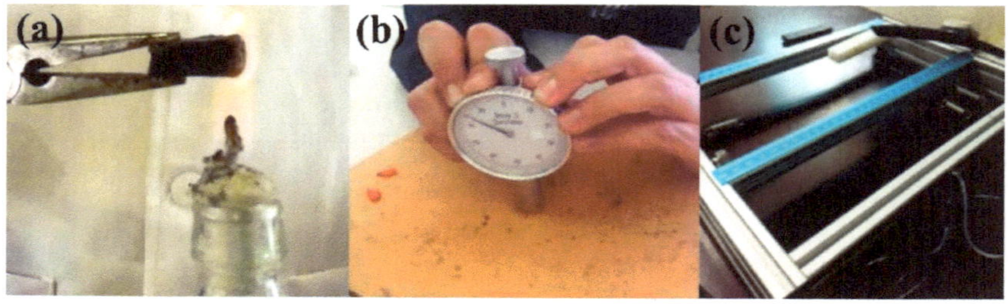

Figure 2. (**a**) Flame test, (**b**) hardness Shore D test, and (**c**) RFDA setup.

For the flame test, there is no standard as a reference. Therefore, it was carried out by exposing a piece of a purge of approximately 1 cm^2 to the flame of an alcohol burner, observing and taking note of all the changes in the material for about 3 min (Figure 2a).

The hardness is evaluated according to the method indicated in the ISO 868 standard [47], with a portable Shore D hardness tester TT 0.90 HD 0.25 mm. The measurement value is taken after 5 s and the result is expressed as "HSD hardness Shore D" (Figure 2b).

The Young's modulus was determined by the impulse excitation technique (IET) using an IMCE brand RFDA equipment. The samples for the RFDA analyses were obtained by cutting the purges using a band saw and polishing them to obtain smooth surfaces for rectangular parallelepiped pieces.

First, the bending vibration frequency was measured and Young's modulus was calculated using the mass and dimensions of the sample according to different measurements and standard ASTM E1876-22 [48]. Next, the samples are mechanically struck by hand with a small flexible hammer. Then, the induced vibration signal is detected with a USB microphone (10 Hz–16 kHz), and later, the elastic properties are calculated using the software RFDA-MF basic v1.2.0. The calculation basis is the following relationship:

$$E = 0.9465\left(\frac{mF_f^2}{w}\right)\left(\frac{L^3}{t^3}\right)T \quad (1)$$

where, E is Young's modulus (GPa), F_f is the bending frequency, m is the mass (g), w is the width (mm), L is the length (mm), t is the thickness (mm), and T is a correction factor. Figure 2c shows the experimental setup with the device bearing one sample.

Thermogravimetric analysis (TGA) was carried out to know the temperatures at which the material undergoes some degradation, so that the material does not exceed this temperature when softening. The test is carried out on a TA-Instruments (New Castle, DE, USA) SDTQ600 unit. The record produced by the analysis is a curve representing a mass variation as a function of temperature. The norm that was taken as a reference was ISO-11358 Plastics [49]. The DSC/TGA analyses were conducted using polymeric powders extracted from the purges. A hand-held electric drill was used to obtain a powder from each type of polymeric purge. The powder was treated using a Krups coffee grinder with stainless steel blades. The powder sieving was performed using a No. 400 mesh (38 µm).

Fourier transform infrared spectroscopy FTIR was carried out to determine the type of bonds and functional groups present, considering that the purges are a mixture of more than one polymer. The samples were analyzed using Bruker (Ettlingen, Germany) Tensor 37 FTIR equipment, using an ATR, in transmittance mode with 32 scans and a resolution of 1 cm^{-1}. Plates of the purges (1 × 1 cm) were cut into different sections of each purge sample. They were analyzed with FTIR with coupled ATR, obtaining several spectra (1 averaged from 30 spectra per point) at different sections of the purge, and the spectra with the highest occurrence were presented.

Except for flame test, all the other tests were carried out before and after softening the purges with heating. The conditions were determined through a series of tests on samples of size around 10 × 5 × 2 cm to determine the time and temperature at which the material softens. When this happens, the pieces are cut to determine that the cut can be made easily in softening conditions. The heating was carried out in a model FE340 muffle (Felisa, Fabricantes Feligneo S.A. de C.V., Zapopan, Mexico), voltage 120 Vac, power 1500 W, frequency 60 Hz.

2.2. Determination of the Processes, Design, and Manufacture of the Prototype to Soften and Cut the Purges

Using the information obtained in the characterization, the Six Sigma methodology is used to design and manufacture the prototype and the processes. Several analysis tools were used to ensure the correct design: Functional Block Diagram "DBF", diagram "Diagram "P", Design of experiments "DOE", ANOVA, Hypothesis Test T, and others. The software used for data analysis were SolidWorks® V2018 Modeling software and MINITAB V14 for statistical data analysis.

3. Results and Discussion

3.1. Higher Frequency Purges and Flame Test

Table 1 shows the most frequent purges in Victoria recycling, their content, description, and observations made during the flame test.

Table 1. Some representative industrial purges from polymer injection.

Polymers	Physical Description	Observations	Image
PP-ABS	Gray with many marked lines	The material quickly deforms, the flame is yellow and blue, bubbles, and becomes rubbery but does not drip, and carbonizes in its walls. As a result, it can be stretched and makes almost no smoke.	
PC-ABS-PP	Two shades of black (one glossy and one opaque)	Quickly begins to burn and deform. The flame is yellow, it causes fumes of black smoke, and has a marked odor. It becomes rubbery, but drips cannot be stretched and carbonized.	
Nylon 66 yellow	Yellow with many marked lines	It takes time to begin to burn and change its form. Then it bubbles, causes a small amount of fumes, the flame is yellow and blue, becomes rubbery but does not drip, can be stretched, and gets carbonized at the end.	
Acetal (polyoxymethylene, POM)	White with many grains	Quickly begins to burn and deform. The flame is yellow and blue, and it causes a small amount of fumes. It has a marked odor. It bubbles a lot but does not drip, can be stretched, becomes rubbery, and gets carbonized at the end.	
Nylon 66/black with fillers	Spaghetti-shaped black	It takes time to begin to burn and change its form. It bubbles but does not drip, making almost no smoke. The flame is yellow and blue, hardly stretched, and nearly no carbonization at the end.	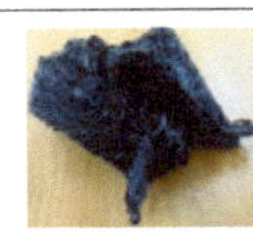

The most representative purges in Victoria recycling were polypropylene/acrylonitrile butadiene styrene (PP-ABS), polycarbonate with ABS and polypropylene (PC-ABS PP), yellow Nylon 66, white acetal (polyoxymethylene, POM), and black Nylon 66 with fillers: their main physical characteristics are described. The separation of the purges that contain mixtures of materials is not always evident.

The flame tests were conducted to determine the general behavior of the material when heat was applied. Table 1, in its third column, shows the observations made. It is noticeable that all the polymers retain their thermoplasticity.

3.2. Softening and Cutting Initial Tests

The temperature at which each type of purge can be softened was determined, and the temperature was varied to evaluate the status of the materials at different times. Table 2 shows the results obtained by maintaining a constant temperature. Once the material softened, the required time was identified, it was removed from the muffle, and some observations were made about the behavior of the purge related to the cutting process.

Table 2. Softening and cutting tests.

Polymers	Weight (g)	Temperature (°C)	Time (min)	Observations
PP-ABS	16.7	250	20	It can be easily cut, and it can be stretched, but it tends to become rubbery.
PC-ABS-PP	52.99	250	20	It can be easily cut and stretched.
Nylon 66 yellow	6.64	250	20	It cools down very fast and hardens. It can be stretched. Changes to a dark color. It is relatively difficult to cut and has a sandy consistency.
Acetal (polyoxymethylene, POM)	29.88	300	45	It cools down very fast and hardens. It can be stretched. Changes to a dark color. Difficult to cut.
Nylon 66/black with fillers	19.63	300	50	It cools down very fast and hardens. It can be stretched but has a dusty consistency.

Table 2 shows that the time for the purge containing nylon is similar to those with a higher amount, even when the quantity is less. This is because this material takes longer to soften. However, the polymers softened and cut easily in most cases.

3.3. Process Design and Equipment

Although the characterization tests were carried out before designing the processes to soften and cut the purges, the results are shown afterward to highlight the differences once the purges have been cut.

The overall design goal is to create processes and devices to soften and diminish the sizes of polymer purges using renewable energy sources, without exceeding the maximum temperature to avoid degradation of the polymers and keeping the design as simple as possible to facilitate its operation.

Figure 3 is a diagram, or process map. It shows the necessary processes, functions, and interactions to achieve the design objective.

Figure 3 is a process scheme showing the necessary stages, functions, and interactions to achieve the design objective. The three stages of the process identified are as follows.

(1) Heat collection system ("solar collector"). The purpose is to convert solar radiation into heat.
(2) Softening zone. The purpose is to soften the purges and maintain the temperature reached for this purpose.
(3) Cutting. Once the purges have been softened, the purpose of this last stage is to cut them off.

The design of each of these stages is explained in the following sub-sections.

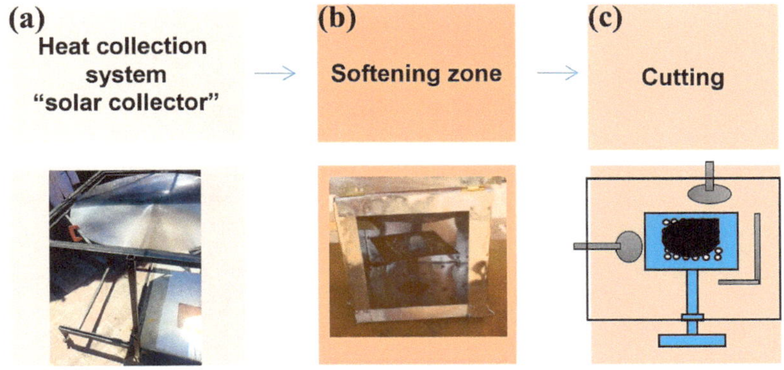

Figure 3. Three stages in softening plastic purges: (**a**) Concentrated solar power collector, (**b**) Heating and softening chamber, (**c**) Segmenting the pieces.

3.4. Heat Collection System ("Solar Collector")

3.4.1. System Modeling and Construction

The types of solar collectors were analyzed, and according to the temperature required to soften without degrading (180–280 °C), a Fresnel lens solar concentrator was used. The auxiliary components are the lens and the support. Table 3 shows the requirements and the mechanism to achieve them. Figure 4 shows the final design.

Table 3. Requirements and the mechanism to achieve them.

Requirements	Answer
Increase or decrease the diameter of the focus	Height adjustment mechanism
Adjusted to the position of the sun	Angle adjustment mechanism
A softening zone	Sample tray
Easy to move	Drive wheels and brake
No risk of the lens falling off	Padded retainer
Samples can be inserted and removed without risk of burns	Handle—endless screw system or rails

Figure 4a shows the three stages of the process (the cutting one is inside the softening zone). Figure 4b shows the attachment designed for the solar collector. The pieces that compose it are the following.

(1) Lens support consists of two frames. The inner frame supports the lens and modifies its inclination angle, referencing the position perpendicular to the height posts. The outer frame gives the lens sun-tracking capabilities.
(2) The design comprises two lateral posts, each comprising two rectangular tubes (one inside the other) that allow movement to achieve different lens heights relative to the sample. In addition, adjustment holes are spaced one inch apart for a total of twenty-six. Thus, a wide range of heights can be achieved.
(3) Plate where the softening zone was located.
(4) This plate incorporates a manual sliding system that allows it to be removed from the focus for user safety.
(5) The device supports have swivel wheels with a braking system that moves and stops the device.

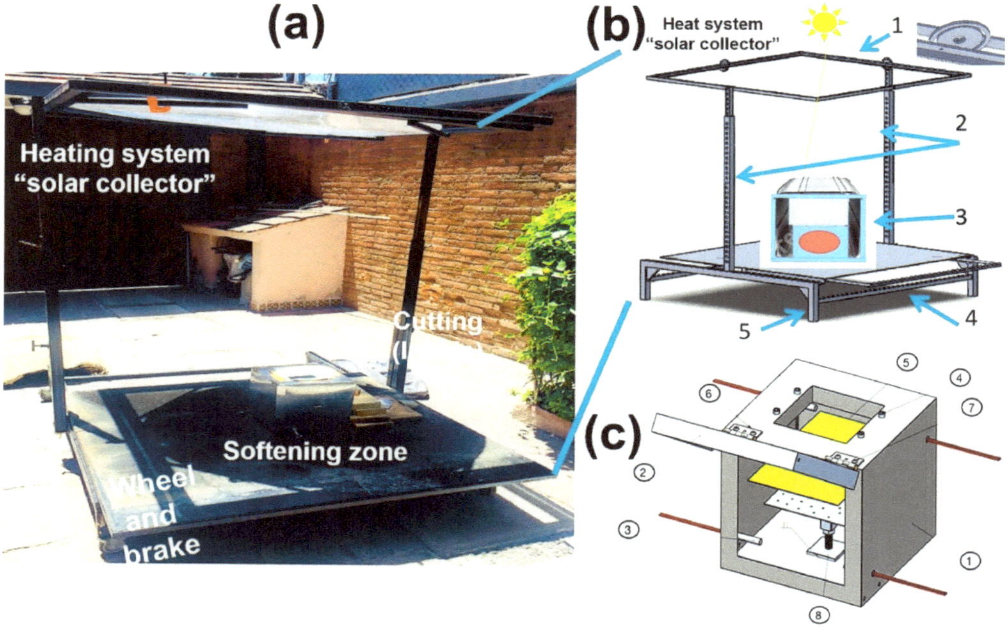

Figure 4. Design of the complete device. (**a**) Solar collector indicating the corresponding parts, (**b**) the collector subsystem, and (**c**) the softening zone.

3.4.2. Softening Zone

The first thing that was determined for this thread was that the radiation would not be directly on the purge due to the solarization of the polymers (breaking of polymeric chains by high-intensity aces) and the difficulty in controlling the temperature. This implies the need for a means to move the heat achieved by the collector toward the softening zone. In the state of the art, the use of added thermofluids with carbon particles and air as heat transport was found [50–54]. This alternative was chosen due to the ease of its implementation. The study for the design of this process was carried out in stages.

Figure 4c shows the proposed heating system related to an electrical furnace. There is an internal zone in which to put the purge (cubic box). Instead of electrical resistances, they were substituted by a receptor plate for solar radiation. An internal fan was used to better distribute heat. Some other factors were addressed, such as the material and dimensions of the box, fan position, material, size, and position of the plate. Considering the required characteristics, the chosen box and plate materials were aluminum and copper, respectively. The absorbing plate could have many variants, such as using graphite, selective absorbing coatings of black nickel [55,56], black copper [57], black cobalt [58], and graphene oxide [59,60], among others.

Figure 4c shows the proposed system for heating and cutting the purges. The components are #1 cab, #2 cab cover, #3 thermocouple, #4 m6 screws, #5 support plate, #6 hinge, #7 m5 screws, and #8 base. The absorbing plate was on the box (#5). This was fixed using four screws (#4) that adjust the distance between the plate and the plastic purge. The base for the purge (#8) rotates and has holes that allow air to flow throughout. The red lines (#3) indicate the placement of thermocouples to sense temperature in four zones. The box had a door or lid for insulation.

Once the softening zone was manufactured, its performance was evaluated, and adjustments were made. This was carried out in two stages: (1) In conjunction with the solar collection system, (2) In evaluating the most relevant components. The focal point was circle–elliptically shaped with a diameter of about 10 cm, which means an area of about 78.54 cm^2. The local ambient conditions were in the range of 20–27 °C and the relative humidity was about 35–45%.

Correlation studies were carried out between two factors to evaluate the functionality and interaction of the solar collection system and the softening zone. Study (A) investigated the relationship between the distance between the Fresnel lens and the receiving plate (height X), and the diameter of the radiation focus on this plate (Y_1). Study (B) investigated the height (X) and the temperature within the softening zone or base of the samples (Y_2). Although a relationship is expected between these variables X vs. Y_1 and X vs. Y_2, the objective in both studies is to find the mathematical function Y = f(x) that models this relationship and find at which value of X the highest temperature is reached, understanding that this also depends on weather conditions and exposure time (as long as it has an R factor higher than 80%).

Table 4 shows the results obtained. The experiment conditions were the following: ambient temperature: 29 °C, relative humidity: 34.8%, pressure: 1023 hPa, solar power: 1350 W/m^2, wind speed: 29.6 km/h, exposure time: 1 min, receiver material: copper plate, independent variable: X = height, dependent variable: Y_1 = diameter of the focus, Y_2 = temperature. The results were analyzed with the Minitab V14 software.

Table 4. The correlation results.

Height (cm)	Diameter (cm)	Temp. (°C)
71.5	52	40
83.3	45	60
95.3	38	80
107	28	200
119.5	22	200

Figure 5 shows the graphic analysis of the height–temperature correlation and height analysis vs. diameter (X vs. Y_1).

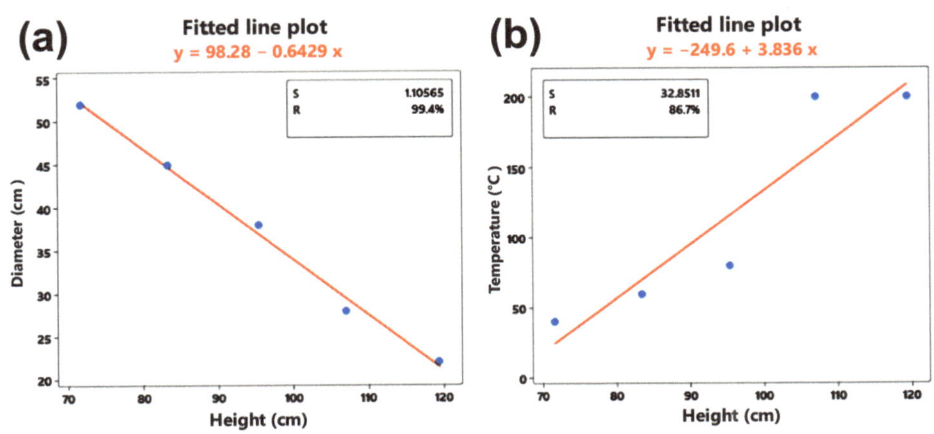

Figure 5. Correlation between (**a**) height and diameter and (**b**) height and temperature.

Figure 5a shows that the greater the height X, the diameter of the radiation focus decreases linearly. Therefore, the mathematical model is valid since the relationship has a factor higher than 80% with an adjustment value of R = 99.2% and a variance of 1.10. Diameter = 98.28 cm and height [cm] = 0.643 cm. Table 5 shows the ANOVA analysis of variance.

Table 5. ANOVA for the Height–Temperature calculus and the Height vs. Diameter.

Source	GL	MC	F	P	GL	SC	MC	F	P
Regression	1	21,082	19.54	0.022	1	592.333	592.333	484.54	0
Height [cm]	1	21,082	19.54	0.022					
Error	3	1079			3	3.667	1.222		
Total	4				4	596			
Source	GL	MC	F	P	3	3.667	1.222		

Considering the value of p, which is less than the value of the statistical error $\alpha = 0.05$, $p = 0.022 < \alpha = 0.05$, it is concluded that the temperature Y_2 depends directly on the height X.

Figure 5b graphically shows the correlation between X and Y_2. It shows that the higher the height X, the more the temperature increases linearly. Therefore, with an adjustment value of R = 82.3% and a variance of 1.10, the mathematical model is valid since the relationship has a factor greater than 80%. Temperature = 249.6 °C and height [cm] = 3.836 cm.

Using the prediction equation, the temperature-maximizing height is about 120 cm. According to the holes that regulate the height, an elevation of about 122 cm was used in tests.

3.4.3. Evaluating the Most Relevant Components of the Softening Zone

Two relevant elements are observed for the scope and temperature control: (1) The fan use. (2) The material of the radiation receptor plate.

Through a design of experiments, these factors are evaluated, determining the impact on the temperature reached in four points of the box and finding the condition that maximizes the temperature in the position of the base or central position (that shortens the time to soften the purge). Table 6 shows the DDE approach.

Table 6. Approach to the design of experiments.

Factors	2	Base design	2, 4	Factor	Level +	Level −
Test runs	8	Replica	2	Fan	With	Without
Blocks	1	Center points (total)	0	Absorbing material	Copper	Steel sheet

Table 6 shows the approach introduced in the Minitab V14 software, DDE: two factors (fan and receptor material), two replicates, without central points, and all in a single block, giving ten experiments. The experimental conditions were as follows. Ambient temperature: 18 °C, relative humidity: 63.6%, pressure: 1025 hPa, solar power: 952 W/m^2, wind speed: 0 km/h, exposure time: 2 min, X_1 = fan, evaluating using it or not, X_2 = material of the absorbing plate, evaluating a copper or steel sheet. The dependent variables were as follows. Y_1 = Upper right position temperature, Y_2 = Central position temperature, Y_3 = Upper left position temperature, Y_4 = Lower left position temperature, Y_5 = Lower right position temperature.

Figure 6a shows the position of the thermocouples on one of the sides and the center (the other side has the same position), and Figure 6b shows the fan.

Table 7 shows, in the "Test run" column, the order of the experiments (random order), while the "Fan" and "Receiving sheet" columns show the conditions under which the corresponding experiment is carried out. The Y_1–Y_5 columns correspond to the temperature °C obtained under each evaluated condition. Table 7 shows the analysis of variance (ANOVA) for Y_1 = Upper right position temperature.

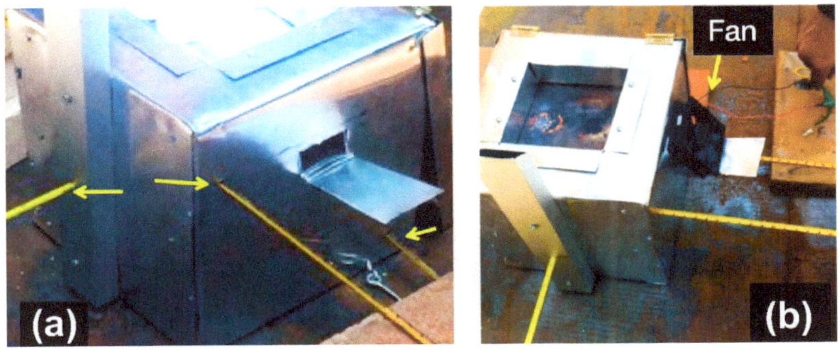

Figure 6. (**a**) Location of the thermocouples (indicated with yellow arrows). (**b**) Location of fan.

Table 7. Shows the test plan with the results obtained.

Test Run	Fan	Receiving Sheet	Y_1	Y_2	Y_3	Y_4	Y_5
1	No	galvanized	74	98	79	84	90
2	Yes	galvanized	74	78	72	72	82
3	No	copper	110	130	104	90	106
4	Yes	galvanized	76	76	76	80	90
5	Yes	copper	92	80	80	72	80
6	Yes	copper	90	82	80	74	80
7	No	copper	120	130	110	88	100
8	No	galvanized	86	104	86	94	100

Table 8 shows the ANOVA analysis of variance. Considering the value of the p statistic, Fan $p = 0.022$, receiver material $p = 0.003$, they are less than the statistical error $\alpha = 0.05$, with which it is concluded that both variables influence linearly the temperature obtained in position Y_1. The interaction of these $p = 0.075$ is above the error; therefore, there is no interaction. The same ANOVA analysis was performed for each thermocouple position (Y_2–Y_5) and the results are summarized in Table 9.

Table 8. ANOVA for Y_1.

Source	DF	Adj SS	Adj MS	F-Value	*p*-Value
Model	3	1901.5	633.83	20.12	0.007
Linear	2	1721	860.5	27.32	0.005
Fan	1	420.5	420.5	13.35	0.022
Receiving sheet	1	1300.5	1300.5	41.29	0.003
2-Way Interactions	1	180.5	180.5	5.73	0.075
Fan * receiving sheet	1	180.5	180.5	5.73	0.075
Error	4	126	31.5		
Total	7	2027.5			

* Interaction between Fan and Receiving sheet.

Table 9. Summary of temperature results in the thermocouples' (Y's) positions.

Y	Fan X_1	Material X_2	Interaction	X with More Influence	Best Combination X_1	X_2
Y_1= Upper right position temperature	Yes	Yes	No	Material	Without	Copper
Y_2 = Central position temperature	Yes	Yes	Yes	Fan	Without	Copper
Y_3 = Upper left position temperature	Yes	Yes	Yes	Fan	Without	Copper
Y_4 = Lower left position temperature	Yes	No	Yes	Fan	Without	Same
Y_5 = Lower right position temperature	Yes	No	No	Fan	Without	Same

The first column, Y, indicates the thermocouple position. The second to fourth indicate if the X factor, or the interaction of these, influences the results of the Y. The fifth indicates which of the X has the greatest influence. The last two columns indicate the combination with which the highest temperature is obtained.

In conclusion to this analysis, we can say that in the central and upper zones, the temperature is affected by the type of receiving material and the use, or not, of the fan; in the lower part, it only affects the use, or not, of the fan. To evaluate the configuration that maximizes the temperature, only position Y_2 is considered; that is, the central area where the sample is placed. The result is "Copper receptor material" and "Without fan".

3.5. Hardness Testing

Thirty hardness measurements were made in different areas of the purge to determine material homogeneity before and after material softening. The t-student statistic test was applied with a 95% confidence level and a hypothesis approach. Null hypothesis H0: $\mu 1 - \mu 2 = 0$, alternative hypothesis H1: $\mu 1 - \mu 2 \neq 0$.

The hypothesis test is carried out with the test statistic of the *p* value to evaluate if there is a statistical difference in the mean hardness values before and after the heat treatment of the sample. An error value of 0.05 is considered to accept or reject the null hypothesis, assuming normal data for this number of measurements. On the other hand, the Levene statistical test is applied to evaluate the impact on the variation in the results obtained in both conditions for each of the purges. Table 10 shows the results obtained and the conclusion of the statistical analysis.

Table 10. Hardness analyses (Shore D) of five polymer purge types with thirty measured samples.

Polymers	Average Hardness before Heating	Average Hardness after Heating–Cooling	Conclusions of the Average Analyses	Conclusions of the Variance Analyses
PP-ABS	59.03	60.6	*p*-value = 0.149 No change	*p*-value = 0.115 No change
PC-ABS-PP	71.33	63.23	*p*-value = 0.00 Different after the heat decreases	*p*-value = 0.591 No change
Nylon 66 yellow	69.53	62.63	*p*-value = 0.001 Different after the heat decreases	*p*-value = 0.364 No change
Acetal	66.37	68.67	*p*-value = 0.005 Different after the heat increases	*p*-value = 0.218 No change
Nylon 66/black with fillers	59.8	64.43	*p*-value = 0.00 Different after the heat increases	*p*-value = 0.701 No change

In the conclusions of the averages analysis, we can see no changes in PP-ABS, PC-ABS-PP, and Nylon 66 yellow. Their hardness decreases after softening, while Acetal and Nylon 66/black with fillers hardness increases. Variation does not change in any of the samples. Figure 7 shows how the data are distributed before and after softening each purge type.

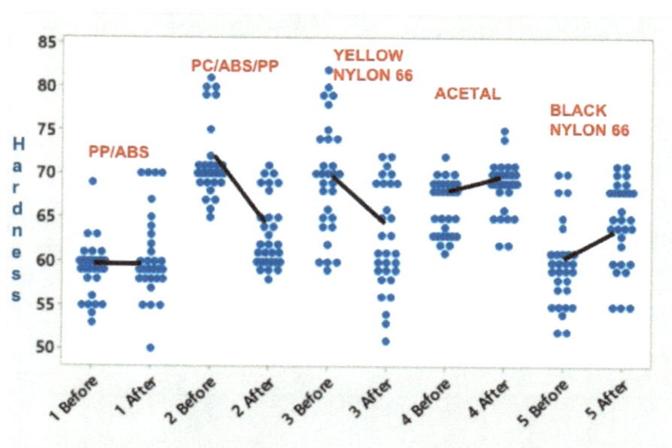

Figure 7. Distribution and averages of hardness values before and after softening.

Figure 7 shows that the data have an equal distribution before and after the softening of the material. There was a slight difference between the means in all purges.

3.6. Young's Module

Table 11 shows the results of the measurements made before and after softening the material.

Table 11. Young's modulus (GPa).

Polymers	YM before	YM after	Comparative
PP-ABS	3.88 (±0.02)	3.17 (±0.01)	↓ 18%
PC-ABS-PP	2.38 (±0.001)	1.70 (±0.001)	↓ 28%
Nylon 66 yellow	3.94 (±0.01)	4.28 (±0.01)	↑ 9%
Acetal (polyoxymethylene, POM)	2.45 (±0.001)	3.04 (±0.01)	↑ 24%
Nylon 66/black with fillers	3.08 (±0.001)	3.45 (±0.005)	↑ 12%

Arrow down: YM decreases, Arrow up: YM increases.

Table 11 shows the average of three data values taken for each purge before and after softening and cutting. The last column indicates with an arrow if the average increased or decreased, and by what percentage. For example, in PP-ABS, Young's modulus decreases by 18% after smoothing and cutting. In general, considering that the purges are a recycling material, the variation is slight in all the samples. Figure 8 shows how the test device reports results.

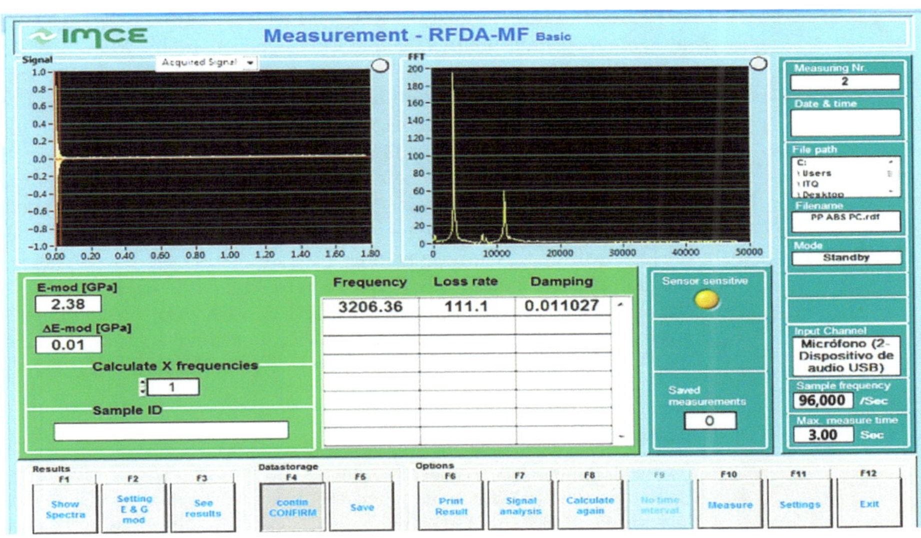

Figure 8. RFDA test set results screen.

Figure 8 shows the graph of the noise signal in the upper part. The lower left part shows Young's modulus. The rest of the information shows the data the software uses for the calculation. Figure 9 shows the results in a bar graph.

Figure 9 shows the results in a bar graph, showing the difference between the average values before and after softening and cutting each purge.

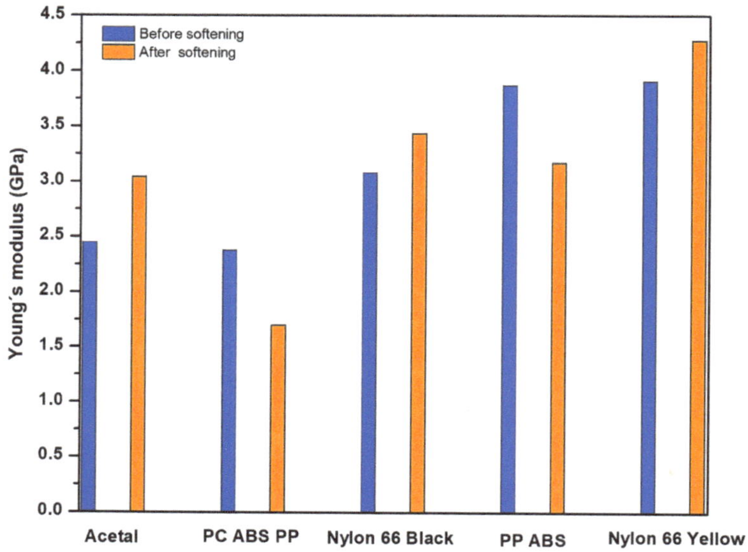

Figure 9. Comparison of Young's modulus average before and after softening.

3.7. DSC and TGA Analyses

The glass transition temperature (T_g) for ABS typically ranges from 105 to 125 °C, depending on the proportions of the three components and the molecular weights of the polymeric chains. Variations in the T_g depend on specific formulation and processing conditions in any polymer. Still, they are even more frequent in industrial polymer purges that could have plasticizers, colorants, or broader molecular weights.

The T_g of PC is in the range of 145–150 °C. Nylon 66, PP, and Acetal (polyoxymethylene) are semi-crystalline polymers in which the T_g is the transition from a glassy to a rubbery state in the range 45–50 °C, −10 to −5 °C, and −50 to −80 °C, respectively.

DSC/TGA calorimetry analyses were performed to determine the maximum temperatures for heating. As long as neither the DSC graphs show irreversible events, nor the TGA graphs show weight loss, heating can be conducted while maintaining the polymer's integrity and original polymer characteristics. For example, in Figure 10a PP ABS, Figure 10b Acetal, Figure 10c Nylon 66 black, Figure 10d Nylon 66 yellow, and Figure 10e PC ABS PP, the proposed maximum temperature could be about 245, 240, 250, 330, and 270 °C, respectively.

3.8. FT-IR Analyses

The five polymers have functional groups with characteristic vibrational bands in their FT-IR spectra (Figure 11). Some of the representative vibrational bands are outlined below.

Polypropylene (PP) has C-H stretching and bending vibrations and C-C stretching in the ranges of 2800–3000, 1350–1470, and 960–970 cm^{-1}, respectively. Also, PP has bands related to its isotactic configuration, such as isotactic CH_3 bending vibrations in the region of 875–885 cm^{-1} (Figure 11a,e).

The terpolymer ABS (Acrylonitrile–Butadiene–Styrene, 3–7: 1–6: 8–12) has a C-H backbone with stretching and bending vibrations in the 2900–3000 and 1300–1400 cm^{-1} (Figure 11a,e) [61–63] regions. Acrylonitrile cyano group's C≡N stretching vibrations are around 2240–2260 cm^{-1}. Butadiene C=C stretching and CH_2 deformation vibrations are in the regions of 1660–1600 and 1450–1375 cm^{-1}, respectively. Styrene vibrations of the aromatic ring deformations are in the range of 700–750 cm^{-1}.

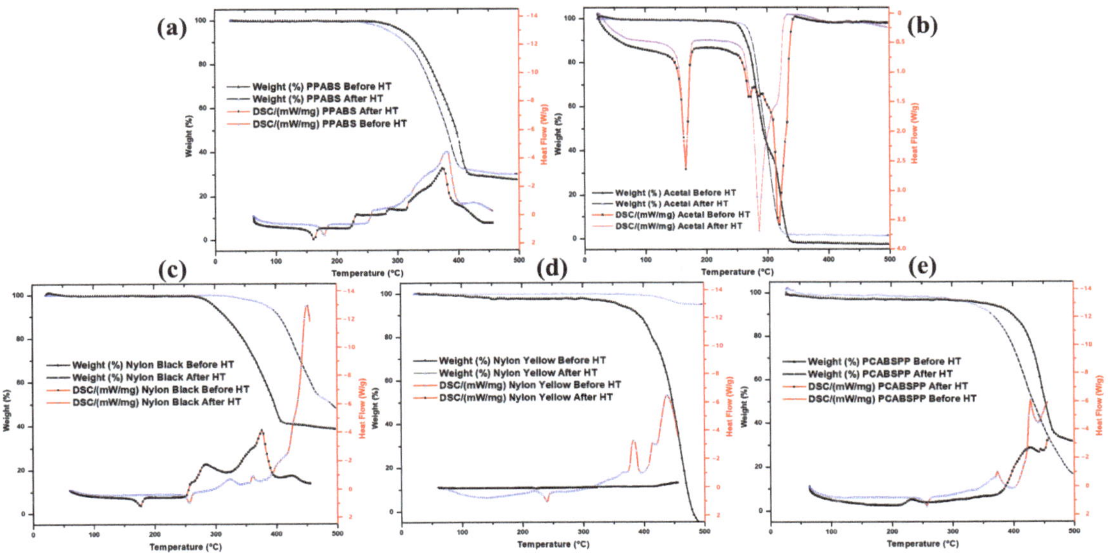

Figure 10. DSC and TGA analyses for the five polymers studied, showing their graphs before and after the heat treatments by indirect sunlight heating (**a**) PP ABS, (**b**) Acetal, (**c**) Nylon 66 black, (**d**) Nylon 66 yellow, and (**e**) PC ABS PP.

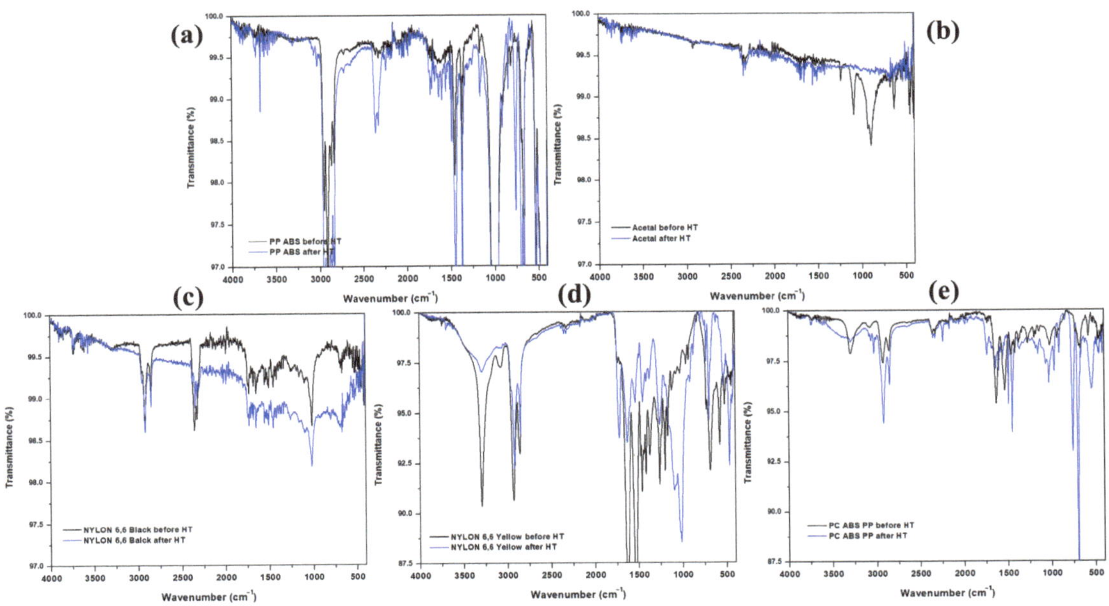

Figure 11. FT-IR spectra for the five polymers studied, showing their vibrational bands before and after the heat treatments by indirect sunlight heating (**a**) PP ABS, (**b**) Acetal, (**c**) Nylon 66 black, (**d**) Nylon 66 yellow, and (**e**) PC ABS PP.

Acetal (polyoxymethylene) has C-H, C=O, C-O, and O-CH_2-O stretching vibrations in the regions of 2900–3000, 1720–1740, 1050–1150, and 1150–1250 cm^{-1} associated with CH_2 groups, acetal groups, polymer backbone, and ether linkage, respectively (Figure 11b).

Nylon 66 (polyhexamethylene adipamide) has N-H bending vibrations and N-H, C=O, and C-N stretching vibrations. There are bands of Amide A, Amide B, Amide I, Amide II, and Amide III in the 3200–3500, 1550–1650, 1620–1660, 1500–1550, and 1200–1300 cm^{-1} ranges, respectively (Figure 11c,d).

Polycarbonate (PC) has the C-H stretching and carbonate ester stretching and bending vibrations in the ranges of 2900–3100, 1740–1760, and 700–800 cm^{-1}, respectively (Figure 11e).

The solar-assisted thermal softening treatment performed in the polymers decreased the intensity of some vibrational bands. The proposed treatment is intended to affect the functional groups and the polymeric chain lengths as little as possible. That was the reason for conducting the DSC/TGA calorimetry analyses shown above to determine the limit temperatures for heating. The heating can be carried out and repeated as long as the original polymer characteristics can be maintained barely unchanged.

4. Conclusions

Plastic injection purges have many sizes and irregular shapes that require reductive changes to be recycled and reintegrated into the polymer supply chains. For this, it is important to know their characteristics, properties, behavior, and grouping by polymer type. Frequency injection purges are blends of processed polymers that are difficult to recycle. This work proposes a process and a prototype device for softening and cutting the injection purges using solar heating. In this study, the starting tested samples were about 30 × 25 × 20 cm in size, and the final size was set to about 10 × 5 × 2 cm.

One of the problems with purges is the combination of materials, which does not allow incorporation into the manufacturing process. Other issues are the complex shapes and properties of the purges, which reduce the possibility of re-insertion into the process. The softening of the material and the feasibility of cutting the pieces with minimal impact on the polymer structure do allow the circularity of the material. There is less waste of materials with reintegration into production, which is the basis of the circular economy.

The thermal properties were evaluated using the flame test, TGA, and DSC techniques. The conclusive results are that none of the purges leak during heating, the five types of studied polymers can stretch, and their softening temperatures are between 250 °C and 280 °C. Once all the evaluated purges softened, they were cut relatively easily. Visually, no smoke was observed when reaching a soft condition. Using this information, a process and a device that operates with solar energy were designed to cut the purges without exceeding the maximum temperature, which varied depending on the type and quantity of polymers. Finally, the mechanical properties of hardness and Young's modulus were evaluated, and the FTIR tests were carried out before and after softening and cutting. Therefore, any change in the polymers could be related to reticulation or chain breakage, which are not associated with reaching stability but with deterioration. Knowing the calorimetry graphs, the maximum limit for the temperature was proposed for each polymer where the TGA graph deviates from a horizontal line. PP ABS, Acetal, Nylon 66 black, Nylon 66 yellow, and PC ABS PP could have, as proposed maximum temperatures for softening and cutting into smaller pieces, about 245, 240, 250, 330, and 270 °C, respectively.

Comparisons of the mechanical properties of the purges before and after treatments showed slight affectation on most of them. Generally, the main functional groups did not change. Further improvements are advisable in both the process and the device. Still, this work shows potential viability for solar-assisted heating for softening and cutting injection purges that constitute a fast-increasing residue of the plastics industry, making them a byproduct to be reprocessed.

This work differs from other research or industry solutions because it aims to soften the pieces to obtain smaller, specific-size elements rather than use them as fillers, composites, or recycling by reprocessing to obtain products such as bottles or fabrics. Considering that the main concern is keeping polymer integrity but softening it, this process can be extended to most organic polymers. Nonetheless, the size of the pieces is important for the system characteristics. Previous experiences advise specializing such a system for a single type of polymer, but if this is not possible, a technical procedure could be elaborated to change the temperature and keep the process clean when changing the materials to be processed.

In the proposed system, each part has a different expected lifespan. This largely depends on factors such as materials, process operation ranges, maintenance, surrounding climatic conditions, and the observation of safety rules during process operation, among others.

Author Contributions: Conceptualization, M.L.M.L.; methodology, M.G.P., M.L.M.L., J.d.J.P.B., J.P.M. and A.X.M.P.; validation, M.G.P., M.L.M.L. and A.X.M.P.; formal analysis, M.G.P., M.L.M.L., J.d.J.P.B. and J.P.M.; investigation, M.G.P., M.L.M.L., J.d.J.P.B. and J.P.M.; resources, M.G.P., M.L.M.L. and J.d.J.P.B.; data curation, M.G.P., M.L.M.L., J.d.J.P.B. and A.X.M.P.; writing—original draft preparation, M.G.P., M.L.M.L., J.d.J.P.B. and J.P.M.; writing—review and editing, M.G.P., M.L.M.L., J.d.J.P.B. and J.P.M.; visualization, M.G.P., M.L.M.L. and J.d.J.P.B.; supervision, M.G.P. and M.L.M.L.; project administration, M.G.P., M.L.M.L. and J.d.J.P.B.; funding acquisition, M.L.M.L. and J.d.J.P.B. All authors have read and agreed to the published version of the manuscript.

Funding: The authors are grateful for the financial support granted by the Tecnológico Nacional de México through project 8096.20-P "Aplicación de aleaciones ternarias en aislamiento acústico" and 9274.20-P "Diseño y construcción de dispositivo empleando lentes Fresnel para la concentración de energía solar".

Institutional Review Board Statement: Not applicable.

Data Availability Statement: Data are contained within the article.

Acknowledgments: The authors express gratitude to Guillermo Eduardo Mejía Hernández, Juan Carlos Méndez Méndez, and Jesús Guerrero Orduña for their technical assistance.

Conflicts of Interest: The authors declare that they have no known competing financial interests or personal relationships that could have appeared to influence the work reported in this paper.

References

1. Gu, F.; Guo, J.; Zhang, W.; Summers, P.A.; Hall, P. From waste plastics to industrial raw materials: A life cycle assessment of mechanical plastic recycling practice based on a real-world case study. *Sci. Total Environ.* **2017**, *601–602*, 1192–1207. [CrossRef]
2. Dorigato, A. Recycling of polymer blends. *Adv. Ind. Eng. Polym. Res.* **2021**, *4*, 53–69. [CrossRef]
3. Fredi, G.; Dorigato, A. Recycling of bioplastic waste: A review. *Adv. Ind. Eng. Polym. Res.* **2021**, *4*, 159–177. [CrossRef]
4. Chanda, M. Chemical aspects of polymer recycling. *Adv. Ind. Eng. Polym. Res.* **2021**, *4*, 133–150. [CrossRef]
5. Zhao, X.; Korey, M.; Li, K.; Copenhaver, K.; Tekinalp, H.; Celik, S.; Kalaitzidou, K.; Ruan, R.; Ragauskas, A.J.; Ozcan, S. Plastic waste upcycling toward a circular economy. *Chem. Eng. J.* **2022**, *428*, 131928. [CrossRef]
6. Johansen, M.R.; Christensen, T.B.; Ramos, T.M.; Syberg, K. A review of the plastic value chain from a circular economy perspective. *J. Environ. Manag.* **2022**, *302*, 113975. [CrossRef]
7. Mostofi Sarkari, N.; Ayar, P.; Hatefi Oskouei, M.; Karimian Khosrowshahi, F.; Mohseni, M. Silane crosslinkable polyethylene waste as bitumen modifier: A new fortunate destiny by in time recycling of thermoplastic waste before conversion to thermoset end-of-life unrecyclable polymer. *Constr. Build. Mater.* **2021**, *287*, 122999. [CrossRef]
8. Martínez-López, M.; Martínez-Barrera, G.; Salgado-Delgado, R.; Gencel, O. Recycling polypropylene and polyethylene wastes in production of polyester based polymer mortars. *Constr. Build. Mater.* **2021**, *274*, 121487. [CrossRef]
9. Eskander, S.B.; Saleh, H.M.; Tawfik, M.E.; Bayoumi, T.A. Towards potential applications of cement-polymer composites based on recycled polystyrene foam wastes on construction fields: Impact of exposure to water ecologies. *Case Stud. Constr. Mater.* **2021**, *15*, e00664. [CrossRef]
10. Jelcic Rukavina, M.; Baricevic, A.; Serdar, M.; Grubor, M. Study on the post-fire properties of concrete with recycled tyre polymer fibres. *Cem. Concr. Compos.* **2021**, *123*, 104184. [CrossRef]
11. Suresh, M.; Pal, M. Utilization of recycled concrete wastes and latex polymer for sustainable road construction. *Mater. Today Proc.* **2021**, *47*, 4171–4176. [CrossRef]
12. Wathiq Hammodat, W. Investigate road performance using polymer modified concrete. *Mater. Today Proc.* **2021**, *42*, 2089–2094. [CrossRef]

13. Varun, B.K.; Anila Kumar, C.P. Strength characteristics of polymer modified high volume fly ash concrete. *Mater. Today Proc.* **2021**, *46*, 285–288. [CrossRef]
14. Viscione, N.; Veropalumbo, R.; Oreto, C.; Biancardo, S.A.; Abbondati, F.; Russo, F. Additional procedures for characterizing the performance of recycled polymer modified asphalt mixtures. *Measurement* **2022**, *187*, 110238. [CrossRef]
15. Kazemi, M.; Fini, E.H. State of the art in the application of functionalized waste polymers in the built environment. *Resour. Conserv. Recycl.* **2022**, *177*, 105967. [CrossRef]
16. Gu, L.; Ozbakkaloglu, T. Use of recycled plastics in concrete: A critical review. *Waste Manag.* **2016**, *51*, 19–42. [CrossRef]
17. Ahmad, J.; Majdi, A.; Babeker Elhag, A.; Deifalla, A.F.; Soomro, M.; Isleem, H.F.; Qaidi, S. A Step towards Sustainable Concrete with Substitution of Plastic Waste in Concrete: Overview on Mechanical, Durability and Microstructure Analysis. *Crystals* **2022**, *12*, 944. [CrossRef]
18. Qaidi, S.; Al-Kamaki, Y.; Hakeem, I.; Dulaimi, A.F.; Özkılıç, Y.; Sabri, M.; Sergeev, V. Investigation of the physical-mechanical properties and durability of high-strength concrete with recycled PET as a partial replacement for fine aggregates. *Front. Mater.* **2023**, *10*, 1101146. [CrossRef]
19. Vila-Cortavitarte, M.; Lastra-González, P.; Calzada-Pérez, M.Á.; Indacoechea-Vega, I. Analysis of the influence of using recycled polystyrene as a substitute for bitumen in the behaviour of asphalt concrete mixtures. *J. Clean. Prod.* **2018**, *170*, 1279–1287. [CrossRef]
20. Chaukura, N.; Gwenzi, W.; Bunhu, T.; Ruziwa, D.T.; Pumure, I. Potential uses and value-added products derived from waste polystyrene in developing countries: A review. *Resour. Conserv. Recycl.* **2016**, *107*, 157–165. [CrossRef]
21. Liguori, F.; Moreno-Marrodán, C.; Barbaro, P. Valorisation of plastic waste via metal-catalysed depolymerisation. *Beilstein J. Org. Chem.* **2021**, *17*, 589–621. [CrossRef]
22. Kosloski-Oh, S.C.; Wood, Z.A.; Manjarrez, Y.; de los Rios, J.P.; Fieser, M.E. Catalytic methods for chemical recycling or upcycling of commercial polymers. *Mater. Horiz.* **2021**, *8*, 1084–1129. [CrossRef] [PubMed]
23. Zhang, F.; Zhao, Y.; Wang, D.; Yan, M.; Zhang, J.; Zhang, P.; Ding, T.; Chen, L.; Chen, C. Current technologies for plastic waste treatment: A review. *J. Clean. Prod.* **2021**, *282*, 124523. [CrossRef]
24. Singh, N.; Hui, D.; Singh, R.; Ahuja, I.P.S.; Feo, L.; Fraternali, F. Recycling of plastic solid waste: A state of art review and future applications. *Compos. Part B Eng.* **2017**, *115*, 409–422. [CrossRef]
25. Chohan, J.S.; Singh, R. Thermosetting Polymers: A Review on Primary, Secondary, Tertiary and Quaternary Recycling. In *Encyclopedia of Materials: Plastics and Polymers*; Elsevier: Amsterdam, The Netherlands, 2022; pp. 603–610.
26. Kumar, S.; Singh, R. Primary and Secondary Recycling of Thermoplastics: A Review. In *Encyclopedia of Materials: Plastics and Polymers*; Elsevier: Amsterdam, The Netherlands, 2022; pp. 471–477.
27. Delva, L.; Hubo, S.; Cardon, L.; Ragaert, K. On the role of flame retardants in mechanical recycling of solid plastic waste. *Waste Manag.* **2018**, *82*, 198–206. [CrossRef] [PubMed]
28. Huang, J.; Veksha, A.; Chan, W.P.; Lisak, G. Support effects on thermocatalytic pyrolysis-reforming of polyethylene over impregnated Ni catalysts. *Appl. Catal. A Gen.* **2021**, *622*, 118222. [CrossRef]
29. Li, N.; Liu, H.; Cheng, Z.; Yan, B.; Chen, G.; Wang, S. Conversion of plastic waste into fuels: A critical review. *J. Hazard. Mater.* **2022**, *424*, 127460. [CrossRef] [PubMed]
30. Garcia, J.M.; Robertson, M.L. The future of plastics recycling. *Science* **2017**, *358*, 870–872. [CrossRef] [PubMed]
31. Vollmer, I.; Jenks, M.J.F.; Roelands, M.C.P.; White, R.J.; Harmelen, T.; Wild, P.; Laan, G.P.; Meirer, F.; Keurentjes, J.T.F.; Weckhuysen, B.M. Beyond Mechanical Recycling: Giving New Life to Plastic Waste. *Angew. Chem. Int. Ed.* **2020**, *59*, 15402–15423. [CrossRef]
32. Arandes, J.; Bilbao, J.; López, D. Reciclado de residuos plásticos. *Rev. Iberoam. Polímeros* **2004**, *5*, 28–45.
33. Ragaert, K.; Delva, L.; Van Geem, K. Mechanical and chemical recycling of solid plastic waste. *Waste Manag.* **2017**, *69*, 24–58. [CrossRef] [PubMed]
34. Thakur, S.; Verma, A.; Sharma, B.; Chaudhary, J.; Tamulevicius, S.; Thakur, V.K. Recent developments in recycling of polystyrene based plastics. *Curr. Opin. Green Sustain. Chem.* **2018**, *13*, 32–38. [CrossRef]
35. Yang, Z.; Lü, F.; Zhang, H.; Wang, W.; Shao, L.; Ye, J.; He, P. Is incineration the terminator of plastics and microplastics? *J. Hazard. Mater.* **2021**, *401*, 123429. [CrossRef] [PubMed]
36. Shen, M.; Hu, T.; Huang, W.; Song, B.; Qin, M.; Yi, H.; Zeng, G.; Zhang, Y. Can incineration completely eliminate plastic wastes? An investigation of microplastics and heavy metals in the bottom ash and fly ash from an incineration plant. *Sci. Total Environ.* **2021**, *779*, 146528. [CrossRef]
37. Lai, Y.; Dong, L.; Sheng, X.; Li, Q.; Li, P.; Hao, Z.; Yu, S.; Liu, J. Swelling-Induced Fragmentation and Polymer Leakage of Nanoplastics in Seawater. *Environ. Sci. Technol.* **2022**, *56*, 17694–17701. [CrossRef]
38. Guo, Y.; Xia, X.; Ruan, J.; Wang, Y.; Zhang, J.; LeBlanc, G.A.; An, L. Ignored microplastic sources from plastic bottle recycling. *Sci. Total Environ.* **2022**, *838*, 156038. [CrossRef]
39. Huang, Q.; Cheng, Z.; Yang, C.; Wang, H.; Zhu, N.; Cao, X.; Lou, Z. Booming microplastics generation in landfill: An exponential evolution process under temporal pattern. *Water Res.* **2022**, *223*, 119035. [CrossRef]
40. Thakur, S.; Chaudhary, J.; Singh, P.; Alsanie, W.F.; Grammatikos, S.A.; Thakur, V.K. Synthesis of Bio-based monomers and polymers using microbes for a sustainable bioeconomy. *Bioresour. Technol.* **2022**, *344*, 126156. [CrossRef]
41. Miri, S.; Saini, R.; Davoodi, S.M.; Pulicharla, R.; Brar, S.K.; Magdouli, S. Biodegradation of microplastics: Better late than never. *Chemosphere* **2022**, *286*, 131670. [CrossRef]

42. Zhang, Q.; Deng, Y.; Shi, C.-Y.; Feringa, B.L.; Tian, H.; Qu, D.-H. Dual closed-loop chemical recycling of synthetic polymers by intrinsically reconfigurable poly(disulfides). *Matter* **2021**, *4*, 1352–1364. [CrossRef]
43. Vimala, P.P.; Mathew, L. Biodegradation of Polyethylene Using Bacillus Subtilis. *Procedia Technol.* **2016**, *24*, 232–239. [CrossRef]
44. Restrepo-Flórez, J.-M.; Bassi, A.; Thompson, M.R. Microbial degradation and deterioration of polyethylene—A review. *Int. Biodeterior. Biodegrad.* **2014**, *88*, 83–90. [CrossRef]
45. Nowak, B.; Pająk, J.; Drozd-Bratkowicz, M.; Rymarz, G. Microorganisms participating in the biodegradation of modified polyethylene films in different soils under laboratory conditions. *Int. Biodeterior. Biodegrad.* **2011**, *65*, 757–767. [CrossRef]
46. Mohan, A.J.; Sekhar, V.C.; Bhaskar, T.; Nampoothiri, K.M. Microbial assisted High Impact Polystyrene (HIPS) degradation. *Bioresour. Technol.* **2016**, *213*, 204–207. [CrossRef] [PubMed]
47. *ISO 868:2003*; Plastics and Ebonite, Determination of Indentation Hardness by Means of a Durometer (Shore Hardness). International Organization for Standardization (ISO): Geneva, Switzerland, 2003.
48. *ASTM E1876-22*; Standard Test Method for Dynamic Young's Modulus, Shear Modulus, and Poisson's Ratio by Impulse Excitation of Vibration. ASTM International: West Conshohocken, PA, USA, 2022; pp. 1–19.
49. *ISO 11358-1:2002*; Plastics, Thermogravimetry (TG) of Polymers, Part 1: General Principles. International Organization for Standardization (ISO): Geneva, Switzerland, 2022.
50. Muraleedharan, M.; Singh, H.; Udayakumar, M.; Suresh, S. Modified active solar distillation system employing directly absorbing Therminol 55–Al_2O_3 nano heat transfer fluid and Fresnel lens concentrator. *Desalination* **2019**, *457*, 32–38. [CrossRef]
51. Borode, A.; Ahmed, N.; Olubambi, P. A review of solar collectors using carbon-based nanofluids. *J. Clean. Prod.* **2019**, *241*, 118311. [CrossRef]
52. Borode, A.O.; Ahmed, N.A.; Olubambi, P.A. Surfactant-aided dispersion of carbon nanomaterials in aqueous solution. *Phys. Fluids* **2019**, *31*, 071301. [CrossRef]
53. Yi, Y.; Salonitis, K.; Tsoutsanis, P.; Litos, L.; Patsavelas, J. Improving the Curing Cycle Time through the Numerical Modeling of air Flow in Industrial Continuous Convection Ovens. *Procedia CIRP* **2017**, *63*, 499–504. [CrossRef]
54. Suman, S.; Khan, M.K.; Pathak, M. Performance enhancement of solar collectors—A review. *Renew. Sustain. Energy Rev.* **2015**, *49*, 192–210. [CrossRef]
55. Sosa Domínguez, A.; Pérez Bueno, J.J.J.; Zamudio Torres, I.; Mendoza López, M.L.L.; Sosa Domínguez, A. Corrosion Study of Electroless Deposited Nickel-Phosphorus Solar Absorber Coatings on Carbon Steel. *Int. J. Electrochem. Sci.* **2017**, *12*, 2987–3000. [CrossRef]
56. Sosa Domínguez, A.; Pérez Bueno, J.J.J.; Zamudio Torres, I.; Mendoza López, M.L.L. Characterization and corrosion resistance of electroless black Ni-P coatings of double black layer on carbon steel. *Surf. Coat. Technol.* **2017**, *326*, 192–199. [CrossRef]
57. Bueno, J.d.J.P.; Rojas, E.R.; Hernandez, J.M.; López, M.L.M.; Esquivel, R.A.C. Copper Black Coatings for the Absorption of Solar Concentration with an APPJ SiO_2 Super-Hydrophobic Protection. *Int. J. Environ. Sustain. Green Technol.* **2020**, *12*, 76–87. [CrossRef]
58. Herrera-Zamora, D.M.M.; Lizama-Tzec, F.I.I.; Santos-González, I.; Rodríguez-Carvajal, R.A.A.; García-Valladares, O.; Arés-Muzio, O.; Oskam, G. Electrodeposited black cobalt selective coatings for application in solar thermal collectors: Fabrication, characterization, and stability. *Sol. Energy* **2020**, *207*, 1132–1145. [CrossRef]
59. Wang, X.; Liu, D.; Liu, X.; He, Y. Graphene oxide stabilized carbon nanotube-water nanofluids for direct absorption solar collectors. *IOP Conf. Ser. Mater. Sci. Eng.* **2019**, *556*, 012037. [CrossRef]
60. Uscanga Olea, M.A.; Pérez Bueno, J.d.J.; Pérez Maldonado, A.X. Nanometric and surface properties of semiconductors correlated to photocatalysis and photoelectrocatalysis applied to organic pollutants—A review. *J. Environ. Chem. Eng.* **2021**, *9*, 106480. [CrossRef]
61. Cacho, L.M.; de Jesús Pérez Bueno, J.; Vong, Y.M.; Stremsdoerfer, G. Effect of swelling of chemical reagents and the sulfuric-chromic acid bath on surface texturizing of poly(acrylonitrile-butadiene-styrene). *Acta Chim. Slov.* **2019**, *66*, 638–647. [CrossRef]
62. Magallón Cacho, L.; Pérez Bueno, J.J.; Meas Vong, Y.; Stremsdoerfer, G.; Espinoza Beltrán, F.J.; Martínez Vega, J. Novel green process to modify ABS surface before its metallization: Optophysic treatment. *J. Coat. Technol. Res.* **2015**, *12*, 313–323. [CrossRef]
63. Magallón-Cacho, L.; Pérez-Bueno, J.J.; Meas-Vong, Y.; Stremsdoerfer, G.; Espinoza-Beltrán, F.J. Surface modification of acrylonitrile-butadiene-styrene (ABS) with heterogeneous photocatalysis (TiO_2) for the substitution of the etching stage in the electroless process. *Surf. Coat. Technol.* **2011**, *206*, 1410–1415. [CrossRef]

polymers

MDPI

Article

The Effect of Various Ti6Al4V Powders on the Behavior of Particle-Reinforced Polyester Matrix Composites

Erdoğan Teke [1], Elif Alyamaç Seydibeyoğlu [2] and Mehmet Özgür Seydibeyoğlu [3,*]

1 Program of Materials Science and Engineering, Graduate School of Natural and Applied Science, Izmir Kâtip Celebi University, 35620 Izmir, Turkey
2 Department of Petroleum and Natural Gas Engineering, Izmir Kâtip Çelebi University, 35620 Izmir, Turkey
3 Department of Materials Science and Engineering, Izmir Kâtip Çelebi University, 35620 Izmir, Turkey
* Correspondence: seydibey@gmail.com; Tel.: +90-532-5475217

Abstract: In this study, recycled and commercial Ti6Al4V powder was added to polyester composites at various filling ratios by weight fraction. Three weight fractions of Ti6Al4V particles, 0%, 5%, and 10%, were chosen for study. By examining the mechanical and microstructural properties of polyester composites, the effects of the Ti6Al4V powder proportion by weight fraction and particle size parameters were investigated. With the filler loading, a 39% increase in the tensile strength of the composites was achieved. A minor decrease in flexural strength was observed at 5% filler weight fraction. The addition of the recycled Ti6Al4V powder to the polyester matrix slightly reduced the thermal conductivity of the composite over that of the neat polymer. However, the incorporation of the commercial Ti6Al4V powder fillers in the polyester matrix considerably increased the thermal conductivity of the composites, suggesting several potential uses. The presence of high levels of oxygen in the powder led to reduced thermal conductivity of the composites due to the reduction in phonon scattering.

Keywords: polyester composite; Ti6Al4V powder; thermal conductivity; mechanical strength

Citation: Teke, E.; Seydibeyoğlu, E.A.; Seydibeyoğlu, M.Ö. The Effect of Various Ti6Al4V Powders on the Behavior of Particle-Reinforced Polyester Matrix Composites. *Polymers* **2023**, *15*, 2904. https://doi.org/10.3390/polym15132904

Academic Editors: Cristina Cazan and Mihaela Cosnita

Received: 13 April 2023
Revised: 12 June 2023
Accepted: 12 June 2023
Published: 30 June 2023

1. Introduction

Composite materials first emerged in the middle of the 20th century and are now one of the most important areas of materials and manufacturing technology [1,2]. They are ideal for a variety of applications in the industrial sectors of aerospace, automotive, construction, sports, and bio-medicine, among many others, due to their excellent properties such as being lightweight and corrosion-resistant [3,4].

Modern technologies frequently require materials with very complex property combinations that are not achievable with more traditional metal alloys, ceramics, and polymeric materials. This is particularly true for materials used in transportation, undersea, and aircraft applications. For instance, structural materials with low density, strength, stiffness, resistance to abrasion and impact, and resistance to corrosion are increasingly sought after by aircraft engineers. The combination of these qualities is ideal in terms of physical strength. Strong materials are usually relatively thick, and increasing strength or stiffness typically causes a reduction in impact strength [5]. These materials have exceptional mechanical and structural qualities, including a high strength-to-weight ratio, resistance to fire, chemicals, corrosion, and wear, as well as being inexpensive to produce [6,7].

Composites are classified based on their morphology of reinforcement (fiber, particulate-reinforced, and laminate composites) or their matrix material (metal, ceramic, or polymer matrix composite). In a polymer composite material, the matrix is a resin and the reinforcement is in the form of dispersed particles, which act as the second phase. A wide range of microstructures can be obtained using combinations of matrix materials and reinforcement. The relationship between the properties and the structure of two-phase materials has been studied by many authors. In commercial production, low-cost particulate fillers are added to plastics for primarily economical reasons, as well as to improve molding characteristics. It is not only the material properties of the two components or the volume fraction of the filler which governs the deformation behavior, but also their shape, size, orientation, and the state of adhesion between the filler and the matrix [8].

Polyester resins and other thermosetting plastics are frequently utilized as matrix materials because they evenly distribute stresses in all directions and can withstand shocks and vibration. The most frequently utilized resin system is polyester resin, notably in the marine industry. The liquid mixtures of low molar mass reactants, such as monomers, which are used to make thermosetting plastics, polymerize to form strongly cross-linked network polymers [9–11]. The qualities of the particle filler dispersed in the composite are enhanced. Polymer composites manufactured in this manner have found numerous applications as lightweight, high-strength materials. Polymer matrix composites can be reinforced with fibers (synthetic or natural), whiskers, and particulate materials. Natural fillers include minerals such as calcium carbonate, mica, talc, and some agricultural byproducts of synthetic fillers, including processed mineral products such as carbon black, fumed silica, and aluminum hydroxide [12,13]. The sizes of particulate fillers range from 0.1 μm to approximately 2 mm. It can be inferred from the literature that the research on Ti6Al4V-powder-reinforced polymer composites is relatively limited [14].

In recent years, metal particles have been viewed as prospective reinforcing materials for various composites [15]. It is stated in the literature that many metal particles such as aluminum, copper, zinc, stainless steel, silver, and nickel increase the electrical and thermal conductivity of metal-filled polymer composites [16].

Yaman and Taga [17] investigated the thermal and electrical conductivity of copper-filled polyester composites. According to their study, both the thermal and electrical conductivity of polyester composites increase with increasing filler particle sizes and filler contents. As a result, the particle size distribution and content of fillers play significant roles in determining the thermal properties of composites. In another similar study, the electrical behavior of aluminum-powder-reinforced polyester composites was investigated. In the study by Berger and McCullough, an increase in electrical conductivity was achieved as the filler content increased [18].

Alam et al. added multi-walled carbon nanotubes (MWCNTs) to a polyester matrix at different compositions (0.1, 0.3, and 0.5 wt.%). They concluded that 0.3 wt.% MWCNT was the optimum amount of filler to mechanically and thermally improve the polyester matrix [19].

In another study, Krishnasamy et al. investigated the effect of aluminum and copper wire mesh on the thermal and mechanical properties of epoxy composites. It was concluded that the tensile strength of copper- and aluminum-reinforced epoxy composites increased [20].

With composites, it is possible to reduce the overall production cost by embedding waste metal particles from other manufacturing processes or recycled materials in a polymer.

This study concerns the evaluation of mechanical properties such as tensile and flexural strength for different weight ratios of Ti6Al4V-particulate-filled polyester composite. Titanium alloys, which are especially preferred in aerospace applications, exhibit tensile strength, creep and fatigue strength, and fracture toughness; they contain α and β stabilizing elements to achieve the necessary mechanical properties such as fatigue cracking, stress-abrasion cracking, and oxidation resistance. These properties classify Ti6Al4V as a precious metal. Recycling chip materials in different forms will provide both environmental and economic benefits. The recycling of Ti6Al4V chips generated during machining is limited due to difficulties in the recycling process. In this study, Ti6Al4V chips were converted into powder form using mechanical milling, and the potential use of the obtained Ti6Al4V powder as a reinforcement material in a polyester matrix was investigated. This article focuses on the effect of the addition of Ti6Al4V particles on the mechanical properties, and thermal stability of the obtained polyester composite.

2. Materials and Methods

2.1. Materials

2.1.1. Preparation of Ti6Al4V Powder

Ti6Al4V scraps obtained in chip form following machining served as the starting materials. The machining chip was supplied by Metrosan Company in Manisa, Turkey. Ti6Al4V alloy powder was utilized as a reinforcement; it was recycling powder obtained from Ti6Al4V chips following mechanical milling. The milling process of the Ti6Al4V chip was performed using the Retsch RS200 vibratory disc mill. In the milling process of the Ti6Al4V chips, a milling duration of 60 min was selected and a milling speed of 1500 rpm was used. In this study, irregularly shaped recycled Ti6Al4V powder obtained as a result of the milling process was used. The characteristics of Ti6Al4V powder used as reinforcement are given in detail in Table 1.

Table 1. The oxygen concentration in the recycled A and B powders and the commercial Ti6Al4V powder.

Powder	d(10) µm	d(50) µm	d(90) µm	Oxygen (ppm)
A	4.4	15.4	35.7	34,840
B	14.4	34.6	60.7	20,966
C (Commercial)	-	15–45	-	2565

Ti6Al4V powder obtained from the milling process was subjected to 325 and 625 mesh sieve analysis, and particle size classification was performed. Two types of recycled Ti6Al4V powder obtained after sieve analysis were coded as A and B according to their particle size distribution. The commercial Ti6Al4V powder was coded as C. The particle size distributions of the recycled powders, are dispersed by a range of dry units, were determined using a laser diffraction particle size measurement instrument (Malvern Panalytical Mastersizer 3000, Malvern, UK). Two kinds of recycled Ti6Al4V powder fillers, coded as A and B, were utilized, with mean particle sizes of 15.4 µm and 34.6 µm, respectively. Particle size distribution histograms of powders A and B are given in Figures 1 and 2, respectively. Commercial Ti6Al4V powder with a particle size of 15–45 µm was also used for comparison in the same parameters. The commercial Ti6Al4V powder was obtained from the Nanokar Company (Istanbul, Turkey).

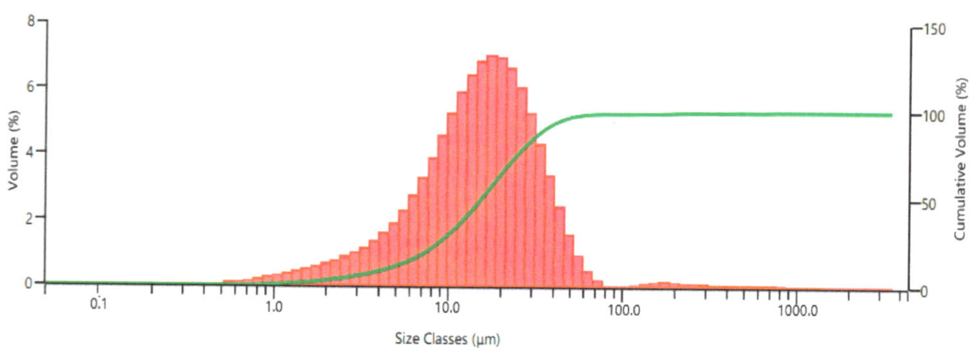

Figure 1. Particle size distribution of powder A.

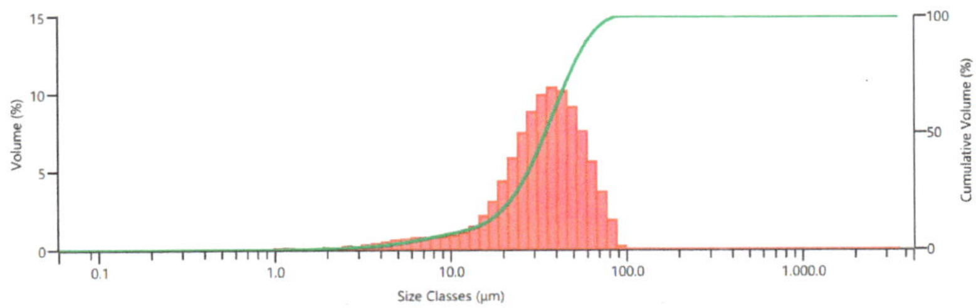

Figure 2. Particle size distribution of powder B.

Table 1 details the oxygen concentration in the recycled A and B powders and the commercial Ti6Al4V powder. The oxygen ratios in the powders used in the study were measured using the LECO TC400 Nitrogen/Oxygen determinator, USA. The increase in oxygen content of recycled powder is due to the milling atmosphere and the increased particle surface area (particle size is decreased). Additionally, this situation is attributable to surface contamination of the starting chip and the jar's temperature.

SEM images of the commercial and recycled Ti6Al4V powders are given in Figure 3. The recycled powders exhibit a rounded morphology and satellite nanoparticles on their particles. The shape of the commercial Ti6Al4V particles is angular and their morphology is smoother than that of recycled Ti6Al4V particles. It is known from the literature that the morphology of the particles has an effect on the mechanical and thermal properties.

Figure 3. *Cont.*

(c)

Figure 3. Particle morphology of powder A (**a**), powder B (**b**), and the commercial powder, C (**c**).

2.1.2. Polymer Matrix

A commercially available Polipol 353 casting-type polyester resin was used as the matrix material. This is a thermoset type of polymer matrix material. Polipol 353 casting-type polyester resin is an orthophthalic unsaturated polyester resin designed for general-purpose, standard casting applications. The polyester resin with added cobalt accelerator was used together with the hardener; both were purchased from Yücel Kompozit (Izmir, Turkey), as detailed in Table 2. Methyl Ethyl Ketone Peroxide (MEK-P) is used for curing general-purpose unsaturated polyesters at room temperature; it is generally added into a resin at a rate of 1% (by weight).

Table 2. The unsaturated polyester specifications.

Appearance	Transparent
Color	Clear
Density	1.11 g/cm^3
Flash point	32 °C
MEK-P hardener	% 1
Usable life	30 min/25 °C
Cure time	12 h/25 °C

2.2. *Experimental Methods*

2.2.1. Manufacturing of Composites

The polyester resin and MEK-P hardener (%1) were mixed according to the datasheet provided by the supplier. Three formulations, 0, 5, and 10 wt.% Ti6Al4V powder in polyester, were developed, as summarized in Table 3.

Table 3. The composition ratio of Ti6Al4V powder in the polyester composite.

Sample Code	Polyester (wt.%)	Ti6Al4V Powder (wt.%)	Powder Type
Neat	100	0	-
PA5	95	5	Recycled
PA10	90	10	Recycled
PB5	95	5	Recycled
PB10	90	10	Recycled
PC5	95	5	Commercial
PC10	90	10	Commercial

A specially designed and fabricated silicon mold was used for this purpose to avoid sticking during curing. The Ti6Al4V powder was oven-dried for 24 h at 80 °C. Ti6Al4V powder was mixed with polyester at a low speed by stirring mechanically to avoid bubbles, and the hardener was mixed into the Ti6Al4V powder/polyester resin mixture. For the synthesis of reinforced polyester composites, the mass ratio of polyester to reinforcement was modified to yield a total of 40 g.

In the case of the particulate-reinforced composites, Ti6Al4V powder was added to the polyester resin and mixed well in a continuous stirring process until a uniform mixture was observed; then, the hardener was added into the Ti6Al4V/polyester mixture. Stirring of the mixture continued for a certain duration depending on the exothermic reaction. The neat polyester resin samples were also prepared under similar processing conditions.

2.2.2. Characterization of Composites

Five specimens of each type were tested. The results were averaged from five tests. The three-point bending test was performed with a constant loading speed of 1 mm/min at room temperature, and a span length of 102.4 mm, as per American Society for Testing and Materials (ASTM) D790, using a Shimadzu universal testing machine. The dimensions of the specimens were 125 mm × 12.7 mm × 3.2 mm with a span of 102.4 mm length, as given in Figure 4. A fixed span-to-depth ratio of 32 was used.

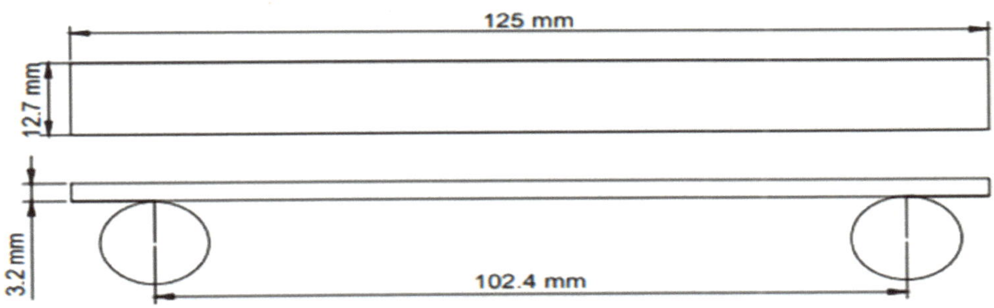

Figure 4. The dimensions of the three-point bending test specimens.

Laboratory tests were performed using the Shimadzu universal tensile strength testing machine (AG-IC 100kN); loading speed of the deformation corresponded to 1 mm/min at room temperature. The tensile test specimen dimensions according to ASTM D638-03 Type I are given in Figure 5.

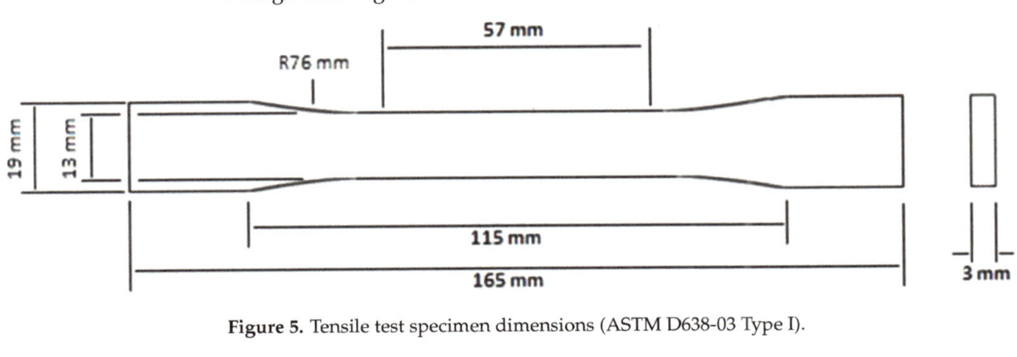

Figure 5. Tensile test specimen dimensions (ASTM D638-03 Type I).

Thermal conductivity measurements were conducted using a C-Therm TCI test machine, in accordance with ASTM D 5470. The cylindrical testing samples had a diameter of 20 mm and a height of 3 mm. The thermal conductivity of various proportions of Ti6Al4V-filled polyester composites was determined.

Morphological investigations of the powders and composites were carried out using Carl Zeiss 300VP scanning electron microscopy (SEM) apparatus.

3. Results

All of the samples, neat and PA-PB-PC, for all cases of filler weight fractions and sample dimensions, were successfully produced with no visible defects or other issues. The standard deviation values of tensile and bending test results were high because each sample was produced by hand.

3.1. Tensile Test Results

After the tensile strength analysis, the fractured regions are given in Figure 6. Moreover, Table 4 shows the effect of adding different wt.% and particle sizes of Ti6Al4V powder on the tensile strength of polyester composites.

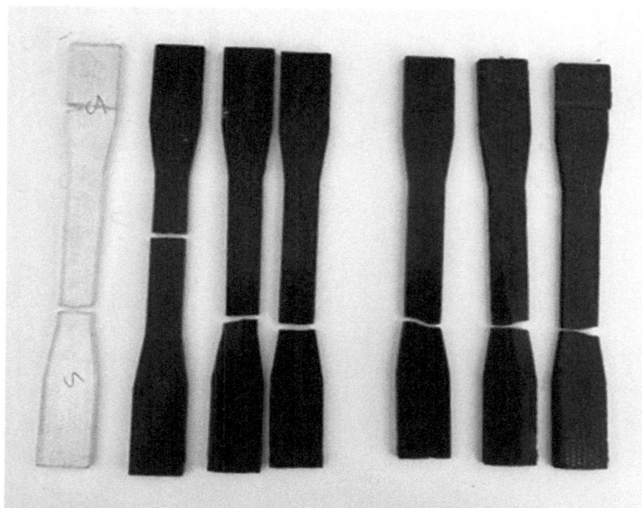

Figure 6. Fractured regions of the samples after the tensile test.

Table 4. Tensile strength of polyester reinforced with different wt.% of Ti6Al4V powder.

Sample	Tensile Strength (MPa)	Young's Modulus (GPa)
Neat	44.83 ± 2.46	1.21 ± 0.03
PA5	47.42 ± 4.20	1.17 ± 0.13
PA10	62.17 ± 3.16	1.92 ± 0.10
PB5	49.35 ± 4.02	1.35 ± 0.13
PB10	55.38 ± 3.34	1.39 ± 0.09
PC5	43.58 ± 2.22	0.99 ± 0.08
PC10	52.67 ± 2.71	1.48 ± 0.05

The mechanical properties of particulate–polymer composites depend on the particle loading, particle size distribution, and particle–matrix interface adhesion [21]. The tensile strength of Ti6Al4V-powder-reinforced polyester composites, except for the PC5 composite, increased when compared with the neat polyester composite. It is clear that the tensile strength is significantly dependent on particle proportion.

The results from the Shimadzu (AG-IC 100 kN) machine and the Young's modulus values of the composites are given in Table 4.

The surface area of the particles increased as the particle size decreased. Therefore, the PA10 composite, which had the largest particle surface area, had higher tensile strength. In addition, PA and PB composites had higher tensile strength compared with PC composites. The morphologies of particles A and B were more irregular than particle C. The irregular form of the particles provided good correlation, such as interlocking between the polymer matrix and the filler.

In Figure 7, the polyester composites show decreasing strain (ductility) as the Ti6Al4V concentration increases. The lowest strain values were obtained with the PA10 sample. The proportion of filled particles in polymer matrix composites (PMCs) was significant for the general mechanical behavior of the composite. On the other hand, extreme particle adding can easily cause agglomeration of the fillers, therefore introducing defects to the matrix [22]. The tensile strength increased slightly when 5% by weight of the particles was added, while the tensile strength increased significantly when 10% by weight of the particles was added to polyester resin. An improvement in tensile strength as the particle proportion increased, indicated strong interfacial bonding between the particle and the matrix. The particle size distribution was also a parameter that affected the mechanical properties of the filled matrix [23].

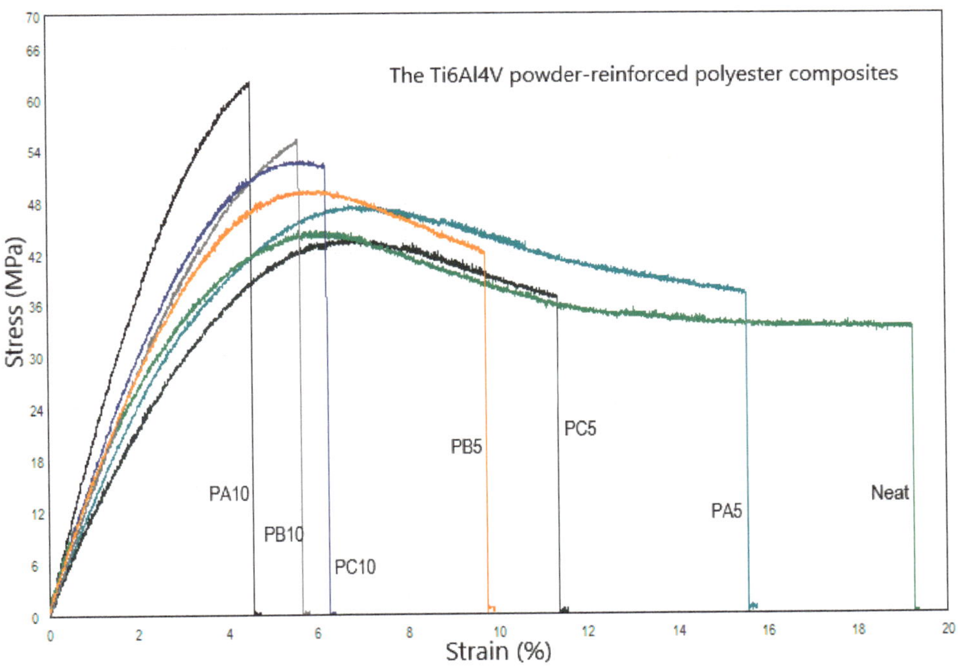

Figure 7. Tensile stress–strain curves of the 'Ti6Al4V-powder-reinforced' polyester composites.

The tensile strength also increased as the particle size decreased for the same weight fraction of the fillers due to the increased interfacial area between the particle surface and the polymer matrix. The commercial Ti6Al4V powder was used as reinforcement in PC5 and PC10 composites produced with the same parameters. The tensile strength of the commercial Ti6Al4V-powder-reinforced composites is lower than that of the recycled Ti6Al4V-powder-reinforced polyester composites. It is known that the particle shape affects the mechanical behavior in PMCs [21]. The commercial Ti6Al4V powder had a more regular morphology than the recycled Ti6Al4V powder. The irregularly shaped reinforcement connected with the matrix in an interlocking manner [24].

3.2. Three-Point Bending Test Results

The flexural strength of a particle-reinforced polymer matrix composite is affected by factors such as the bonding strength between the particles and the matrix, particle size distribution, and particle loading. The bending test is a simple method for testing supports that involves a three-point bending test, in which a loading pin is lowered from above at a constant rate. The specimen is placed on two supporting pins at a set distance from the specimen, and the fixture is mounted on a universal testing machine at crosshead speed and load was gradually applied until breakage of the specimen occurred. The test results are affected by details of the test preparation, conditioning, and load rate. An average of five specimens was taken. Table 5 shows the effect of adding Ti6Al4V powder with different weight fractions on the flexural strength and flexural modulus of polyester composites.

Table 5. Flexural strength of the polyester composites.

Sample	Flexural Strength (MPa)	Flexural Modulus (MPa)
Neat	106.61 ± 2.44	2.280 ± 0.05
PA5	103.92 ± 4.58	2.525 ± 0.11
PA10	118.33 ± 3.71	2.945 ± 0.09
PB5	101.78 ± 4.40	1.973 ± 0.10
PB10	135.85 ± 3.03	3.098 ± 0.08
PC5	93.54 ± 3.82	1.956 ± 0.08
PC10	104.09 ± 2.87	2.774 ± 0.06

The results indicated that the flexural strength decreased when the particle content was 5 wt.% and then increased with the particle content increasing up to 10 wt.%. For the PA5-PB5-PC5 composites with a reinforcement ratio of 5 wt.%, the decreased flexural strength led to a reduction in bending stress. This may be a result of poor bonding of the particles with the matrix and the inhomogeneous filling of reinforcement in matrices (poor particle distribution) that diminished the support to bending stress. The particles easily detached when the composite was subjected to load, which could lead to the formation of large voids [25].

The flexural strength values of the PA10-PB10-PC10 composites were significantly enhanced compared with the unfilled polyester composite due to good particle dispersion and strong polymer/filler interface adhesion for effective stress transfer [26,27]. This implies that the particle proportion has a significant effect on composite strength. For the particulate composites, the flexural strength depends on the transfer of stress between the matrix and particles [28]. The greatest improvement in flexural strength was obtained at a 10 wt.% content of PB10. The lowest flexural strength of the composites was obtained with a 10% particle proportion in the PC10 composite.

3.3. Thermal Conductivity Test Results

The experimental results of the thermal conductivity are presented in Table 6. The effective thermal conductivity of a composite material composed of one type of filler introduced into a polymer matrix depends on the thermal conductivity of the components, the filler's shape, size, concentration, their dispersion into the polymer, and the thermal interfacial resistance [29].

Table 6. The experimental results of the thermal conductivity measurements.

Sample	Thermal Conductivity (W/mK)
Neat	0.2502 ± 0.001
PA5	0.2271 ± 0.003
PA10	0.2317 ± 0.003
PB5	0.2121 ± 0.001
PB10	0.2344 ± 0.003
PC5	0.6079 ± 0.020
PC10	0.6319 ± 0.010

The addition of the recycled Ti6Al4V powder to the polyester matrix slightly reduced the thermal conductivity of the composite over that of the neat polymer. However, the incorporation of the commercial Ti6Al4V powder filler in the polyester matrix significantly increased the thermal conductivity of the PC5-PC10 composites. In the previous sections, the properties of Ti6Al4V powders used as fillers were detailed. The morphology of the commercial Ti6Al4V powder was more regular compared with that of the recycled Ti6Al4V powder. Thus, the Ti6Al4V filler particle shape and the distribution of filler particles are the critical factors impacting the thermal properties of metal-filled composites. The result is most likely due to the different particle shapes between the commercial and recycled types of filler, since the particle shape has a substantial effect on thermal conductivity [30]. The presence of oxygen in the A and B powders used as fillers acts as a vacancy and negatively affects the thermal conductivity. Therefore, the decrease in the thermal conductivity of PA and PB can be attributed to the high oxygen content. However, the incorporation of C powder fillers with a lower oxygen content in the polyester matrix significantly increased the thermal conductivity of the composites.

3.4. Morphology of the Composites

The fracture surfaces of neat polyester and the Ti6Al4V-particle-reinforced composite samples were investigated for various fracture features under a scanning electron microscope. According to Figure 8, the results of the experiment also show the irregular stratification of filler micro particles in the matrix. According to this observation, the quality of powder-matrix mixing process was not sufficient and better mixing may improve the powder dispersion in the matrix. The brittle-type fracture mode was observed in all polyester composite samples. However, micron-sized voids formed by the particles were observed on the fracture surfaces. During the tensile test, cracks formed at the applied stress. The lines around the particles indicated that it caused the extra stress concentration. When adding to the particles to the matrix, the wetting of the particles with the matrix was very important [31]. The presence of particle traces on the fracture surfaces indicated that the particles were held in place by the matrix material.

Figure 8. *Cont.*

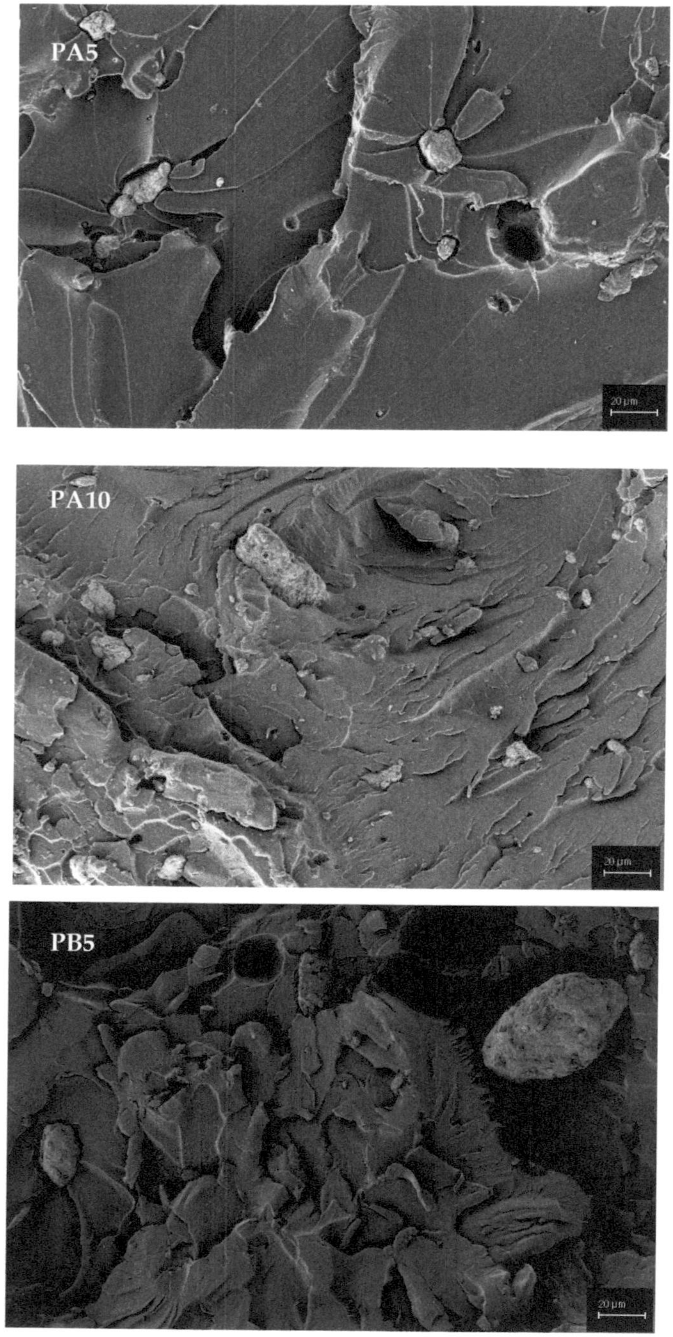

Figure 8. *Cont.*

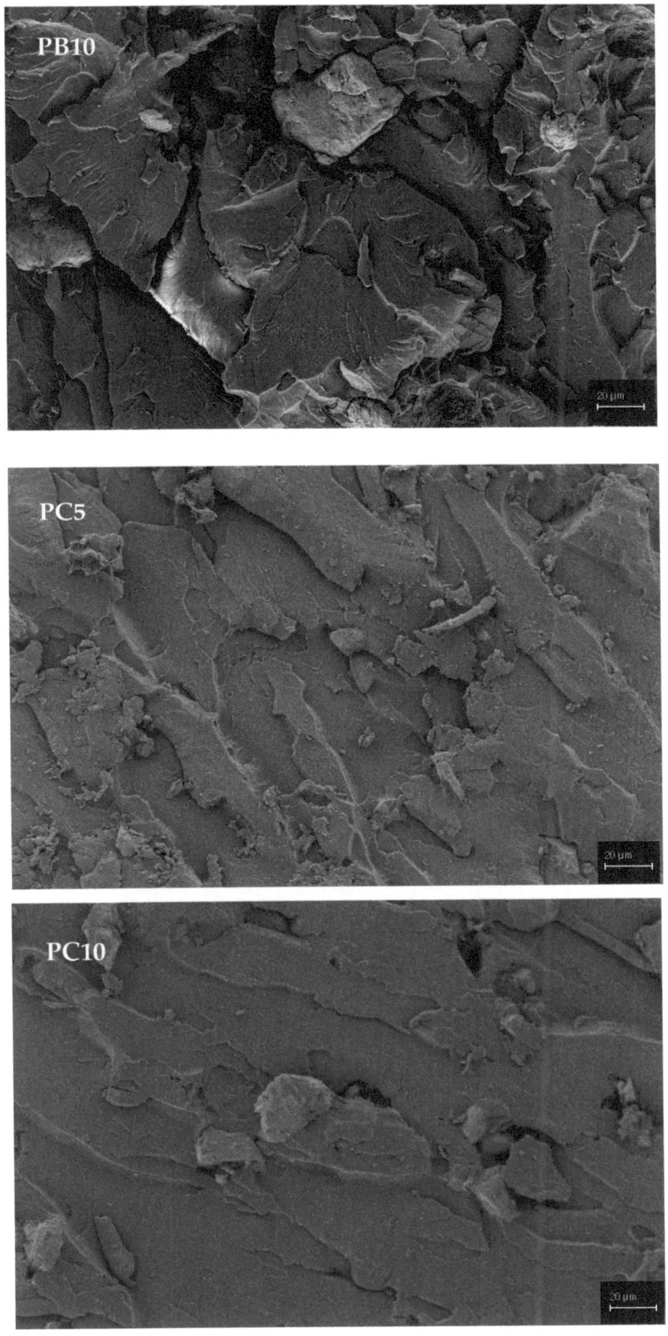

Figure 8. Morphology of the polyester composite specimens at different magnifications.

3.5. *Effect of Powder Oxygen Concentration on the Mechanical and Thermal Properties of Composites*

We focused on the effects of oxygen on the mechanical properties in the polyester composites. The measured oxygen concentrations in the A, B, and commercial Ti6Al4V powder were 34,840 ppm, 20,966 ppm, and 2565 ppm, respectively. The results show that the powder oxygen concentration had a significant effect on the mechanical properties of the Ti6Al4V-powder-reinforced polyester composites. The tensile and flexural strength of PC samples decreased compared with PA-PB samples with high oxygen concentrations. High oxidation may contribute to achieving strong particle–matrix adhesive bonding. In general, high oxide concentrations can provide strong particle–matrix interfacial bonding [32,33]. The oxidation level of the particles used as filler also affects the thermal properties of the composites. Although a high oxygen content improves the mechanical properties, it has a negative effect on thermal conductivity. Thermal conductivity decreased in PA and PB composites due to the high oxygen contents of powders A and B. However, a significant increase in thermal conductivity was observed in PC samples with low oxygen contents. The presence of high oxygen reduced thermal conductivity due to the reduction in phonon scattering. Phonon scattering from grain boundaries and internal defects are significant factors affecting thermal conductivity. The presence of high oxygen reduced thermal conductivity due to reduced phonon scattering. As a result, the thermal conductivity determined at high oxidation levels was lower than that measured at low oxidation levels [34–36].

4. Conclusions

Circular economy and recycling studies are critical in today's materials world. In addition to recycling, upcycling processes that convert waste streams to value-added products are a vital subject of study for humanity and the environment.

In this study, we have investigated another waste product obtained from the medical industry. The novelty of this work is that the effect of the recycled and commercial Ti6Al4V powder on the properties of the polyester composite was studied. In this study, the effects of particle size, shape, and filler content on the mechanical behavior of polymer matrix composites reinforced with the recycled and commercial Ti6Al4V powder were investigated.

With the filler loading, an increase in the tensile strength was achieved due to the increased surface contact area between the particles and the resin. A small decrease in bending strength with a filler weight fraction of 5% was observed. This may be because a 5% amount of filler is not sufficient to transmit the stress transfer from the matrix to the filler. However, an increase in flexural strength was observed with a 10% filler weight fraction. The addition of the recycled Ti6Al4V powder to the polyester matrix slightly reduced the thermal conductivity of the composite compared with that of the neat polymer.

Author Contributions: Conceptualization, E.T. and E.A.S.; methodology, E.T.; investigation, E.T.; data curation, E.A.S.; writing—original draft preparation, E.T. and E.A.S.; writing—review and editing, M.Ö.S.; supervision, M.Ö.S. All authors have read and agreed to the published version of the manuscript.

Funding: This research received no external funding.

Institutional Review Board Statement: Not applicable.

Data Availability Statement: All the data is presented are new. This article is based on Dr. Erdogan Teke's PhD thesis that was finished in March 2023.

Conflicts of Interest: The authors declare that they have no known competing financial interests or personal relationships that could have appeared to influence the work reported in this paper.

References

1. Dvorkin, L.; Konkol, J.; Marchuk, V.; Huts, A. Effectiveness of Polymer Additives in Concrete for 3D Concrete Printing Using Fly Ash. *Polymers* **2022**, *14*, 5467. [CrossRef]
2. Siraj, S.; Al-Marzouqi, A.H.; Iqbal, M.Z.; Ahmed, W. Impact of Micro Silica Filler Particle Size on Mechanical Properties of Polymeric Based Composite Material. *Polymers* **2022**, *14*, 4830. [CrossRef] [PubMed]
3. Zurowski, W.; Zepchlo, J.; Cep, R.; Cepova, L.; Rucki, M.; Krzysiak, Z.; Caban, J.; Samociuk, W. The Effect of Powder and Emulsion Binders on the Tribological Properties of Particulate Filled Glass Fiber Reinforced Polymer Composites. *Polymers* **2023**, *15*, 245. [CrossRef] [PubMed]
4. Nukala, S.G.; Kong, I.; Patel, V.I.; Kakarla, A.B.; Kong, W.; Buddrick, O. Development of Biodegradable Composites Using Polycaprolactone and Bamboo Powder. *Polymers* **2022**, *14*, 4169. [CrossRef] [PubMed]
5. Callister, W.D.; Rethwisch, D.G. *Materials Science and Engineering: An Introduction*; Wiley: New York, NY, USA, 2018; Volume 9.
6. Yan, D.-X.; Ren, P.-G.; Pang, H.; Fu, Q.; Yang, M.-B.; Li, Z.-M. Efficient electromagnetic interference shielding of lightweight graphene/polystyrene composite. *J. Mater. Chem.* **2012**, *22*, 18772–18774. [CrossRef]
7. Sahmaran, M.; Li, V.C.; Andrade, C. Corrosion resistance performance of steel-reinforced engineered cementitious composite beams. *ACI Mater. J.* **2008**, *105*, 243.
8. Rajak, D.K.; Pagar, D.D.; Kumar, R.; Pruncu, C.I. Recent progress of reinforcement materials: A comprehensive overview of composite materials. *J. Mater. Res. Technol.* **2019**, *8*, 6354–6374. [CrossRef]
9. Mazuki, A.A.M.; Akil, H.; Safiee, S.; Ishak, Z.A.M.; Abu Bakar, A. Degradation of dynamic mechanical properties of pultruded kenaf fiber reinforced composites after immersion in various solutions. *Compos. Part B Eng.* **2011**, *42*, 71–76. [CrossRef]
10. Velmurugan, G.; Venkatesan, S.P.; Prakash, P.V.; Sathishkumar, N.; Vijayakumar, N. Mechanical testing of hybrid composite material (Sisal and Coir). *Int. J. Sci. Res. Publ.* **2014**, *4*, 1–6.
11. Selvam, R.; Ravi, S.; Raja, R. Fabrication of SiC particulate reinforced polyester matrix composite and investigation. In *IOP Conference Series: Materials Science and Engineering*; IOP Publishing: Chennai, India, 2017; Volume 197.
12. Njoku, R.; Okon, A.; Ikpaki, T. Effects of variation of particle size and weight fraction on the tensile strength and modulus of periwinkle shell reinforced polyester composite. *Niger. J. Technol.* **2011**, *30*, 87–93.
13. Teh, P.L. Development Of Core Layer Materials Using Particulate Filled Epoxy Composites [TK7870. 15. T261 2008 f rb]. Ph.D. Thesis, Universiti Sains Malaysia, Pulau Pinang, Malaysia, 2008.
14. Solberg, D.; Wågberg, L. On the mechanism of GCC filler retention during dewatering-New techniques and initial findings. *J. Pulp Pap. Sci. JPPS* **2002**, *28*, 183–188.
15. Vakharia, V.S.; Kuentz, L.; Salem, A.; Halbig, M.C.; Salem, J.A.; Singh, M. Additive Manufacturing and Characterization of Metal Particulate Reinforced Polylactic Acid (PLA) Polymer Composites. *Polymers* **2021**, *13*, 3545. [CrossRef]
16. Ilyas, R.A.; Sapuan, S.M.; Asyraf, M.R.M.; Dayana, D.A.Z.N.; Amelia, J.J.N.; Rani, M.S.A.; Norrrahim, M.N.F.; Nurazzi, N.M.; Aisyah, H.A.; Sharma, S.; et al. Polymer Composites Filled with Metal Derivatives: A Review of Flame Retardants. *Polymers* **2021**, *13*, 1701. [CrossRef]
17. Yaman, K.; Taga, Ö. Thermal and Electrical Conductivity of Unsaturated Polyester Resin Filled with Copper Filler Composites. *Int. J. Polym. Sci.* **2018**, *2018*, 8190190. [CrossRef]
18. Berger, M.A.; McCullough, R.L. Characterization and analysis of the electrical properties of a metal-filled polymer. *Compos. Sci. Technol.* **1985**, *22*, 81–106. [CrossRef]
19. Alam, A.M.; Beg, M.; Yunus, R. Microstructure and fractography of multiwalled carbon nanotube reinforced unsaturated polyester nanocomposites. *Polym. Compos.* **2017**, *38*, E462–E471. [CrossRef]
20. Krishnasamy, P.; Rajamurugan, G.; Thirumurugan, M. Performance of fiber metal laminate composites embedded with AL and CU wire mesh. *J. Ind. Text.* **2022**, *51*, 6884S–6901S. [CrossRef]
21. Fu, S.-Y.; Lauke, B. Characterization of tensile behaviour of hybrid short glass fibre/calcite particle/ABS composites. *Compos. Part A Appl. Sci. Manuf.* **1998**, *29*, 575–583. [CrossRef]
22. Fu, S.; Lauke, B. Analysis of mechanical properties of injection molded short glass fibre (SGF)/calcite/ABS composites. *J. Mater. Sci. Technol.* **1997**, *13*, 389.
23. Tjong, S.; Xu, S. Ternary polymer composites: PA6, 6/maleated SEBS/glass beads. *J. Appl. Polym. Sci.* **2001**, *81*, 3231–3237. [CrossRef]
24. Lee, C.H.; Khalina, A.; Lee, S.H. Importance of Interfacial Adhesion Condition on Characterization of Plant-Fiber-Reinforced Polymer Composites: A Review. *Polymers* **2021**, *13*, 438. [CrossRef] [PubMed]
25. Peng, X.; Fan, M.; Hartley, J.; Al-Zubaidy, M. Properties of natural fiber composites made by pultrusion process. *J. Compos. Mater.* **2012**, *46*, 237–246. [CrossRef]
26. Jain, K.; Shit, S.; Jain, S. Evaluation of mechanical & thermal properties of polypropylene—Palm kernel nut shell powder composites for green roof technology. *J. Inf. Knowl. Res. Mech. Eng.* **2013**, *2*, 456–459.
27. Singha, A.; Rana, A.K.; Jarial, R. Mechanical, dielectric and thermal properties of Grewia optiva fibers reinforced unsaturated polyester matrix based composites. *Mater. Des.* **2013**, *51*, 924–934. [CrossRef]
28. Uyor, U.O.; Popoola, A.P.I.; Popoola, O.M.; Aigbodion, V.S. Enhanced Thermal and Mechanical Properties of Polymer Reinforced with Slightly Functionalized Graphene Nanoplatelets. *J. Test. Evaluation* **2019**, *47*, 2681–2692. [CrossRef]

29. Nakamura, Y.; Yamaguchi, M.; Okubo, M.; Matsumoto, T. Effects of particle size on mechanical and impact properties of epoxy resin filled with spherical silica. *J. Appl. Polym. Sci.* **1992**, *45*, 1281–1289. [CrossRef]
30. Mamunya, Y.; Davydenko, V.; Pissis, P.; Lebedev, E. Electrical and thermal conductivity of polymers filled with metal powders. *Eur. Polym. J.* **2002**, *38*, 1887–1897. [CrossRef]
31. Shahrezaei, M.A.S.; Goharpey, F.; Foudazi, R. Effect of particle-particle and polymer-particle interactions on nanosilica aggregation in polystyrene. *Polym. Compos.* **2018**, *39*, 2904–2914. [CrossRef]
32. Meng, X.; Huang, Y.; Xie, Y.; Li, J.; Guan, M.; Wan, L.; Dong, Z.; Cao, J. Friction self-riveting welding between polymer matrix composites and metals. *Compos. Part A Appl. Sci. Manuf.* **2019**, *127*, 105624. [CrossRef]
33. Li, Y.; Meng, S.; Gong, Q.; Huang, Y.; Gan, J.; Zhao, M.; Liu, B.; Liu, L.; Zou, G.; Zhuang, D. Experimental and Theoretical Investigation of Laser Pretreatment on Strengthening the Heterojunction between Carbon Fiber-Reinforced Plastic and Aluminum Alloy. *ACS Appl. Mater. Interfaces* **2019**, *11*, 22005–22014. [CrossRef]
34. Rumble, J. *CRC Handbook of Chemistry and Physics;* CRC Press: Boca Raton, FL, USA, 2017.
35. BBalandin, A.A. Thermal properties of graphene and nanostructured carbon materials. *Nat. Mater.* **2011**, *10*, 569–581. [CrossRef]
36. Sonvane, Y.; Gupta, S.K.; Raval, P.; Lukačević, I.; Thakor, P.B. Length, width and roughness dependent thermal conductivity of graphene nanoribbons. *Chem. Phys. Lett.* **2015**, *634*, 16–19. [CrossRef]

polymers

MDPI

Article

3D-Printed PLA Molds for Natural Composites: Mechanical Properties of Green Wax-Based Composites

Mihai Alin Pop [1], Mihaela Cosnita [2,*], Cătălin Croitoru [3], Sebastian Marian Zaharia [4], Simona Matei [1] and Cosmin Spîrchez [5]

1 Materials Science Department, Transilvania University of Brasov, 29 Eroilor Ave., 500484 Brasov, Romania; mihai.pop@unitbv.ro (M.A.P.); simona.matei@unitbv.ro (S.M.)
2 Department of Product Design, Mechatronics and Environment, Transilvania University of Brasov, 29 Eroilor Ave., 500484 Brasov, Romania
3 Materials Engineering and Welding Department, Transilvania University of Brasov, 29 Eroilor Ave., 500484 Brasov, Romania; c.croitoru@unitbv.ro
4 Manufacturing Engineering Department, Faculty of Technological Engineering and Industrial Management, Transilvania University of Brasov, 29 Eroilor Ave., 500484 Brasov, Romania; zaharia_sebastian@unitbv.ro
5 Wood Processing and Furniture Design of Wood, Transilvania University of Brasov, 29 Eroilor Ave., 500484 Brasov, Romania; cosmin.spirchez@unitbv.ro
* Correspondence: mihaela.cosnita@unitbv.ro

Abstract: The first part of this paper is dedicated to obtaining 3D-printed molds using poly lactic acid (PLA) incorporating specific patterns, which have the potential to serve as the foundation for sound-absorbing panels for various industries and aviation. The molding production process was utilized to create all-natural environmentally friendly composites. These composites mainly comprise paper, beeswax, and fir resin, including automotive function as the matrices and binders. In addition, fillers, such as fir needles, rice flour, and *Equisetum arvense* (horsetail) powder, were added in varying amounts to achieve the desired properties. The mechanical properties of the resulting green composites, including impact and compressive strength, as well as maximum bending force value, were evaluated. The morphology and internal structure of the fractured samples were analyzed using scanning electron microscopy (SEM) and an optical microscopy. The highest impact strength was measured for the composites with beeswax, fir needles, recyclable paper, and beeswax fir resin and recyclable paper, 19.42 and 19.32 kJ/m^2, respectively, while the highest compressive strength was 4 MPa for the beeswax and horsetail-based green composite. Natural-material-based composites exhibited 60% higher mechanical performance compared to similar commercial products used in the automotive industry.

Keywords: green composites; beeswax; fir needles; recyclable paper; *Equisetum arvense*; mechanical properties

Citation: Pop, M.A.; Cosnita, M.; Croitoru, C.; Zaharia, S.M.; Matei, S.; Spîrchez, C. 3D-Printed PLA Molds for Natural Composites: Mechanical Properties of Green Wax-Based Composites. *Polymers* **2023**, *15*, 2487. https://doi.org/10.3390/polym15112487

Academic Editor: Paul Joseph

Received: 25 April 2023
Revised: 19 May 2023
Accepted: 22 May 2023
Published: 28 May 2023

1. Introduction

The circular (bioresources) economy is considered one of the most sustainable potential solutions to the global challenges of energy crises and environmental pollution. The research community has approached this issue from two main paths: recycling waste to develop novel materials and using natural resources to create new materials. One area of significant attention for researchers is the development of fiber-reinforced composite materials, which have the potential to be used in (semi) structural applications [1]. In the last decade, great attention has been paid by the research community to the developments of fiber (carbon, graphite, graphene, aramid, kevlar, natural, and wood fibers)—polymer (thermoplastic or thermosetting matrices) composites. Fiber-reinforced composite materials (FRPs) are widely used in modern products due to their exceptional combination of properties, including high specific strength and stiffness, low weight, durability, and resistance to creep and fatigue. Due to the special properties in relation to weight, composite materials have found a safe place in top industries, such as aerospace and aeronautics (applications for propellers of torpedoes, submarines, etc.) [2], in the automotive industry (components of transmission, brake discs, engine components) [2–7], in electronics and electrotechnics (manufacture of packages and layers of integrated circuit-plate supports on which electronic circuits are mounted to capacitors, electronic transducers or electronic filters) [8], and so on. However, achieving adequate interfacial strength between the filler and matrix often requires the use of toxic or expensive compatibility agents. Additionally, when synthetic fibers are used, the manufacturing process can become energy intensive due to the high-temperature curing required for the fiber–polymer blend [9,10].

Extensive studies in the specialized literature indicate that many composite materials with natural reinforcements use polymeric matrices that exhibit good or even very good physical–mechanical properties [11,12]. However, the advantage of these materials is diminished by the polluting factor over their lifetime, as well as the high costs and complex technological/chemical processes required for their manufacturing and the limited recycling options available at the end of their product lifecycle [13–15]. The environmental sustainability and cost effectiveness has pushed researchers' focus to be moved from synthetic FRP composites to more environmentally friendly fillers. Natural plant fibers have emerged as a suitable alternative to synthetic fibers due to their lightweight nature, renewability, sustainability, eco-friendliness, carbon neutrality, low cost, biodegradability, and availability. Moreover, these fibers provide good insulation, toughness, vibration damping, flexibility, and high specific strength and modulus [16,17]. Moreover, legislative pressures on the industrial sector have led to an increased demand for environmentally friendly materials. As a result, natural fibers have gained traction as a commonly used alternative to synthetic fibers in composite materials. Flax, hemp, jute, kenaf, banana, pineapple, and sisal are amongst the most commonly used natural plant fibers as reinforcements in composite materials [1,9,11,17–20]. Compared to synthetic carbon and glass fibers, natural plant fibers are less expensive, lighter, biodegradable, easy to produce, and environmentally friendly. Researchers are now exploring the use of natural fibers for higher-load applications by incorporating ceramic fillers and synthetic fibers as reinforcements. This shift towards biologically sourced and recyclable materials is driven by the growing demand for sustainable and environmentally friendly materials in order to diminish the burden on the environment [21,22]. A study on ramie-reinforced PLA composites was reported, with a positive impact on the mechanical properties (tensile strength) of ramie/PLA composites [17,19]. Farid et al. [23] investigated the use of natural fibers, including kenaf, jute, waste cotton, and flax, in blends with PP and polyester binders for sound-absorbing floor coverings. They found that the sound absorption coefficient increased with the frequency of sound [24]. This research highlights the potential of natural fibers in creating sustainable products. In this context, the assimilation of pine needles (PNs) through biotechnology can offer a valuable opportunity to enhance the circular bio-economy. Converting PNs into biomaterials and bio-energy can help reduce the reliance on petroleum products and promote a healthier environment by maintaining the energy–environment network [22–37].

Composite material development by using natural fibers promotes reductions in greenhouse gas emission, biodegradability, job availability increase, energy saving, easy decomposition, lightweight, high specific mechanical properties, less tool wear during processing, and less density, lowering the product cost. However, there are some shortcomings when using natural fiber composites due to their hydrophilic nature and the potential of moisture absorption, debonding the fillers from the matrix [30–34,38]. The CO_2 release during when obtaining natural fiber composites is negligible as compared to synthetic fiber [35]. The composite structure comprises biodegradable natural fibers, which have found significant applications as sound-proofing material for automobile components. Composite laminates with bamboo, cotton, and flax fibers with PLA fibers showed bending stiffness of 2.5 GPa, but the impact strength, another mechanical property required in such applications, was been reported [36,37].

Beeswax is derived from honey combs and is shown as a complex mixture of several chemical compounds mostly used in the food industry (glazing, carrier of additives in food, and texturizing agent in chewing gum production) [38]. Praveen B. et al. [38] blended cellulose triacetate (CTA) microspheres and beeswax in order to evaluate the controlled release of antidiabetic nateglinide [39]. Pine needles and natural materials were used to obtain valuable products, metabolites, and bio-energy [23,24,28]. Converting natural materials into valuable products not only promotes a sustainable energy–environment network but also reduces the use of conventional petroleum products, resulting in a positive impact on the environment and human health [32,34]. While technology has advanced significantly, virtual information storage cannot completely replace readable paper supports, leading to the continued generation of paper waste.

To the best of our knowledge, there are no previous studies on the fabrication of green composites by pouring natural material blends based on beeswax, fir resin, fir needles, and *Equisetum arvense* onto recyclable paper.

This work presents a novel approach for the development of eco-friendly composites, consisting of two stages. Firstly, specific molds were 3D-printed using PLA to mold the natural composite receipts based on recyclable paper, beeswax, fir resin, fir needles, rice, and *Equisetum arvense*. In the second stage, eco-composites were prepared through mold casting (pouring) several natural blends of beeswax, fir resin, with or without fir needle, rice, or *Equisetum* onto recyclable paper. This study aims to contribute to the development of sustainable materials and to reduce waste generation from paper consumption, while also exploring the potential of natural resources for composite material production.

The major advantage of these composite materials is that both the matrix and reinforcement are natural, biodegradable, and recyclable.

This study is part of a larger research program aimed at assessing the feasibility of using the obtained composites for various industrial applications, such as sound insulation, thermal-insulating panels, or shock absorbers. In this particular study, we will focus specifically on the mechanical properties of the composites. By examining the mechanical properties, we can better understand how these materials can be optimized to provide the necessary insulation and damping and to gain a more specific insight into their potential application domains.

2. Experimental Section

2.1. 3D-Printing Molds for Green Composite Development

The research employs a variety of natural materials, such as polymers, resins, waxes, and biomass waste, to produce composite materials with good sound-absorbing properties. These materials, which include beeswax, fir resin, ground fir needles, horsetail (*Equiseti herba*), rice flour, and paper pulp, offer a high degree of recyclability, low cost, and minimal environmental impact.

Natural wax is an excellent candidate for use as an organic matrix in sound-absorbing materials, due to its high content of saturated fatty acids, n-alkanes, and long-chain alcohols. Similarly, fir resin, with its abundance of resin acids, provides an ideal matrix material for these composites. Not only do these materials possess excellent sound-dissipating properties due to their amorphous nature but they are also easy to process, have low softening points, and act as effective binders for filler materials. This combination of properties makes them an attractive option for creating effective sound-absorbing materials at a low cost.

In the pre-testing phase, it was determined that the most suitable filler material for the composite was in ground form. This decision was based on the groundings' relatively large specific surface area, which enables them to bond more effectively with the matrix material and also helps to create a structure that is as loose as possible. This loose structure is particularly beneficial for sound absorption, as it allows sound waves to be effectively dissipated within the material.

Equisetum and fir needles are not only rich in lignocellulosic material but also in phenolic compounds, resin acids, and silicon dioxide (in the case of horsetail), which contribute to the strengthening and thermal stabilizing effect of wax and/or resin. Rice flour, on the other hand, contains starch, waxes, and proteins that are compatible with the chosen matrices, while milled paper (recycled cellulose) has a high specific surface area and absorbency compared to wax and resin. These unique properties of the natural materials make them excellent candidates for the development of eco-friendly composites with enhanced mechanical and thermal properties.

Thus, the following eco-composites were preliminarily analyzed from the point of view of SHORE D hardness:

(a) Beeswax (Ca) and ground horsetail (Cc)—a total of 14 composite materials were obtained. Percent of ground horsetail between 0 and 39.4%.
(b) Beeswax (Ca) and recycled paper (Hr)—a total of 18 composite materials were obtained. Percent of recycled paper was between 0 and 46%.
(c) Beeswax (Ca) and ground rice (Or)—a total of 10 composite materials were obtained. Percent of ground rice was between 0 and 86.5%.
(d) Beeswax (Ca) and ground fir needles (Abr)—a total of 10 composite materials were obtained. Percent of ground fir was between 0 and 86.5%.

A common/household grinding machine was used for grinding and bringing it to the desired size, and to determine the granulation size, a system equipped with mesh sieves between 0.036 mm and 0.5 mm diameter was used.

For each material in powder form, the amount was gradually increased step by step with 0.25 g and then calculated as a percentage of the entire composite.

First of all, all the elements (recycled cellulose, *Equiseti herba*, rice flour, and paper pulp) were chopped to the desired granulation, weighed, and prepared according to the preset recipes for the next step, namely mixing with beeswax.

The second step was to create the composites: beeswax was melted (up to a temperature of 100 °C) and mixed with various reinforcements (presented above) in different proportions until the reinforcer/hardener was maximally moistened.

Last step was to pour the composite while it was still in a pasty state in the mold.

The combinations that produced the best results were identified, and we decided to use these combinations as a basis for developing new composite materials, as follows:

1. Recyclable paper 100% denoted as S1 (as base for comparison with new composites)
2. Beeswax (50%) + fir resin (50%) denoted as S2
3. Beeswax (62.5%) + horsetail (37.5%) denoted as S3
4. Beeswax (55.56%) + recycled paper (44.44%) denoted as S4
5. Beeswax (45.5%) + milled rice (54.5%) denoted as S5
6. Beeswax (61.5%) + milled fir needles (38.5%) denoted as S6
7. Beeswax (31.25%) + fir resin (31.25%) + horsetail (37.5%) denoted as S7
8. Beeswax (27.78%) + fir resin (27.78%) + recycled paper (44.44%) denoted as S8

To assess the mechanical properties of the natural composite materials, including their flexural strength, impact resistance, and compression resistance, specific molds were designed using SolidWorks 2016 software. These molds were specifically tailored to the unique characteristics of the natural composite materials under study. The molds were then created using 3D-printing technology, utilizing the Fused Filament Fabrication (FFF) technique to produce positive molds.

Silicone rubber was subsequently cast into the positive molds, creating negative molds that were then used to shape the natural blends into their final form. These natural blends were poured into the negative molds, which were designed to be made from recyclable paper, enabling a sustainable and environmentally friendly production process. By utilizing this process, precise and accurate measurements of the natural composite materials' mechanical properties were obtained, providing valuable insights into their potential applications.

International Standard ISO 14125—Plastics, Subcommittee SC 13, Composites and reinforcement fibers—was the basis for testing the samples but not the only one; ISO 178 or ASTM D 790 was consulted, for example.

The photo images taken from the 3D-printed molds are presented in Figure 1.

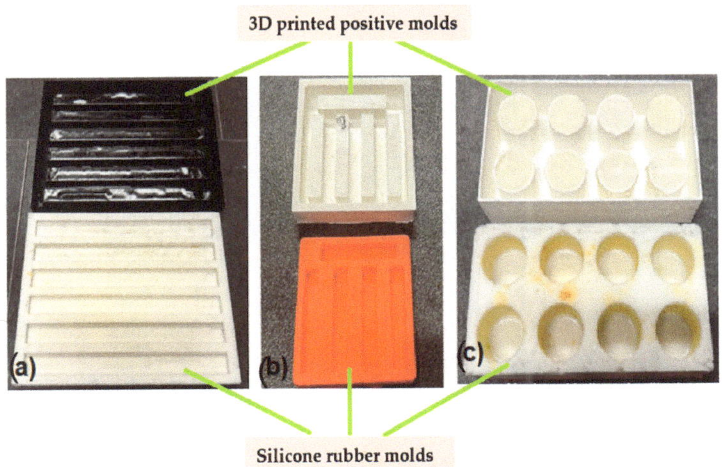

Figure 1. 3D-printed positive and negative molds for mechanical testing of the green composites: (**a**) for bending; (**b**) for impact; (**c**) for compression.

This method of obtaining the molds for testing the mechanical properties of the samples was chosen due to the particularities of these eco-composites: precise, simple, and efficient.

2.2. Materials and Green Composite Preparation

2.2.1. Materials

The green composite materials were developed using natural materials, with beeswax and fir resin serving as the matrix and fir needles, rice, and *Equisetum arvense* as the reinforcing materials. These materials were selected for their biodegradability, recyclability, and sustainable sourcing. Specifically, the fir resin and fir needles were collected during trips to coniferous forests in Brasov, Romania, from fallen branches on the ground. The natural beeswax was purchased from Apsirom SRL Vaslui, Romania, while the alimentary (grocery-grade) rice and *Equisetum arvense* were obtained from a local shop in Brasov, Romania.

The raw materials used for developing green composite materials are presented in Figure 2.

All these natural materials used for green composite preparation were dried and ground before their blending. After grinding the fir needles, the rice and *Equisetum arvense* grain diameter was under 0.5 mm.

2.2.2. Obtaining Composites

The following steps were taken to prepare the green composite materials:

(1) The base natural materials (beeswax and/or fir resin) were melted at a temperature of approximately 80–100 °C.
(2) The reinforcing natural filler powders (fir needles, rice, *Equisetum arvense*) were added in the proportions specified above.

(3) The natural material blend was mixed thoroughly.
(4) The blend was poured into silicone rubber molds, as shown in Figure 1, over the recyclable paper.

Figure 2. Raw natural materials used for the green composites.

The recyclable paper was prepared according to the process outlined in Figure 3.

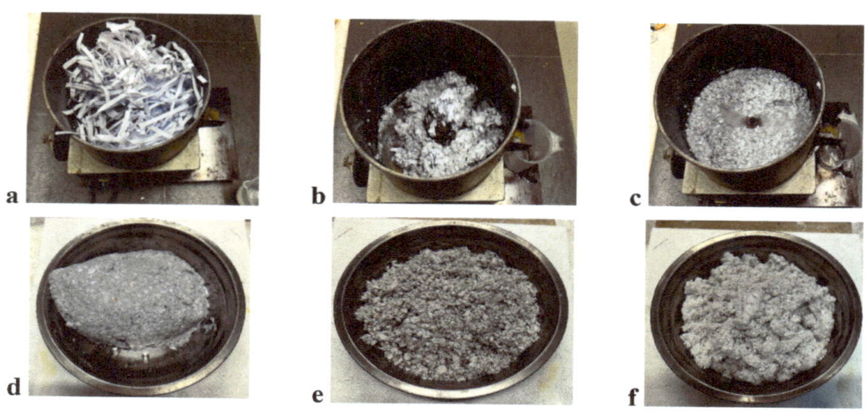

Figure 3. Steps in preparing the recyclable paper: (**a**) shredded paper; (**b**) water addition; (**c**) homogenizing with a blender; (**d**) paste sieving; (**e**) drying the paper paste; (**f**) grinding recyclable paper.

The green composites obtained after pouring the natural material blends onto the recyclable paper can be seen in Figure 4.

(a) (b)

Figure 4. Green composites prepared for mechanical tests; (**a**) casting in the silicon rubber molds; (**b**) samples removed from the molds.

All the natural composition blends were poured onto recyclable paper in the PLA molds.

2.3. Characterization Techniques

2.3.1. Mechanical Tests

Bending and compression strength and Young's modulus (E) for the composite structures (sandwich structures and wing sections) were performed on the WDW-150S universal testing machine with a constant crosshead speed of 10 mm/min for all test (Jinan Testing Equipment IE Corporation, Jinan, China). The impact (resilience) strength of the green composites obtained was measured on a Galdabini Impact 25 equipment, Cardano al Campo, Italia, with 25 kJ maximum energy. Compression and bending tests were performed with a speed of 10 mm/s. The impact energy of the impact hammer was 5.5 J. Five samples were tested for each mechanical test property, and the average values are reported.

2.3.2. Surface Morphology Analysis

Micrographs were obtained by using a scanning electron microscope (SEM), Hitachi, Japan, S3400N, type II, and the images were taken from the impact-fractured composite surface.

2.3.3. Optical Microscope

Images of the fractured composite surfaces were taken using an optical microscope type Leica (Arnhem, The Netherlands), Emspira 3 model.

3. Results and Disscusion

3.1. Mechanical Tests

The mechanical, thermal, and durability properties of a composite are largely determined by the interfacial adhesion zone in the composite system. The strength of this zone depends on several factors, including the physical–chemical properties of each composite component, their size and shape, mass ratio, dispersion grade of fillers into the matrix composite, preparation technique, and technological parameters used for composite preparation. In this study, we evaluated the mechanical properties of the green composites in terms of impact, compression, and bending maximum force, as shown in Figures 5–7.

Figure 5 shows the variation in impact resistance (resilience) for the natural composites. It is worth noting that the addition of the melted beeswax and fir resin positively influences the impact strength of the resulting composites (S2 and S3) compared to S1 (recyclable paper only). The addition of *Equisetum arvense* powder (S4) and ground rice (S5) negatively impacted the resilience value of the beeswax-based samples (S2), likely due to their higher modulus of elasticity, which includes silica-based compounds and unplastified starch.

Conversely, the addition of fir needle powders (S6) and fir resin (S8) led to an increase in shock resistance or resilience. The best resilience value was achieved with the S6 composite, which was based on beeswax and fir resin. Additionally, improvements in impact strength were recorded for S2, S6, S7, and S8, with the highest values found in S6 and S8 at 19.42 and 19.32 kJ/m^2, respectively. These results can be assigned to the mechanical strength provided by fir resin, as well as the viscoelastic properties of the cellulosic fibers, as noted by Jakob et al. [40]. It is widely recognized that cellulosic fibers significantly contribute to the mechanical enhancement of fiber-based composites [40–44]. These findings are consistent with a study conducted by Butnaru et al. [44], which demonstrated the superior thermal and mechanical properties of fir needles compared to fir cone and bark. Furthermore, in addition to their remarkable mechanical and thermal strength, fir needles also exhibit antioxidant properties, which are of great importance in the development of natural composite materials [35].

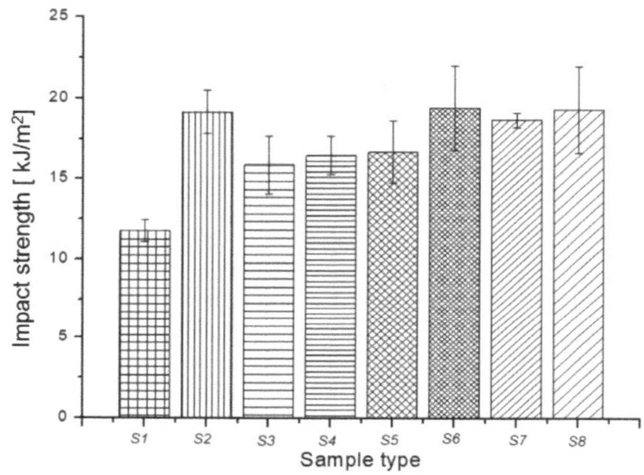

Figure 5. Impact strength variation for green composites.

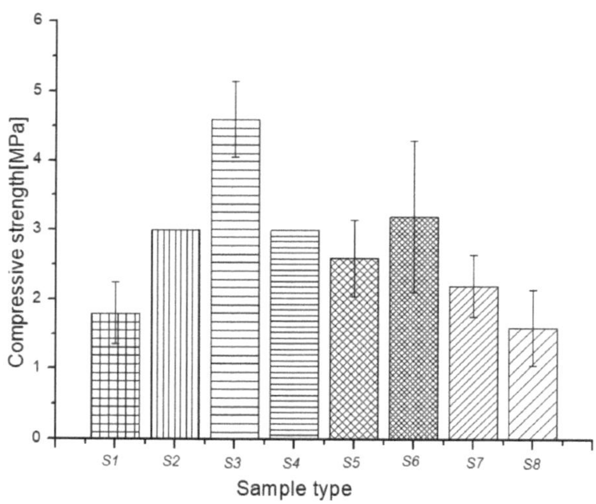

Figure 6. Compressive strength variation for green composites.

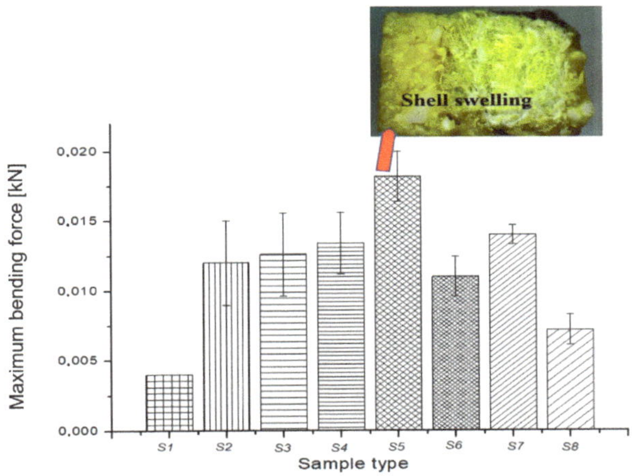

Figure 7. Maximum bending force variation for composites.

Figure 6 shows the compression strength of the natural blends poured onto recyclable paper. Sample S2, based on beeswax poured onto recyclable paper, exhibited a 66.6% increase in compression strength, while sample S3 showed a remarkable increase, exceeding 100%, from 1.8 MPa recorded for S1 (recyclable paper) to 4 MPa recorded for S3. This significant increase in compression strength can be attributed to the insertion of melted fir resin molecules through the paper fibers, as well as the formation of an ordered shell that covers the paper core, as observed in the next section using light microscopy. The hardened fir resin in S3 effectively reinforced the paper fibers, while the shell provided additional support, resulting in a remarkable increase in compression strength.

Ground rice addition caused a slight decrease in compressive strength for the S5 sample, while the addition of fir needle powder caused a slight increase in compressive strength for the S6 sample. The sample based on fir resin (S3) showed the best value for compressive strength. However, the addition of *Equisetum* powder and beeswax (S7) and beeswax alone (S8) had a strong negative effect on the compressive strength of sample S3.

In Figure 7, the maximum bending force of the composites is presented. The addition of ground fir needles to the S2 sample caused a decrease in the maximum bending force for the S6 composite. On the other hand, the addition of *Equisetum* powders (S4), ground rice (S5), and fir resin with *Equisetum* powders (S6) led to an increase in the maximum bending force for sample S2. Among all the samples, composite S5 based on beeswax and ground rice demonstrated the best value for maximum bending strength. This can be attributed to the swelling of the rice fibers, as can be observed in the inset surface fracture image of S5 in Figure 7. The rice fibers covered the paper core, resulting in higher bending force compared to other samples.

In this study, we compared the mechanical properties of the green composites, which were considered as the core materials for the final green products with those of similar commercial products from the automotive sector (see Table 1). We mechanically tested the commercial core product used as a sound-absorbing panel and found that its properties were inferior to those of the core material composites developed in this research. These results indicate that the green composites have the potential to be a superior alternative to the commercial products in terms of mechanical strength and could be applied in various fields, including automotive and sound-absorbing panels.

Table 1. Mechanical properties of the green composites compared to similar commercial products used in the automotive industry.

Mechanical Property/Commercial Automotive Sample/Green Composites	Impact Strength [kJ/m^2]	Compressive Strength [MPa]	Bending Force [N]
	11.41	<1	<3
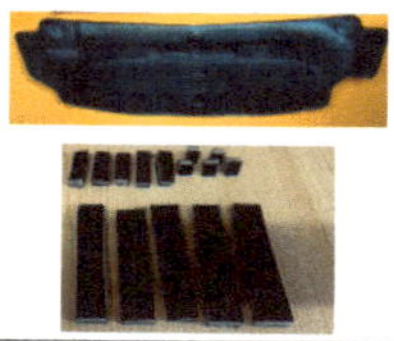	13.26	<1	<3
Green composites' mechanical properties			
	19.42	4.60	18.20

The results obtained from the mechanical tests on the proposed composites indicate that improvements and a different approach are necessary to further enhance these properties without sacrificing their advantages, such as being natural, ecological, recyclable, and sustainable. However, the improved physical–mechanical properties make these composites suitable for various applications beyond the automotive industry, including aeronautics and other fields where sound-absorbing panels are needed.

It was also observed that high mechanical properties are not always required for materials used in the automotive industry, particularly for the lining of the engine hood and linear luggage, where good sound-absorbing properties are more important. Overall, the use of natural, sustainable materials in these applications can provide a more environmentally friendly solution while maintaining or even improving performance.

3.2. Light Microscopy

To observe the effect of the natural material blend incorporation poured onto the recyclable paper matrix, optical images of the fractured green composites' surfaces were analyzed using light microscopy. Figure 8 displays representative images of all samples, where the lowest incorporation of natural material onto the recyclable paper matrix corresponds to sample S3, which was poured with fir resin. The rapid toughening of the fir resin forms a stiff shell that encapsulates the paper matrix, resulting in the highest compression strength of this sample compared to the others, as previously shown in the compression test with a recorded value of 4.6 MPa.

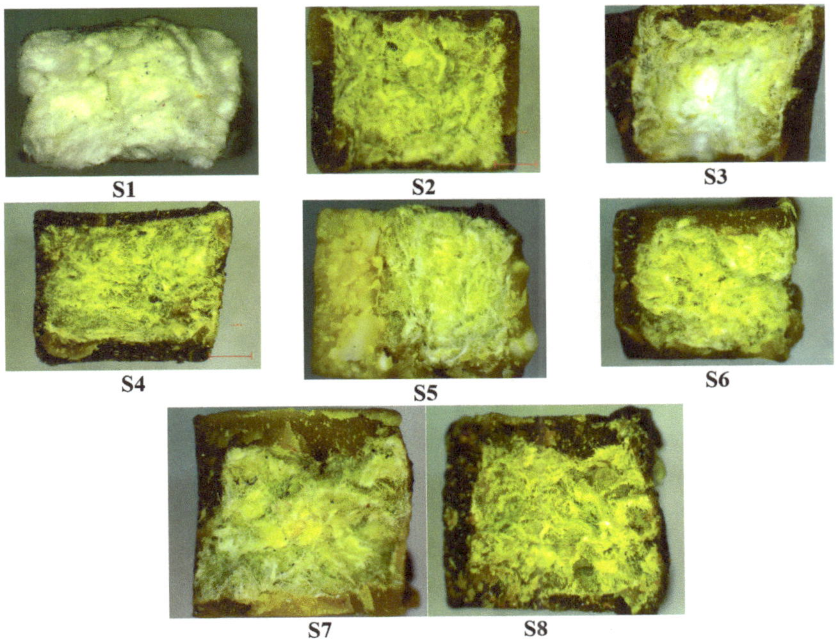

Figure 8. Fractured surfaces' microscope images of the green composites (30× magnification).

The observations from the light microscopy images show that the natural material blends were able to wet the paper core effectively. The dispersion of natural blends was remarkable for samples S2, S6, S7, and S8, with the blends encapsulating the paper core fibers, thereby adding mechanical strength to the resulting green composites. These findings confirm the results from the mechanical tests, particularly the high impact strength values of 19.16, 19.42, and 19.32 kJ/m^2 for the green composites denoted as S2, S6, and S8, respectively.

3.3. *Surface Morphlogy and Internal Structure*

Scanning electron microscopy was performed to investigate the surface morphology and interface structure of the impact-fractured composite samples with good mechanical strength (S2, S6, S7, and S8), as shown in Figure 9. The SEM images reveal a low-rugosity surface, indicating strong bonding between the composite components and reflecting good interface strength.

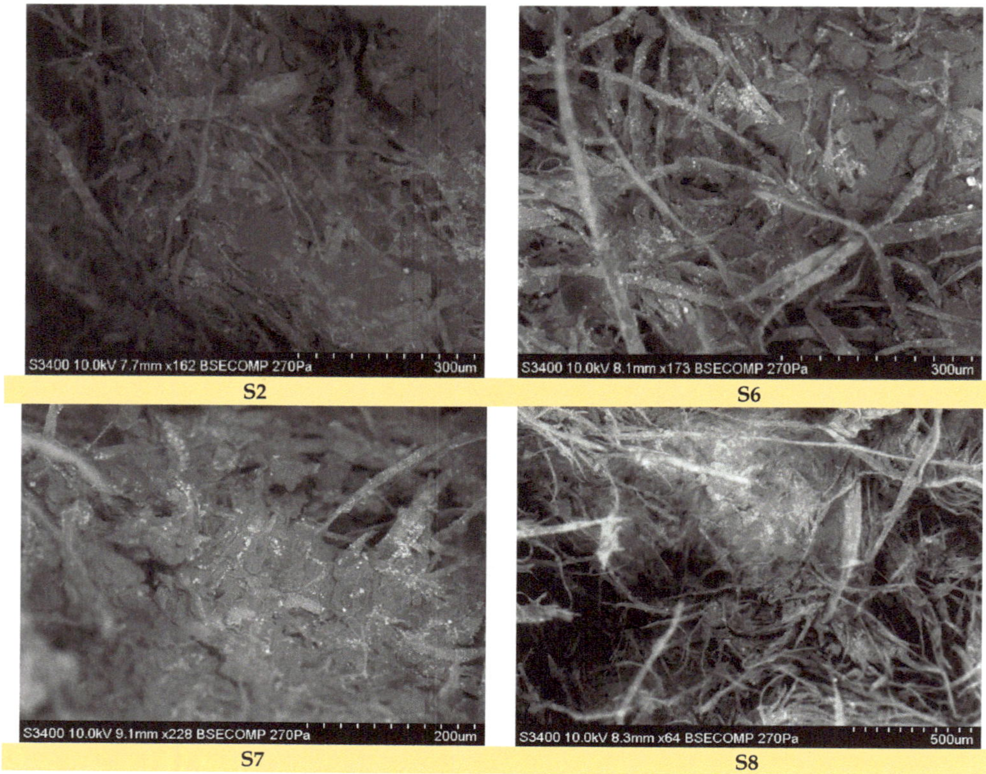

Figure 9. SEM images of fractured green composites' surfaces.

The SEM images obtained from the fractured S2, S6, and S7 samples demonstrate a uniform morphology, indicating a high dispersion of the melted natural materials (beeswax, fir resin with/without fir needles, and *Equisetum* fibers) throughout the internal structure of the paper fiber cores. This behavior, particularly in the case of the S2 sample, confirms its mechanical performance, as the mechanical tests showed an approximately doubled impact strength compared to the recyclable paper-only reference. The interface failure is observed as a matrix failure, with tearing and shearing of the matrix visible. The images in Figure 9 also show that even after composite fracture, the paper core fibers are still covered by the melted natural material blend.

4. Conclusions

- This paper presents the steps involved in obtaining 3D-printed PLA molds for natural composites made of recyclable paper, beeswax, fir resin, needles, rice, and *Equisetum arvense*, and it assesses the mechanical properties of the green composites;
- The highest impact strengths, provided mainly by fir resin and cellulose, were attributed to the samples with beeswax and fir needles S6 and beeswax fir resin and recyclable paper S8, 19.42 and 19.32 kJ/m^2, respectively, while the highest compressive strength was 4 MPa for the S3 sample. The rice fibers positively influenced the maximum bending for the S5 sample;
- The superiority of the green composites' mechanical properties was proved, considering the core materials for the final green products over similar commercial products from the automotive sector;

- The green composites exhibited over 60% higher mechanical properties compared to similar products from the market used as sound-absorbing core materials in the automotive industry;
- The physico-mechanical properties were directly influenced by the wetting degree of the powders for the beeswax and the fir resin and through the interaction between the complementary chemical groups from the fillers and the matrices (possibly silicates, phenolic carboxylic acids, resinic acids, and so on);
- Further research is ongoing to determine the thermal and sound-absorbing properties of these composites, as well as finding new solutions to enhance their physical–mechanical properties. One possible solution is combining the benefits of eco-friendly filament printing, such as polylactic acid, with the sound-absorbing properties of these new materials.

Author Contributions: Conceptualization, M.A.P. and M.C.; methodology, C.C., S.M.Z., M.A.P. and M.C; software, S.M. and C.S.; validation, C.C., investigation, M.A.P., C.C., S.M.Z. and M.C.; writing—original draft preparation, M.C., M.A.P., S.M. and C.S.; project administration, M.A.P. All authors have read and agreed to the published version of the manuscript.

Funding: This work was supported by a grant from the Ministry of Research, Innovation and Digitization, CNCS–UEFISCDI, project number PN-III-P1-1.1-TE-2021-0294, within PNCDI III.

Institutional Review Board Statement: Not applicable.

Data Availability Statement: The data presented in this study are available on request from the corresponding author.

Acknowledgments: This work was supported by a grant from the Ministry of Research, Innovation and Digitization, CNCS–UEFISCDI, project number PN-III-P1-1.1-TE-2021-0294, within PNCDI III. We also acknowledge the structural funds project PRO-DD (POS-CCE O.2.2.1, ID 123, SMIS 2637, ctr. No 11/2009) for partially providing the infrastructure used in this work at the CDI Institute of Transilvania University of Brasov.

Conflicts of Interest: The authors declare no conflict of interest. The funders had no role in the design of the study; in the collection, analyses, or interpretation of data; in the writing of the manuscript; or in the decision to publish the results.

References

1. Hoang, A.T.; Varbanov, P.S.; Nižetić, S.; Sirohi, R.; Pandey, A.; Luque, R.; Ng, K.H.; Pham, V.V. Perspective review on Municipal Solid Waste-to-energy route: Characteristics, management strategy, and role in circular economy. *J. Clean. Prod.* **2022**, *359*, 131897. [CrossRef]
2. Isitman, N.A.; Kaynak, C. Nanoclay and carbon nanotubes as potential synergists of an organophosphorus flame-retardant in poly(methyl methacrylate). *Polym. Degrad. Stab.* **2010**, *95*, 1523–1532. [CrossRef]
3. Gay, D. *Matériaux Composites*; Hermes: Paris, France, 1991.
4. Stefanescu, F.; Neagu, G.; Mihai, A. *Materiale Composite*; Publishing house "Didactica si Pedagogica": Bucuresti, Romania, 1996.
5. Abraham, T.; Bryant, R.W.; Mooney, P.J. The Prospects for Advanced Polymer-, Metal- and Ceramic-Matrix Composites. *JOM* **1988**, *40*, 46–48. [CrossRef]
6. Chawla, K.K. The high-temperature application of ceramic-matrix composites. *JOM* **1995**, *47*, 19–21. [CrossRef]
7. Baltes, L. *Materiale Amorfe—Materiale Compozite*; Publishing House "Transilvania" University: Brasov, Romania, 2003.
8. Sumesh, G.; Mailadil, T.S. Three-phase polymer–ceramic–metal composite for embedded capacitor applications. *Compos. Sci. Technol.* **2009**, *69*, 1298–1302.
9. Usmani, Z.; Sharma, M.; Awasthi, A.K.; Sivakumar, N.; Lukk, T.; Pecoraro, L.; Thakur, V.K.; Roberts, D.; Newbold, J.; Gupta, V.K. Bioprocessing of waste biomass for sustainable product development and minimizing environmental impact. *Bioresour. Technol.* **2021**, *322*, 124548. [CrossRef]
10. Rajak, D.K.; Pagar, D.D.; Menezes, P.L.; Linul, E. Fiber-Reinforced Polymer Composites: Manufacturing, Properties, and Applications. *Polymers* **2019**, *11*, 1667. [CrossRef]
11. Ismail, S.O.; Akpan, E.; Dhakal, H.N. Review on natural plant fibres and their hybrid composites for structural applications: Recent trends and future perspectives. *Compos. Part C Open Access* **2022**, *9*, 100322. [CrossRef]
12. Mueller, D.H.; Krobjilowski, A. New Discovery in the Properties of Composites Reinforced with Natural Fibers. *J. Ind. Text.* **2003**, *33*, 111–130. [CrossRef]

13. Rodríguez, E.; Petrucci, R.; Puglia, D.; Kenny, J.M.; Vázquez, A. Characterization of Composites Based on Natural and Glass Fibers Obtained by Vacuum Infusion. *J. Compos. Mater.* **2005**, *39*, 265–282. [CrossRef]
14. Mann, G.S.; Singh, L.P.; Kumar, P.; Singh, S. Green composites: A review of processing technologies and recent applications. *J. Thermoplast. Compos. Mater.* **2018**, *33*, 1145–1171. [CrossRef]
15. Syduzzaman, M.; Al Faruque, M.A.; Bilisik, K.; Naebe, M. Plant-Based Natural Fibre Reinforced Composites: A Review on Fabrication, Properties and Applications. *Coatings* **2020**, *10*, 973. [CrossRef]
16. Faruk, O.; Bledzki, A.K.; Fink, H.-P.; Sain, M. Progress Report on Natural Fiber Reinforced Composites. *Macromol. Mater. Eng.* **2014**, *299*, 9–26. [CrossRef]
17. Ates, B.; Koytepe, S.; Ulu, A.; Gurses, C.; Thakur, V.K. Chemistry, Structures, and Advanced Applications of Nanocomposites from Biorenewable Resources. *Chem. Rev.* **2020**, *120*, 9304–9362. [CrossRef] [PubMed]
18. Sharath Shekar, H.S.; Ramachandra, M. Green Composites: A Review. *Mater. Today Proc.* **2018**, *5*, 2518–2526. [CrossRef]
19. Islam, M.Z.; Sarker, M.E.; Rahman, M.M.; Islam, M.R.; Ahmed, A.; Mahmud, M.S.; Syduzzaman, M. Green composites from natural fibers and biopolymers: A review on processing, properties, and applications. *J. Reinf. Plast. Compos.* **2022**, *41*, 526–557. [CrossRef]
20. Ntimugura, F.; Vinai, R.; Harper, A.; Walker, P. Mechanical, thermal, hygroscopic and acoustic properties of bio-aggregates—lime and alkali—Activated insulating composite materials: A review of current status and prospects for miscanthus as an innovative resource in the South West of England. *Sustain. Mater. Technol.* **2020**, *26*, e00211. [CrossRef]
21. Núñez-Decap, M.; Wechsler-Pizarro, A.; Vidal-Vega, M. Mechanical, physical, thermal and morphological properties of polypropylene composite materials developed with particles of peach and cherry stones. *Sustain. Mater. Technol.* **2021**, *29*, e00300. [CrossRef]
22. Kumar, A.; Gupta, V.; Singh, S.; Saini, S.; Gaikwad, K.K. Pine needles lignocellulosic ethylene scavenging paper impregnated with nanozeolite for active packaging applications. *Ind. Crop. Prod.* **2021**, *170*, 113752. [CrossRef]
23. Farid, M.; Purniawan, A.; Rasyida, A.; Ramadhani, M.; Komariyah, S. Improvement of acoustical characteristics: Wideband bamboo based polymer composite. *IOP Conf. Ser. Mater. Sci. Eng.* **2017**, *223*, 012021. [CrossRef]
24. Paul, P.; Ahirwar, M.; Behera, B.K. Acoustic behaviour of needle punched nonwoven structures produced from various natural and synthetic fibers. *Appl. Acoust.* **2022**, *199*, 109043. [CrossRef]
25. Atabani, A.; Ali, I.; Naqvi, S.R.; Badruddin, I.A.; Aslam, M.; Mahmoud, E.; Almomani, F.; Juchelková, D.; Atelge, M.; Khan, T.Y. A state-of-the-art review on spent coffee ground (SCG) pyrolysis for future biorefinery. *Chemosphere* **2022**, *286*, 131730. [CrossRef] [PubMed]
26. Bhatia, S.K.; Palai, A.K.; Kumar, A.; Bhatia, R.K.; Patel, A.K.; Thakur, V.K.; Yang, Y.-H. Trends in renewable energy production employing biomass-based biochar. *Bioresour. Technol.* **2021**, *340*, 125644. [CrossRef]
27. Moscariello, C.; Matassa, S.; Pirozzi, F.; Esposito, G.; Papirio, S. Valorisation of industrial hemp (*Cannabis sativa* L.) biomass residues through acidogenic fermentation and co-fermentation for volatile fatty acids production. *Bioresour. Technol.* **2022**, *355*, 127289. [CrossRef]
28. Rana, A.K.; Guleria, S.; Gupta, V.K.; Thakur, V.K. Cellulosic pine needles-based biorefinery for a circular bioeconomy. *Bioresour. Technol.* **2023**, *367*, 128255. [CrossRef]
29. Sealy, C. Recyclable graphite-cellulose composite stronger than steel. *Mater. Today* **2019**, *28*, 7–8. [CrossRef]
30. Dong, C. Review of natural fibre-reinforced hybrid composites. *J. Reinf. Plast. Compos.* **2018**, *37*, 331–348. [CrossRef]
31. Khalid, M.Y.; Arif, Z.U.; Ahmed, W.; Arshad, H. Recent trends in recycling and reusing techniques of different plastic polymers and their composite materials. *Sustain. Mater. Technol.* **2022**, *31*, e00382. [CrossRef]
32. Angin, N.; Caylak, S.; Ertas, M.; Cavdar, A.D. Effect of alkyl ketene dimer on chemical and thermal properties of polylactic acid (PLA) hybrid composites. *Sustain. Mater. Technol.* **2022**, *32*, e00386. [CrossRef]
33. Patel, V.K.; Rawat, N. Physico-mechanical properties of sustainable Sagwan-Teak Wood Flour/Polyester Composites with/without gum rosin. *Sustain. Mater. Technol.* **2017**, *13*, 1–8. [CrossRef]
34. AL-Oqla, F.M.; Hayajneh, M.T.; Al-Shrida, M.M. Mechanical performance, thermal stability and morphological analysis of date palm fiber reinforced polypropylene composites toward functional bio-products. *Cellulose* **2022**, *29*, 3293–3309. [CrossRef]
35. Singh, M.K.; Tewari, R.; Zafar, S.; Rangappa, S.M.; Siengchin, S. A comprehensive review of various factors for application feasibility of natural fiber-reinforced polymer composites. *Results Mater.* **2023**, *17*, 100355. [CrossRef]
36. Zhang, J.; Khatibi, A.A.; Castanet, E.; Baum, T.; Komeily-Nia, Z.; Vroman, P.; Wang, X. Effect of natural fibre reinforcement on the sound and vibration damping properties of bio-composites compression moulded by nonwoven mats. *Compos. Commun.* **2019**, *13*, 12–17. [CrossRef]
37. Joseph-Leenose-Helen, J.; Noor, N.; Mushtaq, M.; Gani, A. Ultrasonics as a tool for development of pine-needle extract loaded bee wax edible packaging for value addition of Himalayan cheese. *Ultrason. Sonochem.* **2022**, *82*, 105914. [CrossRef]
38. Kajjari, P.B.; Manjeshwar, L.S.; Aminabhavi, T.M. Novel blend microspheres of cellulose triacetate and bee wax for the controlled release of nateglinide. *J. Ind. Eng. Chem.* **2014**, *20*, 397–404. [CrossRef]
39. Jakob, M.; Mahendran, A.R.; Gindl-Altmutter, W.; Bliem, P.; Konnerth, J.; Müller, U.; Veigel, S. The strength and stiffness of oriented wood and cellulose-fibre materials: A review. *Prog. Mater. Sci.* **2022**, *125*, 100916. [CrossRef]
40. Bazli, M.; Heitzmann, M.; Hernandez, B.V. Durability of fibre-reinforced polymer-wood composite members: An overview. *Compos. Struct.* **2022**, *295*, 115827. [CrossRef]

41. Taurino, R.; Bondioli, F.; Messori, M. Use of different kinds of waste in the construction of new polymer composites: Review. *Mater. Today Sustain.* **2023**, *21*, 100298. [CrossRef]
42. Cosnita, M.; Cazan, C.; Duta, A. The influence of inorganic additive on the water stability and mechanical properties of recycled rubber, polyethylene terephthalate, high density polyethylene and wood composites. *J. Clean. Prod.* **2017**, *165*, 630–636. [CrossRef]
43. Al-Oqla, F.M.; Sapuan, S.M. Natural fiber reinforced polymer composites in industrial applications: Feasibility of date palm fibers for sustainable automotive industry. *J. Clean. Prod.* **2014**, *66*, 347–354. [CrossRef]
44. Butnaru, E.; Brebu, M. The Thermochemical Conversion of Forestry Residues from Silver Fir (*Abies alba* Mill.) by Torrefaction and Pyrolysis. *Energies* **2022**, *15*, 3483. [CrossRef]

polymers

MDPI

Review

Insights into Anthropogenic Micro- and Nanoplastic Accumulation in Drinking Water Sources and Their Potential Effects on Human Health

Maria Râpă [1], Raluca Nicoleta Darie-Niță [2,*], Ecaterina Matei [1], Andra-Mihaela Predescu [1], Andrei-Constantin Berbecaru [1,*] and Cristian Predescu [1]

1 Faculty of Materials Science and Engineering, University Politehnica of Bucharest, 313 Splaiul Independentei, 060042 Bucharest, Romania; maria.rapa@upb.ro (M.R.); ecaterina.matei@upb.ro (E.M.); andra.predescu@upb.ro (A.-M.P.); cristian.predescu@upb.ro (C.P.)
2 Physical Chemistry of Polymers Department, Petru Poni Institute of Macromolecular Chemistry, 41A Grigore Ghica Voda Alley, 700487 Iasi, Romania
* Correspondence: darier@icmpp.ro (R.N.D.-N.); andrei.berbecaru@upb.ro (A.-C.B.)

Abstract: Anthropogenic microplastics (MPs) and nanoplastics (NPs) are ubiquitous pollutants found in aquatic, food, soil and air environments. Recently, drinking water for human consumption has been considered a significant pathway for ingestion of such plastic pollutants. Most of the analytical methods developed for detection and identification of MPs have been established for particles with sizes > 10 µm, but new analytical approaches are required to identify NPs below 1 µm. This review aims to evaluate the most recent information on the release of MPs and NPs in water sources intended for human consumption, specifically tap water and commercial bottled water. The potential effects on human health of dermal exposure, inhalation, and ingestion of these particles were examined. Emerging technologies used to remove MPs and/or NPs from drinking water sources and their advantages and limitations were also assessed. The main findings showed that the MPs with sizes > 10 µm were completely removed from drinking water treatment plants (DWTPs). The smallest NP identified using pyrolysis–gas chromatography–mass spectrometry (Pyr-GC/MS) had a diameter of 58 nm. Contamination with MPs/NPs can occur during the distribution of tap water to consumers, as well as when opening and closing screw caps of bottled water or when using recycled plastic or glass bottles for drinking water. In conclusion, this comprehensive study emphasizes the importance of a unified approach to detect MPs and NPs in drinking water, as well as raising the awareness of regulators, policymakers and the public about the impact of these pollutants, which pose a human health risk.

Keywords: drinking water treatment plant; microplastics; nanoplastics; tap water; bottled water; quantification; toxicological effect

Citation: Râpă, M.; Darie-Niță, R.N.; Matei, E.; Predescu, A.-M.; Berbecaru, A.-C.; Predescu, C. Insights into Anthropogenic Micro- and Nanoplastic Accumulation in Drinking Water Sources and Their Potential Effects on Human Health. *Polymers* **2023**, *15*, 2425. https://doi.org/10.3390/polym15112425

Academic Editor: Graeme Moad

Received: 15 February 2023
Revised: 13 May 2023
Accepted: 19 May 2023
Published: 23 May 2023

1. Introduction

Global plastic production, including fossil-based plastics, post-consumer recycled plastic, and bio-based plastics rose to 390.7 million tons in 2021 [1]. From approximately 300 million tons of plastics manufactured annually [2], it was estimated that 13 million tons of plastic waste enter rivers and oceans [3]. Due to the resistance of plastic waste to degradation, it currently causes a serious pollution for the environment [4]. The management of plastic waste is very deficient, for example, 9% of global plastic waste is recycled, 12% incinerated, and 79% disposed in landfills, dumps, and oceans [5]. Only 5.5 million tons of post-consumer recycled plastics were reintroduced to the European economy in 2021 [1].

In the aquatic environment, due to the continuous abiotic degradation, plastic items (macroplastics) eventually break down into smaller fragments less than 5 mm in size to a few nanometers, typically known as microplastics (MPs) and nanoplastics (NPs) [3,6–10].

The recognized size ranges for MPs and NPs are still confusing. According to ISO/TR 21960:2020, "microplastic" stands for "any solid plastic particle insoluble in water with any dimension between 1 µm and 1000 µm (=1 mm)," and "nanoplastic" is defined as "plastic particles smaller than 1 µm" [11]. According to the Committee for Risk Assessment (RAC) and Committee for Socio-Economic Analysis (SEAC), "microplastic" means "particles containing solid polymer, to which additives or other substances may have been added, and where $\geq$1% *w/w* of particles have (i) all dimensions 0.1 µm $\leq$ x $\leq$ 5 mm, or (ii) a length of 0.3 µm $\leq$ x $\leq$ 15 mm and length to diameter ratio of >3" [12]. The term "particles" is "a minute piece of matter with defined physical boundaries; a defined physical boundary is an interface" [12]. Some authors defined the dimension of MPs according to the ISO definition [6,8], whilst others reported a classification between <5 mm and 1 µm [3,9,13]. Other authors categorized NPs as plastic particles with dimensions < 100 nm [6,7,10]. In this paper, we assumed that plastic particles with dimensions ranging between <5 mm and 1 µm are MPs, while those measuring below 1 µm are NPs. The "microplastic" term should not be used for natural polymers that have not undergone chemically modification, except in the case of hydrolysis [12]. In accordance with their origin, MPs could be classified as primary (resulting from anthropogenic activities, cosmetics, textiles, personal care products) and secondary (derived from fragmentation of primary ones) [14]. MPs possess a hydrophobic nature and a morphology of microbeads, fibers, foils, pellets or fragments, while NPs have colloidal behavior [15].

The presence of MPs and NPs has been documented in the aquatic environment [16–22], food [23–26], soil [27–29] and air media worldwide [30,31]. Artificial turf used in sports fields and the degradation of larger plastic pieces from commercial packaging waste are the primary contributors to the presence of MP pollution in European waters [32]. The increase number in the number of scientific publications on the occurrence, detection, characterization and impact of MPs and NPs on aquatic organisms and human health indicates the true importance of these pollutants [33]. It is expected that MPs and NPs may have unique reactivity and bioavailability for aquatic organisms [34]. MPs and NPs remain in the environment for a long time due to their chemical stability after entering water compartments and greatly affect the function of aqueous systems, being considered ubiquitous pollutants [35,36]. However, in a natural aquatic environment, the microorganisms immediately colonize the surface of MPs and NPs, which forms biofilm that alters not only the physicochemical characteristics of MPs/NPs but also their mobility, stability, bioreactivity, settlement, and fate in the environment [37–39]. The mechanism of biofilm formation on the surface of MPs in a water environment involves (i) the attachment of microorganisms to the surface of MPs, (ii) secretion of extracellular polymers (EPS) by microorganisms, and, (iii) multiplication of microorganisms [39]. This biofilm occurs on the surface of MPs and leads to their degradation through fragmentation.

Biodegradable polymers are plastic materials with high molecular weights that break down into H_2O, CO_2, and microbial biomass final products over time with the help of naturally occurring microorganisms [40]. It is important to note that there is a distinction between biodegradable plastics and bio-based plastics, which are obtained from renewable resources as an alternative to petroleum resources but may not be biodegradable [40]. Studies showed that if the conditions for assuring biodegradability are not met, the biodegradable plastics contribute to the generation of MPs and NPs like conventional plastic materials [41–43]. A promising approach for the total removal of MPs from aqueous media could be the use of microorganisms [44,45]. Unfortunately, 100% degradation of biodegradable materials cannot be reached under natural environments, so the occurrence of biodegradable MPs could be an additional threat to the environment [40,46,47]. According to Wei et al. [48], biodegradable polymers produce more MPs in aqueous environments compared to conventional polymers because they hydrolyze faster in basic environments.

Both conventional and biodegradable polymers can contain chemical contaminants such as UV filters [49], preservatives, per- and polyfluoroalkyl substances (PFAS) [50] or flame retardants [51–53], which are considered potential carriers of MPs due to their strong hydrophobicity [54]. The migration and deposition of chemical contaminants can occur due to continuous fragmentation and an abundance of aged MPs [55] increasing the potential risk to human health. In addition, MPs can bind or adsorb emerging pollutants from the water environment, such as pharmaceuticals and personal care products [13,56–58], heavy metals, polycyclic aromatic hydrocarbons [54], and antibiotics [59,60], based on the specific polymer type, chemical and physical properties of MPs. Compared to MPs, NPs are considered to pose an increasing potential risk to ecosystems and human populations due to their higher specific surface-area-to-volume ratio, thus increasing the potential source of chemical contaminants [61].

According to the World Health Organization (WHO), people should consume between 3.7 L and 2.7 L of liquids per day depending on their body weight [62]. Most of these liquids comes from tap water or drinks made with tap water. A research paper reported average concentrations of 94.37 MPs/L in bottled water and 4.23 MPs/L in tap water [63]. Recently, WHO has called for a more advanced assessment of plastic pollution in the environment, following the fact that small plastic particles have been identified in 90% of bottled water [64]. This includes the identification of the sources and pathways of MPs, as well as the development of effective strategies to reduce plastic pollution and mitigate its impact on human health. Human ingestion of MPs from water bottles can occur through the use of disposable water bottles, bottles made from recycled plastic, or glass bottles [26]. The US Food and Drug Administration (FDA) has proposed tolerable levels of ingestible contaminant from recycled plastic to be less than 1.5 μg/person/day [10]. However, the potential daily ingested dose imposed by the FDA was exceeded for children and adults, where intakes of 87.8 mg/kg/body weight and 40.1 mg/kg/body weight, respectively, were reported [65]. According to Cox et al. [63], children and adults consume an average of approximately 79,828 MPs and 97,827 MPs per year through drinking water. Other studies have reported an annual intake of 2550–5100 MPs [62] or even 4.1×10^4 items [66] for the average consumer. The most commonly ingested types of MPs from drinking water were fibers and fragments [63].

The sources of drinking water for human consumption are tap water and bottled drinking water. Tap water is supplied by drinking water treatment plants (DWTPs), which play a vital role in ensuring water safety and meeting the social standards for human consumption [67,68]. Recently, the presence of MPs in DWTP was reported [69–71] and removal methods are still under investigation. From DWTPs, the unremoved MPs can enter water compartments used for daily human drinking water consumption, causing potential toxicological effects through ingestion, dermal exposure and inhalation [72–75]. Despite water being essential for life, few studies have been reported on the removal of MPs and NPs from drinking water sources [75] and their impact on human health. The risk of MPs uptake from drinking water is currently unpredictable. Furthermore, these plastic particles add to the plastic potentially ingested through the consumption of other foods/beverages, including sea salt, beer, food and seafood [76]. Until recently, significant knowledge regarding the detection and identification of these plastic pollutants, especially those in complex matrices present in raw and treated drinking water, was not available. This situation is a consequence of various factors, including the varied chemical composition and surface properties of MPs/NPs [77]. Additionally, distinguishing between NPs and natural matter can be challenging, and the aggregation of NPs can lead to changes in the solution ionic strength, which further complicates advanced characterization [78]. According to reports from the Agency for Toxic Substances and Disease Registry (ATSDR), polystyrene (PS) and polyvinyl chloride (PVC) are among the most carcinogenic contaminants [79]. The toxicological effects associated with MPs/NPs present in drinking water depend on the duration and intensity of exposure, as well as the susceptibility, gender and age of the host [73]. However, the toxicological mechanisms by which MPs/NPs affect human health are still unknown.

Nevertheless, the investigation of naturally occurring MPs and NPs in drinking water coming from treatment plants, tap and commercial water bottles in terms of quantification, toxicological effects on human health and methods of their elimination has not been fully explored been fully explored. In a recent review conducted for the period 2016–2021, it was reported that only nine studies had been devoted to the particle sizes and concentrations of MPs detected in bottled drinking water [73].

The main objectives of this review are: (a) to evaluate the main technologies applicable for removal of MPs/NPs from drinking water sources, such as DWTP, tap and bottled drinking water, (b) to review the innovative analytical methods used for the detection of MPs/NPs occurring in drinking water sources, and (c) to assess the potential toxicological risks associated with the human consumption of MPs/NPs from drinking water sources.

2. Methodology

Our search strategy involved the analysis of the most recent papers published in the last 10 years using the Web of Science database, the search terms being "microplastics/nanoplastics", "microplastics/nanoplastics in drinking water", "microplastics/nanoplastics in bottled water", "toxicological effect of microplastics/nanoplastics". The bibliographic survey was conducted by selecting papers based on their titles and abstracts, and later, analysis of the full-length articles.

Figure 1 shows the overview of the review structure.

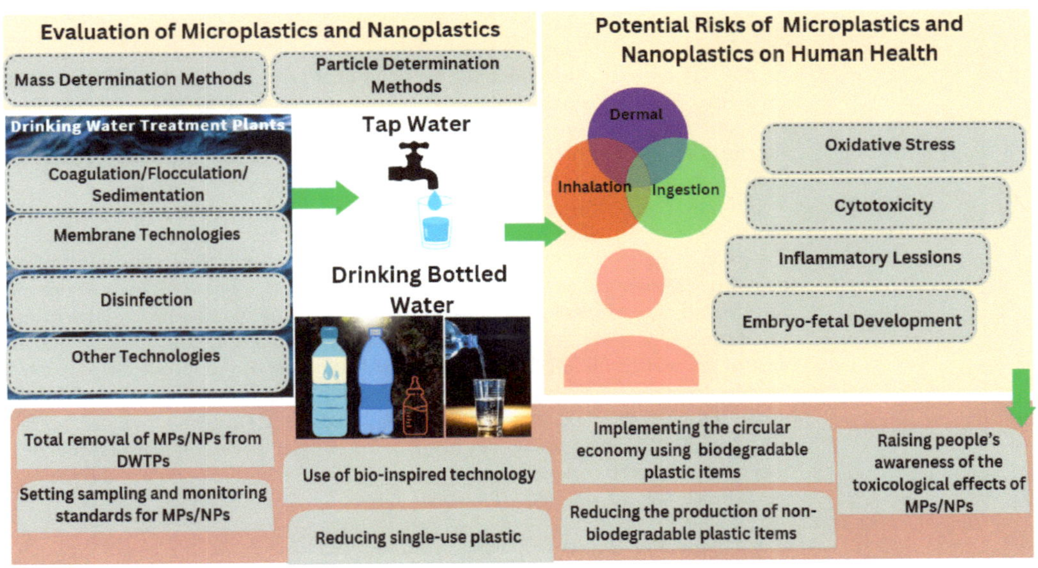

Figure 1. Overview of the review structure.

The first part of the review presents the conventional physicochemical methods used for removal of MPs and NPs from drinking water sources, such as coagulation–flocculation–sedimentation (CFS), filtration, ozonation, membrane filtration technology and adsorption, their performance and limitations, and new approaches for elimination of MPs and NPs.

Next, other physicochemical methods for removal of MPs and NPs, newly introduced or that may be suitably combined depending on the complexity and diversity of MPs/NPs in liquid media, are discussed. Multiple techniques are usually required to obtain information about the concentration, chemical identification, and shape of NPs. In this section, the newest methods based on mass determination and particle measurement are discussed with respect to the concentration, morphology and polymer type of MPs/NPs detected in DWTP, tap water or bottled drinking water. Methods based on mass determination are pyrolysis–gas chromatography–mass spectrometry (Pyr-GC/MS) and thermal desorption–proton transfer reaction–mass spectrometry (TD-PTR/MS). General methods based on particle measurement include micro-Fourier-transform infrared (μ-FTIR) spectroscopy, micro-Raman (μ-Raman) spectroscopy, single particle inductively coupled plasma mass spectrometry (SP-ICP-MS), dynamic light scattering (DLS), scanning electron microscopy–energy dispersive X-ray spectroscopy (SEM-EDX), and transmission electron microscopy (TEM).

Subsequently, the potential toxicological effects of MPs/NPs from drinking water sources on human health are assessed based on the size and concentration of MPs/NPs, as well as other additives incorporated into bottled drinking water. Most studies have been conducted on the exposure of marine organisms to commercial PS NPs purchased from authorized institutions or polyethylene terephthalate (PET) NPs. This is because PS NPs have been detected in the marine environment [80] and PET is the most widely used material for manufacturing bottles and packaging [81,82]. Few studies have been dedicated to investigating the effects of NPs released into drinking water sources on human health.

Finally, this paper highlights the main legal, technical, and social measures necessary to reduce the risks associated with contamination of drinking water sources with MPs/NPs and summarizes the key conclusions.

3. Results and Discussions

3.1. *Physicochemical Methods for Removal of Micro- and Nanoplastics from Drinking Water Sources*

DWTPs are processes with multiple stages that provide safe drinking water for human consumption [83]. Conventional methods used to remove MPs/NPs from DWTPs include coagulation–flocculation–sedimentation (CFS), filtration [3,17,18,84], ozonation [19,85], and chlorination [85]. The treatment process in a DWTP influences the quality of the drinking water produced [72].

In this section, the most frequently used physicochemical methods for removing of MPs/NPs from drinking water sources have been assessed (Table S1). The effectiveness of MPs/NP removal from drinking water sources depends on various factors, such as their physicochemical parameters (concentration, molecular weight, size, shape, age), the adopted treatment method, the type and dosage of coagulant, the type of filtration, water source, as well as the presence of mixture of natural compounds, named natural organic matter (NOM) [72]. Also, the old and worn components from the DWTP could potentially be a potential source of MPs in drinking water [69]. Studies have reported the successful removal of intentionally added MPs/NPs in water [3,17,54,78,85], as well as those present in DWTPs [69–71]. Recently, it was demonstrated that MPs with sizes > 10 μm can be completely removed from DWTPs, and over 80% of the removed MPs had a dimension > 1 μm [86].

3.1.1. Coagulation–Flocculation–Sedimentation (CFS)

Coagulation is an important conventional method used for the removal of pollutants from drinking water sources [54]. Most studies carried out for plastic contaminant removal from drinking water have used spherical engineered particles, especially PS (Table S1), without surface roughness or biofilm. A few studies have dealt with the monitoring of small MPs (<10 μm) in drinking water [3,17,78]. The presence of NPs in drinking water treatments has been reported [3], but their complete removal has not yet been revealed. The mechanism of PS NP removal was attributed to physical retention and straining processes [3].

PE MPs of three sizes (10–20 μm, 45–53 μm, and 106–125 μm), PS NPs (180 nm in size), and PS MPs (1.2 μm in size) were removed from drinking water by conventional treatment using coagulation–flocculation combined with sedimentation (CFS) and granular filtration in the presence of $Al_2(SO_4)_3$ and diallyldimethylammonium chloride (polyDADMAC) as coagulant aid [17]. The results showed that granular filtration was more efficient than the CFS method, resulting in a 99.9% removal rate for polyethylene (PE) MPs with dimensions ranging from 106 to 125 μm [17]. Additionally, the presence of biofilm increased the efficiency of the CFS treatment, improving the removal of MPs from <2.0% to 16.5% [17]. The occurrence of biofilms can significantly modify MP characteristics (size, shape, and density) and subsequently the efficiency of water treatment. Other authors demonstrated that the biofilm formed on the surface of MPs/NPs increased the removal efficiency in column experiments conducted with aged sand from 43% to 77% [17,78]. The explanation consists in the presence of humic substances from NOM [87] facilitating the positively charged coagulants to adsorb NPs with a negative surface charge [88,89].

Regarding the concentration of coagulant, the data showed that 20 ppm is the typical maximum coagulant concentration used for effective drinking water treatment [17]. Aluminum sulfate [$Al_2(SO_4)_3$] or alum, polyaluminum chloride (PACl) [(Al(OH)m Cl(3-m))n], polyDADMAC, aluminum chloride ($AlCl_3$), and iron chloride ($FeCl_3$) are the most commonly employed coagulants in water treatment [17,54,71,88]. The effect of two conventional coagulants, PACl and $FeCl_3$, on NP removal from three bottled mineral waters and Lake Geneva was investigated [88]. It was found that at lower doses required for the coagulation of NPs, PACl was more effective than $FeCl_3$ [88]. PACl is a conventional inorganic coagulant preferred for the treatment of water because it assures a low concentration of residual metal in treated water [83]. The efficiency of PACl was also reported by other authors [3].

The removal efficiency of engineered PE MPs from deionized water containing humic acid in the presence of $FeCl_3{\cdot}6H_2O$ and $AlCl_3{\cdot}6H_2O$ and sodium bicarbonate ($NaHCO_3$) as a buffer followed by the ultrafiltration membrane technology was investigated by Ma et al. [54]. $AlCl_3{\cdot}6H_2O$ was found to be more effective in removing PE MPs than Fe-based salts. Both coagulant agents led to increased removal efficiency with decreased particle sizes of PE MPs. However, a lower removal efficiency of 25.83% ± 2.91% after coagulation and slight membrane fouling were recorded.

Other coagulants and coagulant aisd typically used in the CFS method, combined with filtration, are alum at a concentration of 20 ppm and polyDADMAC at 0.5 ppm [17]. PolyDADMAC is a type of high-molecular-weight polymer coagulant aid that helps in bridging and binding of particles, improving floc strength and achieving optimal floc size, leading to a higher rate of sedimentation. The authors reported a significant difference in the removal efficiency of the polyDADMAC coagulation aid (13.6% ± 6.8% for particle sizes 45–53 μm) compared to raw water and CFS treatment [17]. This behavior was explained by the ability of polyDADMAC to bind the particles, strengthen the floc, and increase the sedimentation rate [17].

Another strategy for removing engineered PE MPs with sizes ranging from 10 to 100 μm from synthetic water was to stain them with Nile red dye and use alum and alum combined with cationic polyamine-coated (PC) sand as coagulants [90]. PC sand (500 mg/L) combined with an alum dose of 20 mg/L showed the highest removal rate (92.7%) compared to using alum alone. The dimension, shape and surface morphology of the MPs played a significant role in the coagulation and flocculation mechanism. The order of MP removal was observed to be elongated-rough (ER) > elongated-smooth (ES) > spherical-rough (SR) > spherical-smooth (SS) based on the results of a flocculation kinetic study [90].

An overall efficiency of 82.1–88.6% was achieved in a DWTP that utilized various processes, such as coagulation–flocculation–sedimentation, sand filtration, ozonation, and granular activated carbon (GAC) filtration, for the removal of MPs [68].

Instead, the use of aluminum salt coagulant combined with sand filtration led to a removal efficiency of 93% ± 5% for MP removal (especially PS and polyester) in a DWTP from Spain [69]. In another study [70], the pre-disinfection with liquid chlorine gas, followed by the addition of alum coagulant (at a flow rate of 3000 L/h), purification with a pulse clarification system, sand filtration, and post-disinfection steps led to 85% removal efficiency in DWTPs.

Velasco et al. [71] compared the effectiveness of coagulant, sand filtration, and activated carbon (AC) for removing MPs such as synthetic fibers (cotton, viscose, and cellulose) from a DWTP. The results indicated that the use of PACl coagulant and AC filter led to a higher removal efficiency of MPs (97 ± 3%). This removal efficiency was compared to a lower rate of 89% when the coagulant was not used. Sand filtration already demonstrated high removal efficiency (95% without coagulant and 92% with coagulant), emphasizing the significance of this step [71]. However, other authors [68] reported that the sand filtration played no significant role in the removal of MPs. They achieved a removal efficiency of only 29.0–44.4% compared to coagulation/sedimentation methods. In terms of the shapes of MPs eliminated, it was found that fibers accounted for 96% when the coagulant, sand, and AC filtration were used [71]. In the case of CFS followed by ozonation integrated with GAC, the percentage of fiber elimination ranged from 51.6% to 78.9% [68].

Disadvantages associated with the coagulation treatments include reduced removal of MPs/NPs, high coagulant consumption [67], and increased presence of residual Al-based salts in the case of Al coagulants, which can have adverse effects on human health [91]. While the CFS technique alone is insufficient for removing plastic pollutants from drinking water, combining it with granular filtration techniques has been successful in removing MPs/NPs larger than 100 μm from DWTPs [17]. Special attention should be given to the configuration of DWTPs and operating parameters for the removal of both MPs and NPs from drinking water.

3.1.2. Disinfection Technologies

During the process of disinfecting drinking water, MPs/NPs may pass through the filter and enter the municipal water supply network, eventually reaching consumers' taps. To address this issue, ozonation and chlorination, which are frequently employed in water treatment plants, have been proposed as potential methods for degrading MPs/NPs [85].

In one study [85], a concentration of 2.5 μg/L PS NPs in water was investigated to determine the effectiveness of ozonation and chlorination technologies for the degradation and mineralization of PS NPs. The average ozone dosage in drinking water was maintained at 4.1 mg/L, while 2.5 mg/L was the concentration of chlorine (in the form of hypochlorite salts) for chlorination. After 30 min, ozonation resulted in the removal of 96.3% of PS NPs, while chlorination only removed 4.2%. This significant difference was attributed to the increased hydrophilicity of PS NPs, which occurred due to the introduction of oxygen-containing groups on the surface during the ozonation treatment. It has been demonstrated through Pyr-GC/MS that ozonation is more effective than chlorination for the destruction of PS NPs from DWTPs. Chlorination, on the other hand, resulted in the formation of a shorter macromolecular chain due to a destruction of a small number of C-C bonds [85]. The Pyr-GC/MS spectrum of PS NPs after ozonation showed several new signals at retention times of 2.286 min, 3.393 min, 11.992 min, 12.091 min, 16.294 min, 17.137 min and 20.046 min, which corresponded to acetic acid (CH_3COOH), phenol (C_6H_6O), acetophenone (C_8H_8O), hydroquinone ($C_6H_6O_2$), methylbenzaldehyde (C_8H_8O), dimethyl acetophenone ($C_{10}H_{12}O$), and phenylpropionic acid ($C_9H_{10}O_2$), respectively. In contrast, the spectrum of PS NPs after chlorination was identical to that observed before this treatment (Figure 2) [85].

In another paper, ozonation and three successive filtration media involving rapid sand, AC, and slow sand filtration were used for the removal of palladium (Pd)-labeled NPs from real DWTPs [78]. A minor impact related to transport through columns was observed when the NPs were pretreated with ozone. Slow sand filtration promotes biofilm formation and results in high efficiency for NP removal (99.5%).

Conversely, the amount of MPs slightly increased in the treated water after the ozonation process [68]. The authors explained the rise of MPs particles during ozonation as a result of the destruction of residual organic matter attached to the MPs, as well as the breaking of MPs due to the cutting force of the water flow.

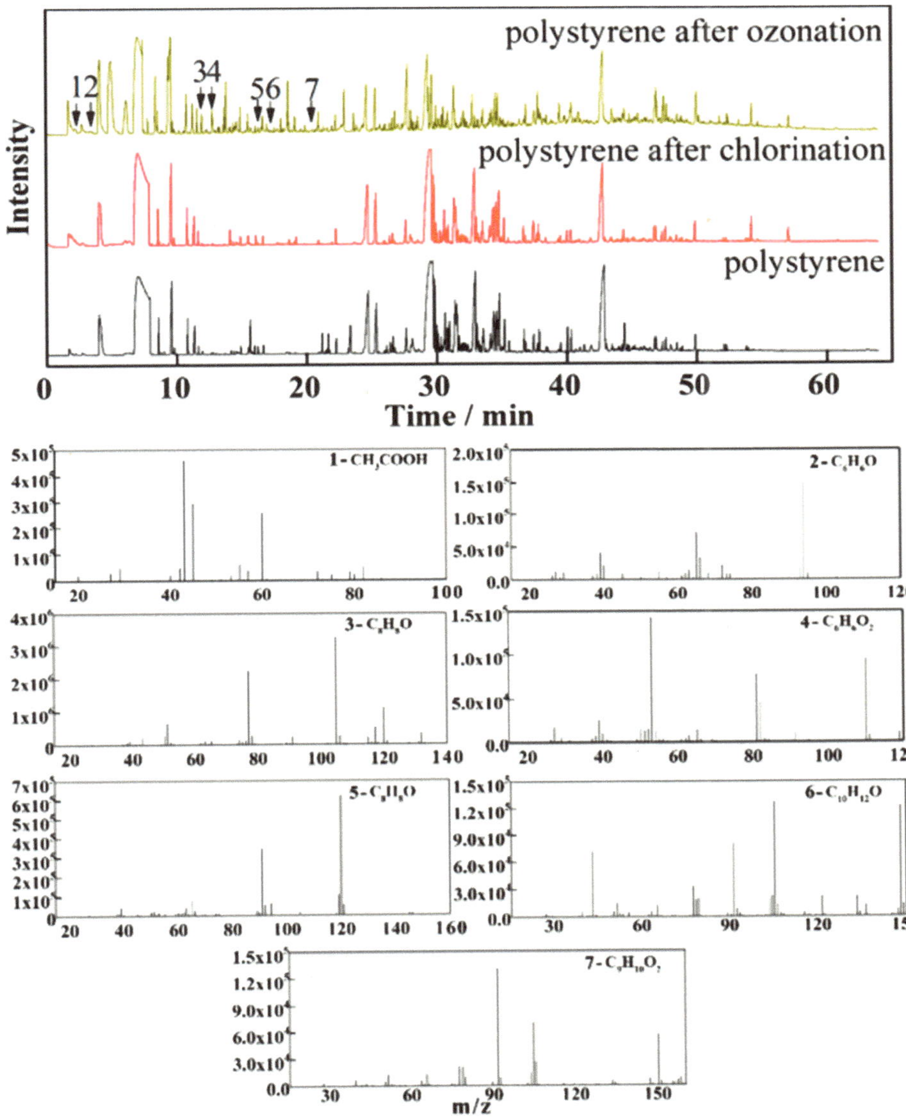

Figure 2. Pyr-GC/MS spectra of PS NPs, PS NPs after ozonation, PS NPs after chlorination, and the seven mass spectra from PS NPs after ozonation. Reprinted with permission from [85]. 1—acetic acid (CH_3COOH), 2—phenol (C_6H_6O), 3—acetophenone (C_8H_8O), 4—hydroquinone ($C_6H_6O_2$), 5—methylbenzaldehyde (C_8H_8O), 6—dimethyl acetophenone ($C_{10}H_{12}O$) and, 7—phenylpropionic acid ($C_9H_{10}O_2$).

3.1.3. Adsorption

Adsorption is widely used for the removal of contaminants from aqueous solutions, being cost-effective and environmentally friendly [92–94]. The possible factors affecting the removal efficiency of MPs/NPs by adsorption include: pH, temperature, adsorbent types, dissolved organic matter (DOM), and ions [72]. pH and temperature are considered the two most important factors. pH affects the adsorption efficiency mainly by influencing the charge on the surface of MPs and the adsorbent, whereas temperature can influence the adsorbate diffusion rate and equilibrium capacity; higher temperatures led to high MPs adsorption [72].

GAC is the most widely used adsorbent material in the drinking water purification process due to its high surface area and porous structure [3,95]. The efficiency of positively charged PS NPs hydrazine was evaluated following sand and GAC filtrations in the main DWTP from Geneva (Switzerland), with and without coagulant agent [3]. An increased removal of PS NPs was found in the case of GAC filtration compared to sand filtration, explained by the adsorption capacity of GAG. By using PACl coagulant, the filtration efficiency increased up to 99.4% ± 1.1%, because the retention of large NPs aggregates was improved and surface charge of PS NPs was reduced, leading to repulsive forces between NPs and filter media, thus improving both their retention and removal.

Another study on the removal of NPs from DWTP by using CFS and sand/GAC filtration revealed a difference in the NP removal mechanism between those two techniques, NPs of 200 nm being more efficiently removed by CFS, while some smaller NPs (50 nm) were better removed by GAC filtration [84].

During the simulation of the dynamic adsorption process of engineered PS NPs by GAC filtration of drinking water, characterized by a surface area of 759 m^2/g and pore volume of 0.357 cm^3/g, a very stable structure of GAC was revealed, its abundant pore structure of 3.09 nm and good regeneration (90% after the first cycle) making this adsorbent effective for adsorption of PS NPs [14]. The addition of Ca^{2+} increased the ionic strength by reducing the electrostatic repulsion between PS NPs. This behavior favors the aggregation and retention of PS NPs in the GAC pores.

Recently, Sajid et al. [96] highlighted the role of metal–organic frameworks, bio-based nanomaterials, carbon-based nanomaterials, and layered double hydroxides as adsorbents for the elimination of MPs from aqueous media. In this vein, Martin et al. [97] exploited the ability of iron oxide nanoparticles (IONPs), synthesized via green chemistry methods and further coated with different hydrophobic or amphiphilic coatings, to separate, concentrate, and remove NPs from water by magnetic separation. Thus, 1000-multifilament yarn of 30 μm diameter, PE nurdles and PE fibers were used as models for the removal of NPs from freshwater and placed in contact with coated iron oxide nanoparticles (IONPs). All PE nurdles and fibers were collected with a simple 2-inch permanent NdFeB magnet [42].

However, at the end of their use, the adsorbents must be regenerated by removing the adsorbed MPs/NPs; otherwise, they present a potential risk of returning to the environment [72].

3.1.4. Membrane Filtration Process

Membrane filtration is currently employed to remove emerging pollutants from contaminated waters [98]. For instance, vacuum-assisted filtration with an inorganic filter membrane (for example, Whatman Anodisc, 47 mm diameter, 0.2 μm pore size) was used to separate NPs from bottled and tap water [62]. Consequently, the total amounts of MPs particles found in tap water and plastic bottled drinking water ranged from 0.99 to 26 MPs/L [62]. In the case of the water deposited in glass bottles, the detected MPs were below the limit of quantification (LOQ) by μ-FTIR analysis. The composition of the detected MPs was: 38% PE, 25% PS, 22% and polypropylene (PP), polyamide (PA), and polyurethane (PU) in small amounts [62]. The most commonly used polymers for production of water bottles are PET for the body [16,99] and high-density polyethylene (HDPE) for the caps [100,101]. However, the source of MPs/NPs contamination depends on the bottle manufacturing technology.

Recently, Cherian et al. [102] reported the performance of three point-of-use (POU) water treatment devices containing GAC, ion exchange (IX), and microfiltration (MF) components used by consumers for optimal MP removal from drinking water. PE and PET as fragments and nylon as fibers were used as spiked MPs. The study showed that the POU device containing GAC and IX did not effectively remove MPs. The POU device containing GAC, IX and filtration membrane with pore size > 1 μm exhibited 78–86% and 94–100% removal for polyvinyl chloride (PVC) and PET, respectively, while, the best performance was reported in the case of the POU device including MF with pore size > 0.2 μm, GAC and IX, when 100% PVC and 94% PET were removed. High nylon fiber removal was observed in all POU devices. This study highlighted the importance of membrane- and small pore-associated membrane filtration process for MP removal.

Ultrafiltration units are often installed as a single filtration stage in newly constructed plants [103]. With pore sizes of about 10 nm, high NP removal efficiencies are expected through such treatment processes. Barbier et al. [104] showed, for the first time, that the nanofiltration process is more efficient than conventional methods (coagulation–flocculation, settling, sand filtration, ozonation, GAC filtration, UV treatment, and chlorination) used to remove MPs from DWTPs. Thus, by applying nanofiltration for MP removal from three DWTPs, the authors reported concentrations ranging from 7.4 to 45.0 MPs/L in the inlet water, while these decreased to 0.260 MPs/L in the case of outlet drinking water. The overall removal rate was >99%, and the MPs identified were PP, PE and PET.

Even though drinking water treatment processes should minimize the chances of MPs entering tap water, thus reducing their risk to consumers, membrane filtration can increase the number of MPs/NPs in some cases. MPs/NPs can be released in the membrane filtration step and further adsorb halogenated by-products resulting from chemical purification, thus becoming a secondary pollution during long-distance water transport [105]. For example, a concentration of MPs/NPs of 67.81 ng/L was found after using membrane filtration compared to 13.23 ng/L detected in tap water [106]. Membrane filtration is considered to exhibit a high potential ecological risk due to the membrane destruction during water filtration and clogged pores [106].

3.1.5. Other Technologies

Another suitable approach for removing low-density polyethylene (LDPE) suspended on the water surface could be dissolved air flotation (DAF) water treatment. The purpose of DAF is to diffuse the air in the form of fine bubbles in the drinking water, followed by flotation of the suspended particles and their final removal by skimming [17]. The removal efficiency of MPs and NPs was evidenced by the analysis of sediment captured by filtration of 250–500 mL of raw water using filter membranes with various pore sizes (pore size of 25 nm for NPs with 180 nm particle size, and pore size of 200 nm for NPs with 1.2 μm, 10–20 μm, 45–53 μm, and 106–125 μm particle sizes) [17]. The filtration removal efficiency was not dependent on the size of the plastic particles. For example, by using five different sizes of fluorescent plastic particles, namely, 106–125 μm, 45–53 μm, 10–20 μm, 1.2 μm, and 180 nm, the corresponding removal efficiencies were reported to be 99.9% ± 0.1%, 97.0% ± 3.0%, 86.9% ± 4.9%, 94.9% ± 0. 4%, and 98.9% ± 0.7%, respectively [17]. Removal efficiency by filtration treatment was affected by multiple mechanisms, such as straining, interception, gravitational sedimentation, diffusion, and particle attachment/detachment.

In addition to the traditional methods (filtration, coagulation, centrifugation, flocculation, and disinfection) used to remove MPs/NPs from drinking water, new effective methods involve the use of microorganism-based degradation, membrane separation with a reactor, and photocatalysis [107].

A membrane bioreactor can be used as an effective technology in the treatment of drinking water. For instance, the removal of PVC with a concentration of 10 particles/L daily and size < 5 μm from synthetic water by using a membrane bioreactor was studied by Li et al. [91]. Besides a removal rate of organic matter and ammonia over 80% and 95%, respectively, higher membrane fouling and irreversible membrane fouling were observed in the case of PVC contamination.

An eco-friendly and sustainable method for removing MPs from water involves using visible light to activate a photocatalytic process [108]. This approach employs glass fiber substrates to capture low-density MPs, such as PP, while also supporting the photocatalyst material. The process uses zinc oxide nanorods immobilized onto the glass fiber substrates in a flow-through system to break down spherical PP MPs suspended in water via visible light irradiation. After two weeks of this treatment, the average particle volume decreased by 65%. Gas chromatography–mass spectrometry was used to identify the primary by-products of photodegradation, which were found to be mostly non-toxic according to the existing literature.

A new approach that could help remove MPs/NPs from contaminated water involves using algal cells as bio-scavengers [109]. These cells bind the particles to their surfaces or incorporate them into their own cells, filtering them from the water. The polluted biomass can then be further processed downstream through microalgal cultivation, together with sustainable biofuel production, ultimately destroying the MPs/NPs.

All the methods used to remove MPs/NPs from DWTPs require special attention, considering the emergence of new MPs/NPs during the distribution network to tap water consumers. Also, it is important to note that these technologies may not remove all MPs from drinking water sources, and a combination of treatment methods may be necessary to achieve the desired level of removal.

3.2. Analytical Methods for Monitoring Micro- and Nanoplastics in Drinking Water Sources

A critical evaluation of applicable methods used for the identification and quantification of MPs/NPs was recently performed by Cella et al. [110], Ivleva et al. [111], Liu et al. [112] and Lee et al. [113]. Criteria used for the quantitative evaluation of the quality of MP concentration data were well reviewed by Koelmans et al. [114]. In this section, the sensitive methods based on mass or particle size needed for identification and quantification of MPs from DWTPs, tap water and bottled drinking water sources have been reviewed.

Table 1 shows the characteristics of MPs/NPs found in drinking water sources in terms of size, concentration, polymer type or morphology monitored by different analytical techniques.

Table 1. Characteristics of MPs/NPs identified in drinking water sources.

Source	Method	Characteristics of MPs/NPs	Polymer Type	Ref.
Three DWTPs	FTIR and Raman	Concentrations of 443 ± 10, 338 ± 76 and 628 ± 28 particles/L	PET, PP, PE	[16]
1200 L to 2500 L water from DWTPs (sampled in 2014 in Germany)	μ-FTIR microscopy coupled to a FPA detector	Four blank samples contained 45 ± 22 fibers, 18% were black and 78% were transparent; DWTPs show 0.7 fibers/m^3, with sizes ranging from 50 to 150 μm	Control: PP, SAN DWTPs: PE, PA, PES, PVC or epoxy resin	[115]
DWTP, Germany	μ-Raman	MPs size varying from 50 to 5000 μm; Shape of MPs: 83.3 % fragments and 16.7% fibers	37.8% PE, 31% PP and 24.4% PS were the most commonly found plastic in the analyzed samples	[116]
38 samples were collected from different tap waters from China	μ-Raman	Concentration of MPs was varying from 0 to 1,247 particles/L Distribution of MPs according to different size classes was: 31.25 to 100% for 1–50 μm; 1.47 to 31.25% for 50–100 μm; 1.72 to 31.25% for 100–300 μm; 1.18 to 7.69% for 300–500 μm; 1.72 to 11.76% for 500–5000 μm	26.8% PE, 24.4% PP, 22% compounds of PE and PP, 7.3% polyphenylene sulfite (PPS), 6.5% PS, 3.3% PET, and 9.8% other	[117]
DWTPs	LDIR, optical microscopy	MPs decreased after pre-treatment (80–99%); Size of MPs ranging from 20 to 500 μm; concentration of 2 MPs/L	PA, PET, PE, rubbers, chlorinated PE	[118]

Table 1. *Cont.*

Source	Method	Characteristics of MPs/NPs	Polymer Type	Ref.
Two DWTPs and ten tap water samples (from Iran)	Density separation techniques, digestion, observation, μ-Raman and FTIR, and SEM	An average of 22–51.8 MPs/m^3 for DWTPs; A high concentration of particles in tap water (85–390 MPs/m^3) compared to those found in DWTPs	PS	[119]
Two tap water samples (collected from Saudi Arabia)	μ-FTIR	First sample: 1.8 MPs/L Second sample: <LOQ	PE	[62]
159 samples of tap water collected between January and April of 2017, from 14 countries	FTIR	Concentration ranging from 0 to 61 MPs/L, with an overall mean of 5.45 MPs/L; 98.3% MPs were identified as fibers, and the remaining particles were identified as fragments or films	Not mentioned	[120]
Tap water sample (from China)	FTIR, AFM-IR and Pyr-GC/MS	The most frequently occurring particles had a size ranging from 58 to 255 nm, and a concentration between 1.67–2.08 μg/L	PE, PP, PS, PVC, PA	[121]
Two brands of bottled water in PET bottles	TD-GC/MS combined with TFU; super-resolution optical nanoscopy with microsphere lens; DLS	Size ranging between 66–605 nm	Degradation products of PET: phthalate derivatives and ethyl *p*-ethoxybenzoate	[122]
Mineral water bottles (0.5 L) consisted of transparent PET, with cap made of white HDPE	SPES and μ-Raman	*Size distribution: fewer than 10% of particles have a dimension of* 0.38 ± 0.03 μm, and *fewer than 9% of particles have a dimension of* 1.04 ± 0.14 μm	HDPE, PET	[123]
Four mineral water bottles	FTIR	Concentrations ranging from <1 MP/L to 317 ± 257 MPs/L, with particle sizes ≥ 11 μm	PE, PP, PS, polyester, PVC, EvOH, and PA	[124]
32 samples were collected from 21 different brands of mineral waters (from Bavarian location)	μ-Raman	Single use PET bottles: 2649 ± 2857 MPs/L Reusable PET bottles: 4889 ± 5432 MPs/L Glass bottles: $6292 \pm 10{,}521$ MPs/L Single and reusable PET: 95% of the plastic particles < 5 μm and 50% < 1.5 μm Glass bottle: ~15% of plastic particles were between 5 μm and 10 μm, and ~7% > 10 μm	PET for PET water bottle; PE (46%), PP (23%) and a styrene-butadiene-copolymer (14%) for glass water bottle	[125]
Ten mineral waters, either still or sparkling, in PET plastic bottles (from Catania, Italy)	SEM, density, statistical analysis	MPs with a mean diameter of 2.44 μm $\pm$ 0.66 μm were detected on PET surface	Not mentioned	[65]
Drinking water stored in PC and PP bottles (from China)	LDIR chemical imaging system, TEM	53 to 393 particles/mL during 100 opening/closing cycles	PC, PP	[126]
63 drinking water samples collected from decentralized refill kiosks in the Mexico City	ATR-FTIR	11 to 860 MPs/L from which: 65% were fibers, 28% fragments, and 7% films	PET, PA, vinyl polymers, polyacetals, cellophane	[127]

Focal plane array (FPA), styrene acrylonitrile (SAN), tangential flow ultrafiltration (TFU), dynamic light scattering (DLS), single particle extinction and scattering (SPES), atomic force microscopy–infrared spectroscopy (AFM-IR), ethylene vinyl alcohol (EvOH), polyamide (PA), polycarbonate (PC), laser direct infrared (LDIR), ATR-FTIR (attenuated total reflectance–Fourier-transform infrared spectroscopy).

3.2.1. Methods Based on Mass Determination

The most used laboratory methods for MP/NP identification based on mass determination are Pyr-GC/MS [85,121,128,129], TD-GS/MS [122], SP-ICP-MS [9], surface-enhanced Raman spectroscopy (SERS) [130], ICP-MS [131], MALDI-ToF/MS [132], TD-PTR/MS [133], quantitative proton nuclear magnetic resonance (q^1H NMR) HPLC [134], and differential scanning calorimetry (DSC) [135]. Scarce results have been reported regarding the detection of MPs in drinking water sources, possible due to a lack of standardized methods, limits of quantification, high cost of analytical tools, as well as variety of plastic pollutants in drinking water sources [76]. The analytical methods developed for the investigation of NPs are still uncertain, mainly due to limitations in laboratory workflow and the low sensitivity of analytical tools, which result in limited signals [136,137].

Funck et al. [128] developed a method for quantification of MPs from 3500 L water collected from DWTP by Pyr-GC/MS analysis using cascade filtration with mesh sizes of 100 μm, 50 μm and 10 μm and a platinum filament having dimensions of 20 mm × 5 mm for sample application. Thus, the quantification limits for PS and PE were 0.03 μg and 1 μg with a relative standard deviation of 11%.

A recent study has reported the quantification of NPs found in bottled water down to 1 nm [122]. Huang et al. [122], developed a new method based on thermal desorption (TD) into a gas chromatography–mass spectrometry (GC/MS) system coupled with tangential flow ultrafiltration (TFU) and evaporation techniques. The principle consisted in the concentration of NPs in retentate fluids by dewatering and desalting, thus preventing the loss of solids. The authors demonstrated the presence of PET-degradation products (from bottles) during a thermal mechanism and exposure to light. The advantages of this TD-GC/MS technique compared with other techniques were a higher resolution analysis without the use of organic solvent, low sample volume for NP determination, and analysis time (27 min per sample) [122].

Li et al. [121] successfully used a sequential filtration with inorganic filters, followed by identification of chemical groups using a micro-zone through atomic force microscopy (AFM) coupled with infrared spectroscopy (AFM-IR), and, finally, employed Pyr-GC/MS for the identification of polymer in tap water [121]. The authors quantified NPs ranging from 58 nm to 255 nm as polyolefins, PS, PVC, PA, and some plastic additives.

A suspension of PS22 model NPs conjugated with AuNPs@gel at a concentration of 1×10^{12} particles/L was used for the development of the SP-ICP-MS method to quantify NPs up to 1 μm and a concentration of 8.4×10^5 NPs/L in drinking water and tap water samples [9]. The strategy of using SP-ICP-MS for the detection and quantification of NPs in water sources is based on the oxidation signature of aged plastic debris, and the conjugation of carboxyl groups from the surface of NPs particles with functionalized positively charged metal (gold)-containing NPs (AuNPs).

Lin et al. [130] confirmed that PE particles were gradually released over time in plastic cup and bottled mineral water samples during irradiation by using of SERS. Analysis of the signal spectrum collected in 15 s revealed that the PE concentrations measured for plastic cups and bottles were 3751 ± 0.19 ng/mL and 1522 ± 0.21 ng/mL, respectively, after 240 min. The detection limit for NPs was 1.6 ng/mL when copper oxide/silver NPs (CuO/Ag NPs) were used as SERS substrate (Figure 3). Similar concentrations of MPs ranging from 1.67–2.08 μg/L were found in tap water by using Pyr-GC/MS [121].

A nanowell-enhanced Raman spectroscopy (NWERS) substrate, composed of self-assembled SiO_2 sputtered with silver films (SiO_2 PC@Ag), was developed for PS NP detection from tap and bottled drinking water, with a size < 200 nm and a limit of detection (LOD) of 5 μg/L [138].

Realistic PET NPs from bottled water labeled with an iridium-containing organic molecular agent were detected in liver, spleen, lung and kidney via inductively coupled plasma mass spectroscopy (ICP-MS) [131].

Asymmetric flow-field flow fractionation (AF4) and multiangle and dynamic light scattering (DLS-MADLS) methodologies were used by Villacorta et al. [139] for monitoring PET NPs from plastic water bottles. To tackle the real NP sample without metal contamination, the authors used diamond burrs to obtain uniform and representative samples with a size of about 100 nm for the investigation of potential health risks.

3.2.2. Methods Based on Particle Determination

The best-known methods used in the laboratory for MP/NP identification based on particle determination are FTIR [16,119,121,124], μ-FTIR [62,115], μ-Raman [116,123,125], SERS [130], SEM [65] and DLS [88]. An LDIR chemical imaging system is also a useful analytical tool based on the μ-FTIR technique for detection of the number of MPs, polymer types and sizes of MPs in the case of diameters > 20 μm [118,126].

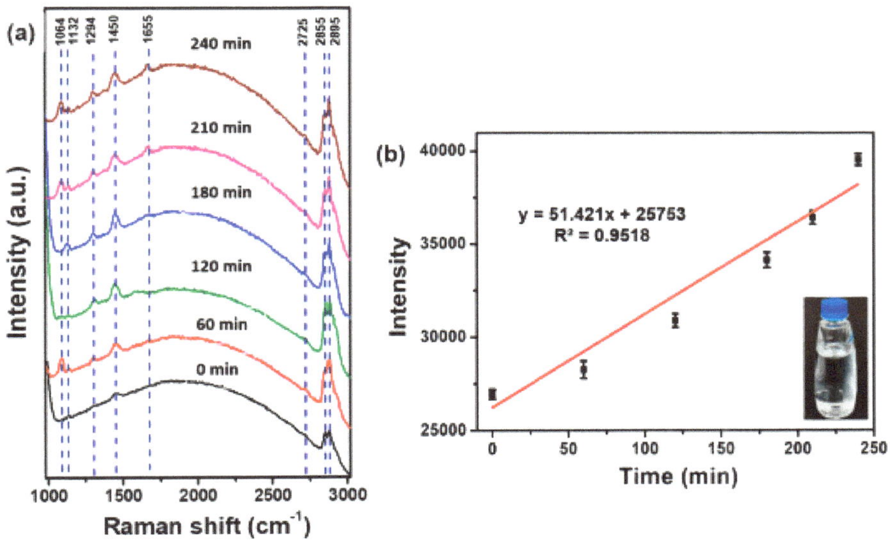

Figure 3. (**a**) SERS spectra of PE particles released from bottled mineral water after four irradiation times: 0, 60, 120, 180, 210 and 240 min. (**b**) Linear regression of the peak absorption band at 2895 cm^{-1} versus time over the full 240 min experiment. Each spectrum was determined three times. Reprinted with permission from [130].

An interesting review on the statistics of MP presence in drinking water sources, published in 2020, reported an average concentration of MPs in conventional water sources of 2.2×10^3 items/m^3, with an identified particle size usually > 50 μm [86].

Pivokonsky et al. [16] studied the removal capacity of plastic particles at three DWTPs and their concentrations in treated water. Plastic fragments and fibers were the most common shapes found in drinking water. These fragments are supposed to occur in drinking water during the breakdown of macroplastics, while fibers could be supplied from the discharge of washing machines into sewage waters.

The aggregation of positively charged PS NPs with a diameter of 15 nm and a specific surface area of 40–60 m^2/g at a concentration of 10 mg/L introduced into three types of commercial bottled mineral waters was observed at a point of zero charge (pzc), pH_{pzc} of 9.9 ± 0.1 [88]. Positively charged PS NPs have been reported to be difficult to coagulate and remove from surface and mineral drinking waters [88].

Various treatments applied in DWTPs, as well as different analytical tools and geographic location, led to different concentrations of MPs. μ-FTIR spectroscopy is a suitable technique for the identification and classification of MPs, with sizes ranging between 25 and 500 μm [62]. Microplastics, such as PE, followed by PS, and PET, were identified using μ-FTIR in 17 out of 30 samples [62]. In treatment sludge from a DWTP (Germany), a volume of 1000 m^3 raw water intake was analyzed by μ-Raman spectroscopy to detect the abundance and type of MPs with sizes > 50 μm over a period of 3 h. A figure of 196 ± 42 MPs/m^3 was found [116]. Similar concentrations (338–628 MPs/L) were reported from another three DWTPs by using FTIR and Raman spectroscopy [16]. μ-Raman and SEM allowed the detection of 6614 ± 1132 particles/L, from which the predominant types were PET, PE, PP, polyacrylamide (PAM), PS and PVC [68]. Interesting was the presence of PAM MPs in the treated water, denoting its use in the removal of suspended particles. However, it is considered that the product derived from PAM is more toxic than to polymers and the dosage of PAM should be according to the regulations in force [83]. Small concentrations of MPs were detected by using LDIR and optical microscopy techniques at various stages of a

DWTP (2 MPs/L) [118], purification and density separation followed by FTIR and SEM revealing concentrations of 0.022–0.051 MPs/L [119]. Additionally, μ-FTIR coupled with an FPA detector detected 0.7 MPs/m^3 [115].

The concentration of MPs in water can vary depending on the source and treatment processes. Tong et al. [117] conducted a study in which particles larger than 300 μm were found in 38 tap water samples. Fragments were found to be the most common morphotype in most tap water samples, followed by fibers and spheres. Plastic particles found in tap water may originate from the use of PE and PP in pipes used for distributing drinking water. It is possible that the water treatment process does not effectively remove or break down larger particles. A comparative study addressing the amount of MPs detected in DWTPs and tap waters showed high concentrations of MPs in tap waters compared to the levels found in DWTPs [119], implying the possible release of MPs from plastic pipes.

According to the research conducted by Zhang et al. [17], 81% of MPs were found in 159 samples of tap water, while 93% of MPs were detected in 259 samples of bottled drinking water from 11 different brands. MPs/NPs can contaminate bottled drinking water when the screw cap is opened and closed [101,123] or due to degradation compounds associated with PET [122]. The larger plastic particles found in bottled drinking water, compared to those found in tap water, are attributed to the types of material used for the bottles and caps [16]. For example, the opening of a plastic bottle can produce 0.46–250 MP particles/cm [140]. Weisser et al. [124] also found that 81% of MPs detected in bottled drinking water were attributed to abrasion of the PE-based cap sealing material. Thick-necked plastic bottles were found to release more MPs than thin-necked glass bottles [126]. Plastic food packaging is also a considerable source of the release of MPs/NPs [22,50,110,141]. On average, isolated MPs weighing 3 mg to 38 mg were detected in each consumer plastic food container [141].

The dimension of plastic particles detected in bottled drinking water varies from 58 to 255 nm [121], 66–605 nm [122], and ≥11 μm [124]. In a remarkable study conducted by Oßmann et al. [125], plastic particles < 5 μm in size were identified in water packaged in single-use PET, reusable PET and glass bottles. The MPs found in glass bottles were explained based on the age of the bottle. The authors prepared the water samples for μ-Raman spectroscopy analysis by mixing an equimolar amount of 250 g/L ethylenediaminetetraacetic acid tetrasodium salt (EDTA) corresponding to Ca^{2+} and Mg^{2+} ions in water with the water sample, for 15 min, by adding 3 mL of 100 g/L sodium dodecyl sulfate (SDS), then vacuum filtration through an aluminum-coated PC membrane filter with pore size of 0.4 μm. Of interest is the high concentration of 384 ± 468 MPs/L found in the blank sample, in which PP, PS, PE and PET were identified.

Single-particle extinction and scattering (SPES) determines the number and size distribution of particles. Quantitative and qualitative analyses of NPs released in drinking water plastic bottles under realistic conditions show that the PE sealing of the bottles released particles with a size distribution ranging from several hundreds of nanometers to about 1 μm, and estimated a mass release in the order of a few tenths of nanograms per opening/closing cycle [123]. The physicochemical characteristics of the produced secondary NPs were influenced by mechanical stress, making their identification difficult. The combination of SPES and μ-Raman represents the minimum set of techniques required for the application of NP quantification and identification methodologies in simple matrices, such as drinking water. The size distribution of NPs released from a package measured using the SPES technique showed 10–90% distribution for population A of $D_{10}\_A = 0.38 \pm 0.03$ μm and $D_{90}\_A = 1.04 \pm 0.14$ μm, respectively [123] (Figure 4).

A novel qualitative characterization of PET MPs derived from PET bottled water was proposed by Asamoah et al. [142], who analyzed the optical surface roughness together with the speckle contrast of the rough MPs. The use of an optical sensor prototype to detect the flat, nearly flat, curved, and rough MPs prepared from commercial PET plastic and PET bottles in water is promising for the development of a portable optical sensor capable of real-time detection of MPs and NPs in an aqueous environment. The optical behavior evaluated by specular reflection technique detected the residence time of MPs in water, the rate of pollutant adsorption, as well as the hydrodynamics and aerodynamics of MPs.

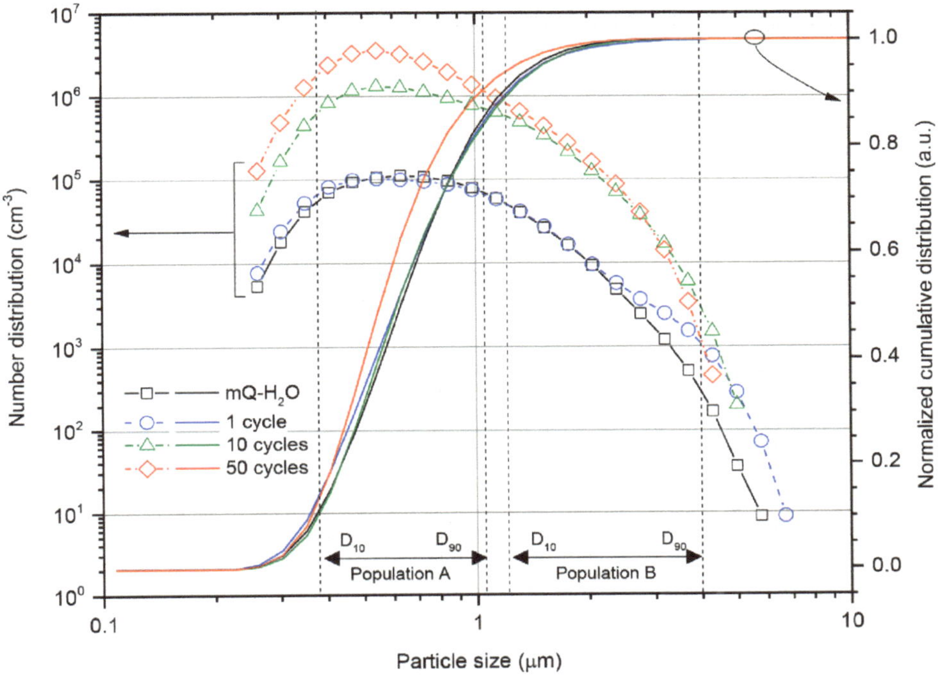

Figure 4. SPES histograms performed for measuring the size of NPs released from three opening–closing cycles—1, 10, and 50 times—at a volume of 15 ± 0.3 mL per sample. Dotted lines are the boundaries of population A and B particle size distributions indicated in terms D_{10} and D_{90}as calculated from SPES histograms [123].

Overall, these methods based on particle determination provide valuable information for identifying and quantifying MPs and NPs in drinking water. However, it is important to use multiple analytical techniques and to validate the results with other methods to ensure accurate and reliable identification.

3.3. Potential Toxicological Effects of Micro- and Nanoplastics from Drinking Water Sources on Human Health

Assessing the clear health effects of MPs/NPs on marine organisms and humans is a huge challenge among researchers [64,79]. The potential human health risks posed by exposure to MPs and NPs are continuously explored, but few published papers have been related to direct human health effects. Most research reports have involved animal studies, mathematical modeling or in vitro cell culture, so there is still a lack of data on direct human exposure and effects.

Since the existence of MPs and NPs has been proven in food, water, air and consumer products, human exposure to MPs/NPs can occur through ingestion (primary route), inhalation and dermal contact [143]. MPs/NPs could be initially ingested by marine species (lower trophic levels) and further bioaccumulate, potentially leading to human exposure. The existence of MPs and NPs in organisms can cause oxidative stress, cytotoxicity, neurotoxicity, inflammatory lesions, increased uptake or translocation, metabolic disturbances, reproductive issues and increased cancer risk in humans [144]. Health risks depend on the concentration, exposure period, route of exposure and the physicochemical properties of the particles [66].

The entrance of MPs and NPs into the human body could additionally involve the introduction of other associated substances incorporated into plastics, such as plasticizers, stabilizers, opacifiers, flame retardants, antistatic, conductive or medical additives or substances considered endocrine disruptors [145], which can migrate from the matrix due to depolymerization and leaching processes, leading to possible cytotoxicity and inflammatory response. A study performed by Tisler and Christensen [146] on reusable plastic (PE and biodegradable PE) and glass sport bottles revealed the migration of >400 plastic-related compounds over 24 h into drinking water, among which plasticizers, antioxidants, and photoinitiators were predominant. Among the toxic chemical additives potentially existing in MPs/NPs that impose high concern for human health are phthalates, bisphenol A (BPA), brominated flame retardants (BFR), triclosan, bisphenone and organotins [147]. BPA is an important chemical used as a monomer for PC [148], antioxidant or plasticizer in PP, PE, PVC, epoxy resins and coating used to line the packaging of food and beverage cans [149,150]. It has been reported that BPA migrates out of PC, epoxy resins and other consumer plastics [151] and may contaminate food products and drinks [152,153], producing adverse effects on human health, such as liver and pancreatic function alternation and respective changes in insulin resistance. BPA leaching also affects the development of offspring in the wombs of pregnant women, causing issues with brain function [154,155] and inhibiting thyroid hormone-mediated transcription by acting as an antagonist [156]. Some authors reported the onset of obesity and cardiovascular disease [157,158].

In respect to the simultaneous assessment of three surfactants, 4-nonylphenol (4-NP), BPA, and triclosan (TCS) in bottled water, the data show that an adult can ingest 340 ng/day of 4-nonylphenol (4-NP), 165 ng/day of BPA and 7 ng/day of TCS [148]. While 4-NP can come from HDPE and PVC containers, BPA can come from PC baby bottles and TCS is widely used as a preservative and antimicrobial agent in personal care products [159]. An adult could ingest 1410 ng/day of 4-NP, 148 ng/day of BPA, and 10 ng/day of TCS when drinking tap water [148]. Daily BPA intake for infants was three times higher than that for adults [148]. According to Directive 2011/8/EU, the import of PC bottles from countries outside of the European Union was prohibited from 1 June 2011, and a total ban on the use of BPA for manufacturing baby bottles was introduced in the European Union on 1 March 2011 [151].

While BPA is not intentionally added during the manufacturing of PET bottles, studies have suggested that the bottle cap, recycled PET or exposure to heat and ultraviolet radiation could lead to its release in water [160]. Another study reported the presence of BPA in PET bottled drinking water, and leaching increased with temperature [151]. For PC bottles exposed to hot water, an increase in the rate of BPA migration up to 55-fold was observed [155]. While there is not a significant risk to human health from exposure to BPA in drinking water, attention should be paid to the cumulative daily dose in the body.

Phthalate esters are employed as plasticizers to improve the flexibility of various plastic materials, much used in manufacturing PVC and plastisol [161] and PET bottles [162]. The potential harmful effects of phthalate esters on human health consist in abnormal sexual development and birth defects [163], while they have also been reported to induce adverse cellular changes in fish [147]. The presence of 17 phthalate esters was reported in drinking water stored in PET bottles without any threat to human health [162]. In other study [164], the phthalate ester identification was assigned to cross-contamination from the laboratory. However, it is recommended that an investigation into the additives released from water plastic bottles from a toxicological perspective should be carried out over the long term.

According to the results of in vitro tests on marine species, MPs have been shown to accumulate in the gills, stomach, and metabolic organs of crabs [165].

MPs larger than 10 μm are unlikely to be transported through an intact intestinal barrier [7,166]. In this regard, Gao et al. [131] suggested that PET NPs produced by mechanical action on bottled water could not be identified in the liver, spleen, lungs, or kidneys of mice, which indicates that they cannot penetrate the intestinal barriers when injected intravenously.

Animal model studies reported the accumulation of NPs in the placenta [167]. Aghaei et al. [74] were the first authors to demonstrate that the MPs and NPs in drinking water pose risks to human pregnancies in late gestation. After exposing fetuses to PS both as MPs and NPs with sizes of 5 μm and 50 nm, respectively, in a concentration of 10^6 ng/L, a 12% decrease in fetal weight was observed. However, the risks to embryo–fetal development associated with unintended ingestion of MPs/NPs require extended investigations.

Once they enter the body, MPs might translocate to distant tissues through the circulatory system, causing a systemic inflammatory response, blood cell cytotoxicity through internalization [168], pulmonary hypertension [169], vascular inflammation or occlusions [170], and decreased organ function, as well as an increased risk of neoplasm due to deoxyribonucleic acid (DNA) damage [171].

In another paper, the ability to induce reactive oxygen species (ROS) and DNA damage were evaluated after the exposure of two human lymphoblastic cell lines to PET NPs removed from drinking water bottles [139]. Preliminary research revealed cellular uptake, but without the induction of significant biological effects, and thus no potential health hazard.

On the contrary, Ji et al. [172] demonstrated that the size of PET NPs (20 nm, 60–80 nm, and 800 nm) obtained by a process of mechanical breakdown and dispersing agents (SDS and bovine serum albumin (BSA)) used for NP stabilization has an important role in in vivo toxicity studies. Thus, PET pieces cut from mineral water bottles have been shown to affect the hatching rate, heart rate, and ROS generation in the development of zebra fish. The BSA-dispersing agent was found to induce a higher level in heart rate abnormalities and more severe oxidative damage from the PET NPs than SDS.

Cell damage caused by extrinsic toxic plastic materials can be evaluated by some typical examinations, such as rupture of the cellular membrane by strong positive charge [173], interference with DNA synthesis, or organelle activities after uptake [174], resulting in cell death due to necrosis or apoptosis [175]. Choi et al. [175] examined the in vitro toxicity of PE MPs of different shapes and sizes, such as 1–100 μm for HDPE and ranging between 25–75 μm and 75–200 μm for LDPE, on cultured cells, including immune cells (human mast cells [HMC-1], peripheral blood mononuclear cells [PBMCs], red blood cells [RBCs], rat basophilic leukemia cells [RBL-2H3]), non-immune cells (cervical cancer cells [HeLa]), and human dermal fibroblasts [HDFs]. Their results showed that the HDPE particles with relatively smooth surfaces did not produce significant cytotoxicity in cells, but induced an immune response in PBMCs and enhanced PBMC differentiation, while the LDPE particles with sharp edges (higher curvature change) caused increased cytotoxicity under direct cell–microplastic interaction, inflammatory response, hemolysis, and ROS production at high concentrations (Figure 5).

Magri et al. [176] mentioned that combining of alternative methodologies, such as metabolomics, with standard biological assays (i.e., cell viability and ROS production) is an important approach to acquire preliminary valuable information on cellular metabolism to facilitate the prediction of potential effects of plastic NPs on human health. Metabolomics is a powerful tool for studying cellular metabolism and can be used to identify changes in metabolite levels in response to exposure to plastic NPs. The authors used pulsed-laser ablation of solid PET films in water to form PET NPs (size distribution in the range of 10 and 80 nm) of similar surface and shape irregularity, broad size distribution, and chemistry to those of particles in the environment [104]. They found a binding capacity of about 3% *w*/*w* NPs for PET NPs with levofloxacin, an antibiotic defined as an emerging contaminant in aqueous environments and demonstrated that these nanoclusters are not toxic to heterogeneous human epithelial colorectal adenocarcinoma cells (Caco-2) in the short term, but affect the cells metabolism as a compensatory response to oxidative stress, suggesting long-term risks.

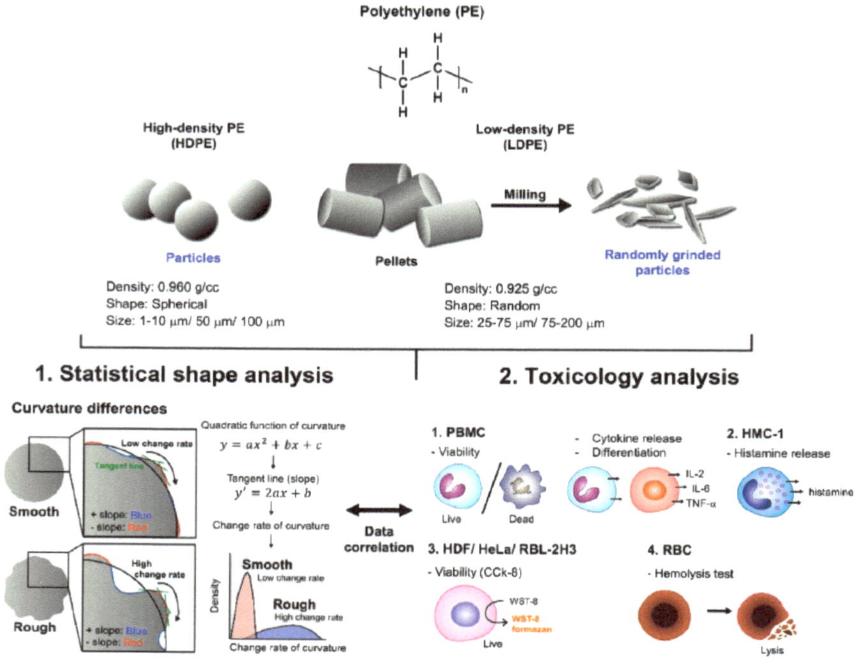

Figure 5. Correlation of physical aspects of PE MPs with statistical data and in vitro toxicity results. CCK-8: cell counting kit-8; WST-8: water-soluble tetrazolium salt-8. Reprinted with permission from [175].

The cytotoxicity of PE and PS MPs was evaluated in vitro in cell lines T98G and HeLa (cerebral and epithelial human cells) exposed for 24–48 h to different contaminant levels of 10 ng/mL to 10 μg/mL under the same conditions [177]. Oxidative stress explains the toxicity of PE and PS MPs at the cell level, being significant in the case of PE MPs for T98G; however, PS showed higher ROS generation in both cell lines.

The effect of engineered aminated PS NPs (model particles) on human liver HepG2 hepatocytes was investigated by Banerjee et al. [178]. They found that the uptake of NPs with sizes of 50 or 100 nm by HepG2 liver cells was higher than 1000 nm particles. Additionally, short-term exposure to aminated PS NPs resulted in cell toxicity, while those with sizes between 500–5000 nm induced apoptosis [178].

Organic matter found in commercial drinking water was analyzed by various analytical methods, such as solid residues, after freeze-drying treatment of drinking water, using analytical methods including dry ash, centrifugation, Raman analysis, and electron microscopy [179]. The study found a concentration of 0.25–2.0 mg/L organic matter. It was proved that the stress applied to the plastic bottle did not significantly increase the concentration of organic matter within one month, and the organic matter did not affect the viability of human intestinal cells [179]. However, the study suggested that prolonged exposure to nanoscale material in drinking water could potentially lead to the accumulation of NOM in the human body over time.

The potential toxicological effects on test organisms, as well as the use of biodegradable MPs as vector for other chemical pollutants and microorganisms, have been reported [43]. However, there is a serious concern regarding the management of bioplastics and MP contamination, particularly with regard to fragmentation, degradability, and toxicity, and their interaction with other chemical contaminants present in aqueous media [41]. Since most MPs/NPs are hardly biodegradable or non-biodegradable, they clearly remain intact inside the living organisms for a long time. The prolonged exposure of the human or other organisms to MPs/NPs may lead to chronic irritation, inflammation, cellular proliferation, and necrosis, and may compromise immune cells [144,180].

Comprehensive studies are needed to establish a clear health-risk assessment for exposure to MPs and NPs, as it is fundamental to set the reference method of investigation to monitor each source of human intake.

4. Conclusions and Future Perspective

The removal efficiency of MPs/NPs from drinking water treatment plants varies depending on the treatment processes applied. The size of MPs detected in DWTPs ranges from 1.2 μm to 2000 μm. Most studies on MPs/NPs have focused on investigating PS NPs models due to their prevalence as debris in water. However, further analytical development is required to monitor real NPs that have been isolated and/or preconcentrated from more complex matrices. Few studies have reported the detection of NPs present in drinking water bottles. A standard that is kept under the same conditions as drinking water is needed to quantify the concentration of NPs in drinking water. Additionally, studies should also investigate the interaction between NPs and other sources, such as food.

In the future, the amount of MPs/NPs in tap water and bottled drinking water are estimated to unfortunately expand as a consequence of the continued degradation and fragmentation of plastics in the environment. The manufacturing technology of reusable drinking bottles (PET, glass) should be carefully examined to decrease the concentration of MPs/NPs in the water packaged in the respective containers. The European Chemicals Agency (ECHA) has proposed far-reaching restrictions on the use of MPs in products marketed in the EU to reduce their release into the environment, as a part of the circular economy plan [35]. The risks associated with MP/NP pollution derived from drinking water sources should be mitigated by applying legal, technical, and social measures [6–8,30,63,73,132], such as:

- setting sampling and monitoring standards for MPs/NPs
- reducing the production of non-biodegradable plastic items
- reducing single-use plastic
- implementing the circular economy using biodegradable plastic items
- using the "refuse, reduce, reuse, and recycle" concept
- innovation for plastics that do not need reusable, recyclable, or compostable materials
- total removal of MPs/NPs from DWTPs
- designing innovative packaging technologies to unscrew bottle caps in other ways, such as easy-to-open caps
- the use of bio-inspired technology related to biomimetics involving the design of advanced systems or devices inspired by nature, where principles from interdisciplinary fields such as engineering, chemistry and biology are applied to the development of materials, synthetic systems or instruments with functions that mimic biological processes [181]
- raising people's awareness of the toxicological effects of MPs/NPs.

In addition to improving multiple interventions and management to prevent the release of plastic related to the removal of MPs/NPs from water infrastructure, other effective measures are necessary to eliminate MPs/NPs directly at the source. A green prevention technology was proposed by the GoJelly Project, which developed a prototype microplastics filter for commercial and public use, employing jellyfish mucus as the main raw material [182]. The use of special household water systems, such as Lint LUV-R and Showerloop, is effective in filtering out microfibers at the domestic level [182]. Synthetic fibers could also be captured in the washing machine by means of laundry balls, such as Cora Ball and Fibre Free [183].

Research is still needed to establish an acceptable upper limit for the concentration of various NPs in drinking water through toxicity assessment, unification of various analytical protocols for NP/MP identification and development of testing standards. In the meantime, it should not be assumed that any level is safe. Toxicity assessment of various MP and NP types of different chemical composition, size and shape should be studied promptly to evaluate concerns regarding human exposure to these particles. Extensive studies are required to establish an explicit health-risk assessment for exposure to MPs and NPs, as a clear distinction of plastic particles entering the human body from different sources (water, air, food, drugs, skin) cannot be achieved. Awareness of MP/NP pollution of drinking water sources should increase among European regulatory bodies, decision-makers, practitioners and researchers.

Supplementary Materials: The following supporting information can be downloaded at https://www.mdpi.com/article/10.3390/polym15112425/s1. Table S1: Physicochemical methods used to remove MPs/NPs from drinking water.

Author Contributions: Conceptualization, M.R. and R.N.D.-N., data curation, A.-M.P. and A.-C.B.; funding acquisition, C.P.; methodology, E.M. and A.-M.P.; project administration, M.R.; validation, E.M.; visualization, C.P. and A.-C.B.; writing—original draft, M.R. and R.N.D.-N.; writing—review and editing, M.R. and R.N.D.-N. All authors have read and agreed to the published version of the manuscript.

Funding: This work was supported by EU Horizon 2020 (InNoPlastic), GA no. 101000612.

Institutional Review Board Statement: Not applicable.

Data Availability Statement: The data presented in this study are available on request from the corresponding author.

Acknowledgments: R.N. Darie-Niță acknowledges the Romanian Academy.

Conflicts of Interest: The authors declare no conflict of interest.

List of Acronyms and Abbreviations

MPs	microplastics
NPs	nanoplastics
DWTPs	drinking water treatment plants
Pyr-GC/MS	pyrolysis–gas chromatography–mass spectrometry
RAC	Committee for Risk Assessment
SEAC	Committee for Socio-Economic Analysis
EPS	extracellular polymers
PFAS	per- and polyfluoroalkyl substances
WHO	World Health Organization
ATSDR	Agency for Toxic Substances and Disease Registry
PS	polystyrene
PVC	polyvinyl chloride
CFS	coagulation–flocculation–sedimentation
TD-PTR/MS	thermal desorption–proton transfer reaction–mass spectrometry
μ-FTIR	micro-Fourier-transform infrared spectroscopy
μ-Raman	micro-Raman
TD-GC/MS	thermal desorption–gas chromatography–mass spectrometry
SP-ICP-MS	single particle inductively coupled plasma mass spectroscopy
SERS	surface-enhanced Raman spectroscopy
ICP-MS	inductively coupled plasma mass spectroscopy
MALDI-ToF/MS	matrix-assisted laser desorption/ionization time-of-flight mass spectrometry
DLS	dynamic light scattering
SEM-EDX	scanning electron microscopy–energy-dispersive X-ray spectroscopy
TEM	transmission electron microscopy
PET	polyethylene terephthalate
NOM	natural organic matter
polyDADMAC	diallyldimethylammonium chloride
PE	polyethylene
PACl	polyaluminum chloride

$Al_2(SO_4)_3$	aluminum sulfate
$AlCl_3$	aluminum chloride
$FeCl_3$	iron chloride
$NaHCO_3$	sodium bicarbonate
PC	polyamine-coated
ER	elongated-rough
ES	elongated-smooth
SR	spherical-rough
SS	spherical-smooth
GAC	granular activated carbon
Pd	palladium
DOM	dissolved organic matter
IONPs	iron oxide nanoparticles
LOQ	*limit of quantification*
AF4	asymmetric flow field flow fractionation
DLS-MADLS	multiangle and dynamic light scattering
PA	polyamide
PU	polyurethane
HDPE	high-density polyethylene
POU	three point-of-use
IX	ion exchange
MF	microfiltration
PVC	polyvinyl chloride
LDPE	low-density polyethylene
DAF	dissolved air flotation
SAN	styrene acrylonitrile
PES	polyester
PPS	polyphenylene sulfite
FPA	focal plane array
TFU	tangential flow ultrafiltration
LOD	limit of detection
EvOH	ethylene vinyl alcohol
LDIR	laser direct infrared
PC	polycarbonate
AFM-IR	atomic force microscope-infrared spectroscopy
ATR-FTIR	attenuated total reflectance Fourier-transform infrared spectroscopy
NWERS	nanowell-enhanced Raman spectroscopy
ILs	ionic liquids
PAM	polyacrylamide
EDTA	ethylenediaminetetraacetic acid tetrasodium salt
SDS	sodium dodecyl sulfate
BPA	bisphenol A
BFR	brominated flame retardants
4-NP	4-nonylphenol
TCS	triclosan
DNA	deoxyribonucleic acid
ROS	reactive oxygen species
BSA	bovine serum albumin
HMC-1	human mast cells
PBMCs	peripheral blood mononuclear cells
RBCs	red blood cells
RBL-2H3	rat basophilic leukemia cells
HeLa	cervical cancer cells
HDFs	human dermal fibroblasts
Caco-2	human epithelial colorectal adenocarcinoma cells
ECHA	European Chemicals Agency
PAN	poly(acrylonitrile)
PMMA	poly(methyl methacrylate)

PVA	poly(vinyl alcohol)
PEVA	ethylene (vinyl acetate) copolymer
AC	activated carbon

References

1. PLASTICS—THE FACTS 2022. Available online: https://plasticseurope.org/knowledge-hub/plastics-the-facts-2022 (accessed on 13 April 2023).
2. Popa, C.L.; Dontu, S.I.; Savastru, D.; Carstea, E.M. Role of Citizen Scientists in Environmental Plastic Litter Research-A Systematic Review. *Sustainability* **2022**, *14*, 13265. [CrossRef]
3. Arenas, L.R.; Gentile, S.R.; Zimmermann, S.; Stoll, S. Fate and removal efficiency of polystyrene nanoplastics in a pilot drinking water treatment plant. *Sci. Total Environ.* **2022**, *813*, 152623. [CrossRef] [PubMed]
4. Ilyas, M.; Ahmad, W.; Khan, H.; Yousaf, S.; Khan, K.; Nazir, S. Plastic waste as a significant threat to environment—A systematic literature review. *Rev. Environ. Health* **2018**, *33*, 383–406. [CrossRef] [PubMed]
5. Reddy, A.S.; Nair, A.T. The fate of microplastics in wastewater treatment plants: An overview of source and remediation technologies. *Environ. Technol. Innov.* **2022**, *28*, 102815. [CrossRef]
6. Ali, I.; Cheng, Q.H.; Ding, T.D.; Qian, Y.G.; Zhang, Y.C.; Sun, H.B.; Peng, C.S.; Naz, I.; Li, J.Y.; Liu, J.F. Micro- and nanoplastics in the environment: Occurrence, detection, characterization and toxicity-A critical review. *J. Clean. Prod.* **2021**, *313*, 127863. [CrossRef]
7. Lee, H.; Shim, J.E.; Park, I.H.; Choo, K.S.; Yeo, M.K. Physical and biomimetic treatment methods to reduce microplastic waste accumulation. *Mol. Cell Toxicol.* **2022**, *19*, 13–25. [CrossRef] [PubMed]
8. Das, R.K.; Sanyal, D.; Kumar, P.; Pulicharla, R.; Brar, S.K. Science-society-policy interface for microplastic and nanoplastic: Environmental and biomedical aspects. *Environ. Pollut.* **2021**, *290*, 117985. [CrossRef] [PubMed]
9. Jimenez-Lamana, J.; Marigliano, L.; Allouche, J.; Grassl, B.; Szpunar, J.; Reynaud, S. A Novel Strategy for the Detection and Quantification of Nanoplastics by Single Particle Inductively Coupled Plasma Mass Spectrometry (ICP-MS). *Anal. Chem.* **2020**, *92*, 11664–11672. [CrossRef]
10. Sorensen, R.M.; Kanwar, R.S.; Jovanovi, B. Past, present, and possible future policies on plastic use in the United States, particularly microplastics and nanoplastics: A review. *Integr. Environ. Assess. Manag.* **2023**, *19*, 474–488. [CrossRef] [PubMed]
11. Plastics—Environmental Aspects—State of Knowledge and Methodologies. Available online: https://www.iso.org/obp/ui/#iso:std:iso:tr:21960:ed-1:v1:en (accessed on 12 March 2023).
12. European Chemicals Agency (ECHA). Opinion on an Annex XV Dossier Proposing Restrictions on Intentionally-Added Microplastics. ECHA/RAC/RES-O-0000006790-71-01/F ECHA/SEAC/. 11 June 2020. Available online: https://echa.europa.eu/documents/10162/23665416/rest_microplastics_opinion_rac_16339_en.pdf/b4d383cd-24fc-82e9-cccf-6d9f66ee9089 (accessed on 13 May 2023).
13. Udenby, F.A.O.; Almuhtaram, H.; McKie, M.J.; Andrews, R.C. Adsorption of fluoranthene and phenanthrene by virgin and weathered polyethylene microplastics in freshwaters. *Chemosphere* **2022**, *307*, 135585. [CrossRef] [PubMed]
14. Ji, H.L.; Liu, Z.Z.; Jiang, W. Transport behavior of nanoplastics in activated carbon column. *Environ. Sci. Pollut. Res.* **2023**, *30*, 26256–26269. [CrossRef] [PubMed]
15. Reynaud, S.; Aynard, A.; Grassl, B.; Gigault, J. Nanoplastics: From model materials to colloidal fate. *Curr. Opin. Colloid Interface Sci.* **2022**, *57*, 101528. [CrossRef]
16. Pivokonsky, M.; Cermakova, L.; Novotna, K.; Peer, P.; Cajthaml, T.; Janda, V. Occurrence of microplastics in raw and treated drinking water. *Sci. Total Environ.* **2018**, *643*, 1644–1651. [CrossRef] [PubMed]
17. Zhang, Y.L.; Diehl, A.; Lewandowski, A.; Gopalakrishnan, K.; Baker, T. Removal efficiency of micro- and nanoplastics (180 nm–125 μm) during drinking water treatment. *Sci. Total Environ.* **2020**, *720*, 137383. [CrossRef] [PubMed]
18. Xue, J.K.; Samaei, S.H.A.; Chen, J.F.; Doucet, A.; Ng, K.T.W. What have we known so far about microplastics in drinking water treatment? A timely review. *Front. Environ. Sci. Eng.* **2022**, *16*, 58. [CrossRef] [PubMed]
19. Jung, J.W.; Kim, S.; Kim, Y.S.; Jeong, S.; Lee, J. Tracing microplastics from raw water to drinking water treatment plants in Busan, South Korea. *Sci. Total Environ.* **2022**, *825*, 154015. [CrossRef] [PubMed]
20. Pojar, I.; Tiron Dutu, L.; Pop, C.I.; Dutu, F. Hydrodynamic observations on microplastic abundances and morphologies in the danube delta, romania. *Agrolife Sci. J.* **2021**, *10*, 142–149. [CrossRef]
21. Zhou, X.X.; He, S.; Gao, Y.; Chi, H.Y.; Wang, D.J.; Li, Z.C.; Yan, B. Quantitative Analysis of Polystyrene and Poly(methyl methacrylate) Nanoplastics in Tissues of Aquatic Animals. *Environ. Sci. Technol.* **2021**, *55*, 3032–3040. [CrossRef] [PubMed]
22. Shruti, V.C.; Perez-Guevara, F.; Elizalde-Martinez, I.; Kutralam-Muniasamy, G. Toward a unified framework for investigating micro(nano)plastics in packaged beverages intended for human consumption. *Environ. Pollut.* **2021**, *268*, 115811. [CrossRef]
23. Sharma, S.; Sharma, B.; Sadhu, S.D. Microplastic profusion in food and drinking water: Are microplastics becoming a macroproblem? *Environ. Sci.-Process. Impacts* **2022**, *24*, 992–1009. [CrossRef]
24. Liu, Q.R.; Chen, Z.; Chen, Y.L.; Yang, F.W.; Yao, W.R.; Xie, Y.F. Microplastics and Nanoplastics: Emerging Contaminants in Food. *J. Agric. Food Chem.* **2021**, *69*, 10450–10468. [CrossRef]
25. Wong, S.L.; Nyakuma, B.B.; Wong, K.Y.; Lee, C.T.; Lee, T.H.; Lee, C.H. Microplastics and nanoplastics in global food webs: A bibliometric analysis (2009–2019). *Mar. Pollut. Bull.* **2020**, *158*, 111432. [CrossRef] [PubMed]

26. Myszograj, M. Microplastic in food and drinking water- environmental monitoring data. *Civ. Environ. Eng. Rep.* **2020**, *30*, 201–209. [CrossRef]
27. Okoffo, E.D.; O'Brien, S.; Ribeiro, F.; Burrows, S.D.; Toapanta, T.; Rauert, C.; O'Brien, J.W.; Tscharke, B.J.; Wang, X.Y.; Thomas, K.V. Plastic particles in soil: State of the knowledge on sources, occurrence and distribution, analytical methods and ecological impacts. *Environ. Sci.-Process. Impacts* **2021**, *23*, 240–274. [CrossRef] [PubMed]
28. Zhang, B.; Yang, X.; Chen, L.; Chao, J.; Teng, J.; Wang, Q. Microplastics in soils: A review of possible sources, analytical methods and ecological impacts. *J. Chem. Technol. Biotechnol.* **2020**, *95*, 2052–2068. [CrossRef]
29. Jia, W.; Karapetrova, A.; Zhang, M.; Xu, L.; Li, K.; Huang, M.; Wang, J.; Huang, Y. Automated identification and quantification of invisible microplastics in agricultural soils. *Sci. Total Environ.* **2022**, *844*, 156853. [CrossRef] [PubMed]
30. Rhodes, C.J. Solving the plastic problem: From cradle to grave, to reincarnation. *Sci. Prog.* **2019**, *102*, 218–248. [CrossRef]
31. Ding, J.F.; Sun, C.J.; He, C.F.; Zheng, L.; Dai, D.J.; Li, F.M. Atmospheric microplastics in the Northwestern Pacific Ocean: Distribution, source, and deposition. *Sci. Total Environ.* **2022**, *829*, 154337. [CrossRef] [PubMed]
32. Antohi, V.M.; Ionescu, R.V.; Zlati, M.L.; Iticescu, C.; Georgescu, P.L.; Calmuc, M. Regional Regression Correlation Model of Microplastic Water Pollution Control Using Circular Economy Tools. *Int. J. Environ. Res. Public Health* **2023**, *20*, 4014. [CrossRef] [PubMed]
33. Krishnan, K. A Systematic Review on the Impact of Micro-Nanoplastics Exposure on Human Health and Diseases. *Biointerface Res. Appl. Chem.* **2023**, *13*, 1–12. [CrossRef]
34. Song, F.H.; Li, T.T.; Hur, J.; Wu, F.C.; Meng, X.G. Eco-Colloidal Layer of Micro/Nanoplastics Increases Complexity and Uncertainty of Their Biotoxicity in Aquatic Environments. *Environ. Sci. Technol.* **2022**, *56*, 10547–10549. [CrossRef] [PubMed]
35. Rubio-Armendariz, C.; Alejandro-Vega, S.; Paz-Montelongo, S.; Gutierrez-Fernandez, A.J.; Carrascosa-Iruzubieta, C.J.; de la Torre, A. Microplastics as Emerging Food Contaminants: A Challenge for Food Safety. *Int. J. Environ. Res. Public Health* **2022**, *19*, 1174. [CrossRef]
36. Cerasa, M.; Teodori, S.; Pietrelli, L. Searching Nanoplastics: From Sampling to Sample Processing. *Polymers* **2021**, *13*, 3658. [CrossRef]
37. Ali, I.; Tan, X.; Li, J.Y.; Peng, C.S.; Naz, I.; Duan, Z.P.; Ruan, Y.L. Interaction of microplastics and nanoplastics with natural organic matter (NOM) and the impact of NOM on the sorption behavior of anthropogenic contaminants—A critical review. *J. Clean. Prod.* **2022**, *376*, 134314. [CrossRef]
38. Ding, R.R.; Tong, L.; Zhang, W.C. Microplastics in Freshwater Environments: Sources, Fates and Toxicity. *Water Air Soil Pollut.* **2021**, *232*, 181. [CrossRef]
39. Sun, X.L.; Xiang, H.; Xiong, H.Q.; Fang, Y.C.; Wang, Y. Bioremediation of microplastics in freshwater environments: A systematic review of biofilm culture, degradation mechanisms, and analytical methods. *Sci. Total Environ.* **2023**, *863*, 160953. [CrossRef] [PubMed]
40. Qin, M.; Chen, C.Y.; Song, B.; Shen, M.C.; Cao, W.C.; Yang, H.L.; Zeng, G.M.; Gong, J.L. A review of biodegradable plastics to biodegradable microplastics: Another ecological threat to soil environments? *J. Clean. Prod.* **2021**, *312*, 127816. [CrossRef]
41. Shruti, V.C.; Kutralam-Muniasamy, G. Bioplastics: Missing link in the era of Microplastics. *Sci. Total Environ.* **2019**, *697*, 134139. [CrossRef] [PubMed]
42. Wang, C.; Yu, J.F.; Lu, Y.; Hua, D.; Wang, X.; Zou, X.H. Biodegradable microplastics (BMPs): A new cause for concern? *Environ. Sci. Pollut. Res.* **2021**, *28*, 66511–66518. [CrossRef] [PubMed]
43. Shen, M.C.; Song, B.; Zeng, G.M.; Zhang, Y.X.; Huang, W.; Wen, X.F.; Tang, W.W. Are biodegradable plastics a promising solution to solve the global plastic pollution? *Environ. Pollut.* **2020**, *263*, 114469. [CrossRef] [PubMed]
44. Miloloza, M.; Cvetnic, M.; Grgic, D.K.; Bulatovic, V.O.; Ukic, S.; Rogosic, M.; Dionysiou, D.D.; Kusic, H.; Bolanca, T. Biotreatment strategies for the removal of microplastics from freshwater systems. A review. *Environ. Chem. Lett.* **2022**, *20*, 1377–1402. [CrossRef]
45. Jeong, Y.; Gong, G.; Lee, H.-J.; Seong, J.; Hong, S.W.; Lee, C. Transformation of microplastics by oxidative water and wastewater treatment processes: A critical review. *J. Hazard. Mater.* **2023**, *443*, 130313. [CrossRef]
46. Wang, Y.; Ding, K.Q.; Ren, L.X.; Peng, A.P.; Zhou, S.D. Biodegradable Microplastics: A Review on the Interaction with Pollutants and Influence to Organisms. *Bull. Environ. Contam. Toxicol.* **2022**, *108*, 1006–1012. [CrossRef] [PubMed]
47. Tong, H.Y.; Zhong, X.C.; Duan, Z.H.; Yi, X.L.; Cheng, F.Q.; Xu, W.P.; Yang, X.J. Micro- and nanoplastics released from biodegradable and conventional plastics during degradation: Formation, aging factors, and toxicity. *Sci. Total Environ.* **2022**, *833*, 155275. [CrossRef] [PubMed]
48. Wei, X.F.; Bohlen, M.; Lindblad, C.; Hedenqvist, M.; Hakonen, A. Microplastics generated from a biodegradable plastic in freshwater and seawater. *Water Res.* **2021**, *198*, 117123. [CrossRef] [PubMed]
49. Do, A.T.N.; Ha, Y.; Kang, H.J.; Kim, J.M.; Kwon, J.H. Equilibrium leaching of selected ultraviolet stabilizers from plastic products. *J. Hazard. Mater.* **2022**, *427*, 128144. [CrossRef]
50. Barhoumi, B.; Sander, S.G.; Tolosa, I. A review on per- and polyfluorinated alkyl substances (PFASs) in microplastic and food-contact materials. *Environ. Res.* **2022**, *206*, 112595. [CrossRef] [PubMed]
51. Jang, M.; Shim, W.J.; Han, G.M.; Rani, M.; Song, Y.K.; Hong, S.H. Widespread detection of a brominated flame retardant, hexabromocyclododecane, in expanded polystyrene marine debris and microplastics from South Korea and the Asia-Pacific coastal region. *Environ. Pollut.* **2017**, *231*, 785–794. [CrossRef] [PubMed]

52. Martin, J.; Santos, J.L.; Aparicio, I.; Alonso, E. Microplastics and associated emerging contaminants in the environment: Analysis, sorption mechanisms and effects of co-exposure. *Trends Environ. Anal. Chem.* **2022**, *35*, e00170. [CrossRef]
53. Oliveira, M.; Almeida, M.; Miguel, I. A micro(nano)plastic boomerang tale: A never ending story? *Trac-Trends Anal. Chem.* **2019**, *112*, 196–200. [CrossRef]
54. Ma, B.W.; Xue, W.J.; Hu, C.Z.; Liu, H.J.; Qu, J.H.; Li, L.L. Characteristics of microplastic removal via coagulation and ultrafiltration during drinking water treatment. *Chem. Eng. J.* **2019**, *359*, 159–167. [CrossRef]
55. Kinigopoulou, V.; Pashalidis, I.; Kalderis, D.; Anastopoulos, I. Microplastics as carriers of inorganic and organic contaminants in the environment: A review of recent progress. *J. Mol. Liq.* **2022**, *350*, 118580. [CrossRef]
56. Wilkinson, J.; Hooda, P.S.; Barker, J.; Barton, S.; Swinden, J. Occurrence, fate and transformation of emerging contaminants in water: An overarching review of the field. *Environ. Pollut.* **2017**, *231*, 954–970. [CrossRef] [PubMed]
57. Caldas, S.S.; Arias, J.L.O.; Rombaldi, C.; Mello, L.L.; Cerqueira, M.B.R.; Martins, A.F.; Primel, E.G. Occurrence of Pesticides and PPCPs in Surface and Drinking Water in Southern Brazil: Data on 4-Year Monitoring. *J. Braz. Chem. Soc.* **2019**, *30*, 71–80. [CrossRef]
58. Zhou, R.R.; Lu, G.H.; Yan, Z.H.; Jiang, R.R.; Bao, X.H.; Lu, P. A review of the in fluences of microplastics on toxicity and transgenerational effects of pharmaceutical and personal care products in aquatic environment. *Sci. Total Environ.* **2020**, *732*, 139222. [CrossRef]
59. Ben, Y.; Hu, M.; Zhang, X.; Wu, S.; Wong, M.H.; Wang, M.; Andrews, C.B.; Zheng, C. Efficient detection and assessment of human exposure to trace antibiotic residues in drinking water. *Water Res.* **2020**, *175*, 115699. [CrossRef] [PubMed]
60. Zhou, Z.Q.; Sun, Y.R.; Wang, Y.Y.; Yu, F.; Ma, J. Adsorption behavior of Cu(II) and Cr(VI) on aged microplastics in antibiotics-heavy metals coexisting system. *Chemosphere* **2022**, *291*, 132794. [CrossRef] [PubMed]
61. Yee, M.S.L.; Hii, L.W.; Looi, C.K.; Lim, W.M.; Wong, S.F.; Kok, Y.Y.; Tan, B.K.; Wong, C.Y.; Leong, C.O. Impact of Microplastics and Nanoplastics on Human Health. *Nanomaterials* **2021**, *11*, 496. [CrossRef] [PubMed]
62. Almaiman, L.; Aljomah, A.; Bineid, M.; Aljeldah, F.M.; Aldawsari, F.; Liebmann, B.; Lomako, I.; Sexlinger, K.; Alarfaj, R. The occurrence and dietary intake related to the presence of microplastics in drinking water in Saudi Arabia. *Environ. Monit. Assess.* **2021**, *193*, 390. [CrossRef] [PubMed]
63. Cox, K.D.; Covernton, G.A.; Davies, H.L.; Dower, J.F.; Juanes, F.; Dudas, S.E. Human Consumption of Microplastics. *Environ. Sci. Technol.* **2019**, *53*, 7068–7074. [CrossRef] [PubMed]
64. Pedersen, A.F.; Meyer, D.N.; Petriv, A.M.V.; Soto, A.L.; Shields, J.N.; Akemann, C.; Baker, B.B.; Tsou, W.L.; Zhang, Y.L.; Baker, T.R. Nanoplastics impact the zebrafish (*Danio rerio*) transcriptome: Associated developmental and neurobehavioral consequences. *Environ. Pollut.* **2020**, *266*, 115090. [CrossRef]
65. Zuccarello, P.; Ferrante, M.; Cristaldi, A.; Copat, C.; Grasso, A.; Sangregorio, D.; Fiore, M.; Conti, G.O. Exposure to microplastics (<10 μm) associated to plastic bottles mineral water consumption: The first quantitative study. *Water Res.* **2019**, *157*, 365–371. [CrossRef] [PubMed]
66. Wu, P.F.; Lin, S.Y.; Cao, G.D.; Wu, J.B.; Jin, H.B.; Wang, C.; Wong, M.H.; Yang, Z.; Cai, Z.W. Absorption, distribution, metabolism, excretion and toxicity of microplastics in the human body and health implications. *J. Hazard. Mater.* **2022**, *437*, 129361. [CrossRef] [PubMed]
67. Shen, M.C.; Song, B.; Zhu, Y.; Zeng, G.M.; Zhang, Y.X.; Yang, Y.Y.; Wen, X.F.; Chen, M.; Yi, H. Removal of microplastics via drinking water treatment: Current knowledge and future directions. *Chemosphere* **2020**, *251*, 126612. [CrossRef] [PubMed]
68. Wang, Z.; Lin, T.; Chen, W. Occurrence and removal of microplastics in an advanced drinking water treatment plant (ADWTP). *Sci. Total Environ.* **2020**, *700*, 134520. [CrossRef] [PubMed]
69. Dalmau-Soler, J.; Ballesteros-Cano, R.; Boleda, M.R.; Paraira, M.; Ferrer, N.; Lacorte, S. Microplastics from headwaters to tap water: Occurrence and removal in a drinking water treatment plant in Barcelona Metropolitan area (Catalonia, NE Spain). *Environ. Sci. Pollut. Res.* **2021**, *28*, 59462–59472. [CrossRef] [PubMed]
70. Sarkar, D.J.; Sarkar, S.D.; Das, B.K.; Praharaj, J.K.; Mahajan, D.K.; Purokait, B.; Mohanty, T.R.; Mohanty, D.; Gogoi, P.; Kumar, V.S.; et al. Microplastics removal efficiency of drinking water treatment plant with pulse clarifier. *J. Hazard. Mater.* **2021**, *413*, 125347. [CrossRef] [PubMed]
71. Velasco, A.N.; Gentile, S.R.; Zimmermann, S.; Stoll, S. Contamination and Removal Efficiency of Microplastics and Synthetic Fibres in a Conventional Drinking Water Treatment Plant. *Front. Water* **2022**, *4*, 34. [CrossRef]
72. Pan, Y.S.; Gao, S.H.; Ge, C.; Gao, Q.; Huang, S.J.; Kang, Y.Y.; Luo, G.Y.; Zhang, Z.Q.; Fan, L.; Zhu, Y.M.; et al. Removing microplastics from aquatic environments: A critical review. *Environ. Sci. Ecotechnol.* **2022**, *13*, 100222. [CrossRef] [PubMed]
73. Gambino, I.; Bagordo, F.; Grassi, T.; Panico, A.; De Donno, A. Occurrence of Microplastics in Tap and Bottled Water: Current Knowledge. *Int. J. Environ. Res. Public Health* **2022**, *19*, 5283. [CrossRef] [PubMed]
74. Aghaei, Z.; Sled, J.G.; Kingdom, J.C.; Baschat, A.A.; Helm, P.A.; Jobst, K.J.; Cahill, L.S. Maternal Exposure to Polystyrene Micro- and Nanoplastics Causes Fetal Growth Restriction in Mice. *Environ. Sci. Technol. Lett.* **2022**, *9*, 426–430. [CrossRef]
75. Jiang, B.R.; Kauffman, A.E.; Li, L.; McFee, W.; Cai, B.; Weinstein, J.; Lead, J.R.; Chatterjee, S.; Scott, G.I.; Xiao, S. Health impacts of environmental contamination of micro- and nanoplastics: A review. *Environ. Health Prev. Med.* **2020**, *25*, 29. [CrossRef] [PubMed]
76. Asmundsdottir, A.M.; Gomiero, A.; Oysaed, K.B. Microplastics and Nanoplastics Occurrence and Composition in Drinking Water from Akureyri Urban Area, Iceland. In Proceedings of the 2nd International Conference on Microplastic Pollution in the

Mediterranean Sea, Capri, Italy, 15–18 September 2019; Springer International Publishing: Berlin/Heidelberg, Germany, 2020; pp. 106–111. [CrossRef]
77. Katare, Y.; Singh, P.; Sankhla, M.S.; Singhal, M.; Jadhav, E.B.; Parihar, K.; Nikalje, B.T.; Trpathi, A.; Bhardwaj, L. Microplastics in Aquatic Environments: Sources, Ecotoxicity, Detection & Remediation. *Biointerface Res. Appl. Chem.* **2022**, *12*, 3407–3428. [CrossRef]
78. Pulido-Reyes, G.; Magherini, L.; Bianco, C.; Sethi, R.; von Gunten, U.; Kaegi, R.; Mitrano, D.M. Nanoplastics removal during drinking water treatment: Laboratory- and pilot-scale experiments and modeling. *J. Hazard. Mater.* **2022**, *436*, 129011. [CrossRef] [PubMed]
79. Kumar, R.; Manna, C.; Padha, S.; Verma, A.; Sharma, P.; Dhar, A.; Ghosh, A.; Bhattacharya, P. Micro(nano)plastics pollution and human health: How plastics can induce carcinogenesis to humans? *Chemosphere* **2022**, *298*, 134267. [CrossRef] [PubMed]
80. Jemec Kokalj, A.; Heinlaan, M.; Novak, S.; Drobne, D.; Kühnel, D. Assessing study quality to evaluate nanoplastics hazard: The case of polystyrene nanoplastics and aquatic invertebrate Daphnia sp. *Nanomaterials* **2022**, *12*, 16.
81. Valh, J.V.; Stopar, D.; Berodia, I.S.; Erjavec, A.; Sauperl, O.; Zemljic, L.F. Economical Chemical Recycling of Complex PET Waste in the Form of Active Packaging Material. *Polymers* **2022**, *14*, 3244. [CrossRef] [PubMed]
82. Kangal, M.O. Selective Flotation Technique for Separation of PET and HDPE Used in Drinking Water Bottles. *Miner. Process. Extr. Metall. Rev.* **2010**, *31*, 214–223. [CrossRef]
83. Garcia-Fayos, B.; Arnal, J.M.; Monforte-Monleon, L.; Sancho, M. Alternatives to the use of synthetic organic coagulant aids in drinking water treatment: Improvements seed. *Desalination Water Treat.* **2015**, *55*, 3635–3645. [CrossRef]
84. Hofman-Caris, C.H.M.; Bauerlein, P.S.; Siegers, W.G.; Mintenig, S.M.; Messina, R.; Dekker, S.C.; Bertelkamp, C.; Cornelissen, E.R.; van Wezel, A.P. Removal of nanoparticles (both inorganic nanoparticles and nanoplastics) in drinking water treatment—Coagulation/flocculation/sedimentation, and sand/granular activated carbon filtration. *Environ. Sci.-Water Res. Technol.* **2022**, *8*, 1675–1686. [CrossRef]
85. Li, Y.; Li, J.; Ding, J.; Song, Z.R.; Yang, B.C.; Zhang, C.M.; Guan, B.H. Degradation of nano-sized polystyrene plastics by ozonation or chlorination in drinking water disinfection processes. *Chem. Eng. J.* **2022**, *427*, 131690. [CrossRef]
86. Li, Y.; Li, W.Y.; Jarvis, P.; Zhou, W.; Zhang, J.P.; Chen, J.P.; Tan, Q.W.; Tian, Y. Occurrence, removal and potential threats associated with microplastics in drinking water sources. *J. Environ. Chem. Eng.* **2020**, *8*, 104527. [CrossRef]
87. Lim, V.H.; Yamashita, Y.; Ogawa, K.; Adachi, Y. Comparison of cationic flocculants with different branching structure for the flocculation of negatively charged particles coexisting with humic substances. *J. Environ. Chem. Eng.* **2022**, *10*, 108478. [CrossRef]
88. Arenas, L.R.; Gentile, S.R.; Zimmermann, S.; Stoll, S. Coagulation of TiO_2, CeO_2 nanoparticles, and polystyrene nanoplastics in bottled mineral and surface waters. Effect of water properties, coagulant type, and dosage. *Water Environ. Res.* **2020**, *92*, 1184–1194. [CrossRef]
89. Ramirez, L.; Gentile, S.R.; Zimmermann, S.; Stoll, S. Behavior of TiO2 and CeO2 Nanoparticles and Polystyrene Nanoplastics in Bottled Mineral, Drinking and Lake Geneva Waters. Impact of Water Hardness and Natural Organic Matter on Nanoparticle Surface Properties and Aggregation. *Water* **2019**, *11*, 721. [CrossRef]
90. Shahi, N.K.; Maeng, M.; Kim, D.; Dockko, S. Removal behavior of microplastics using alum coagulant and its enhancement using polyamine-coated sand. *Process Saf. Environ. Prot.* **2020**, *141*, 9–17. [CrossRef]
91. Li, L.; Liu, D.; Song, K.; Zhou, Y.W. Performance evaluation of MBR in treating microplastics polyvinylchloride contaminated polluted surface water. *Mar. Pollut. Bull.* **2020**, *150*, 110724. [CrossRef] [PubMed]
92. Matei, E.; Predescu, A.M.; Rapa, M.; Tarcea, C.; Pantilimon, C.M.; Favier, L.; Berbecaru, A.C.; Sohaciu, M.; Predescu, C. Removal of Chromium(VI) from Aqueous Solution Using a Novel Green Magnetic Nanoparticle—Chitosan Adsorbent. *Anal. Lett.* **2019**, *52*, 2416–2438. [CrossRef]
93. Râpă, M.; Ţurcanu, A.A.; Matei, E.; Predescu, A.M.; Pantilimon, M.C.; Coman, G.; Predescu, C. Adsorption of copper (II) from aqueous solutions with alginate/clay hybrid materials. *Materials* **2021**, *14*, 7187. [CrossRef]
94. Predescu, A.M.; Matei, E.; Rapa, M.; Pantilimon, C.; Coman, G.; Savin, S.; Popa, E.E.; Predescu, C. Adsorption of Lead(II) from Aqueous Solution Using Chitosan and Polyvinyl Alcohol Blends. *Anal. Lett.* **2019**, *52*, 2365–2392. [CrossRef]
95. Arenas, L.R.; Gentile, S.R.; Zimmermann, S.; Stoll, S. Nanoplastics adsorption and removal efficiency by granular activated carbon used in drinking water treatment process. *Sci. Total Environ.* **2021**, *791*, 148175. [CrossRef]
96. Sajid, M.; Ihsanullah, I.; Khan, M.T.; Baig, N. Nanomaterials-based adsorbents for remediation of microplastics and nanoplastics in aqueous media: A review. *Sep. Purif. Technol.* **2023**, *305*, 22453. [CrossRef]
97. Martin, L.M.A.; Sheng, J.; Zimba, P.V.; Zhu, L.; Fadare, O.O.; Haley, C.; Wang, M.C.; Phillips, T.D.; Conkle, J.; Xu, W. Testing an Iron Oxide Nanoparticle-Based Method for Magnetic Separation of Nanoplastics and Microplastics from Water. *Nanomaterials* **2022**, *12*, 2348. [CrossRef]
98. Matei, E.; Covaliu-Mierla, C.I.; Ţurcanu, A.A.; Râpa, M.; Predescu, A.M.; Predescu, C. Multifunctional Membranes—A Versatile Approach for Emerging Pollutants Removal. *Membranes* **2022**, *12*, 67. [CrossRef]
99. Benavides, P.T.; Dunn, J.B.; Han, J.; Biddy, M.; Markharn, J. Exploring Comparative Energy and Environmental Benefits of Virgin, Recycled, and Bio-Derived PET Bottles. *ACS Sustain. Chem. Eng.* **2018**, *6*, 9725. [CrossRef]
100. Lambre, C.; Baviera, J.M.B.; Bolognesi, C.; Chesson, A.; Cocconcelli, P.S.; Crebelli, R.; Gott, D.M.; Grob, K.; Mengelers, M.; Mortensen, A.; et al. Safety assessment of the process Starlinger recoSTAR HDPE FC 1-PET2PET used to recycle post-consumer HDPE closures into food contact closures. *Efsa J.* **2022**, *20*, e07001. [CrossRef]

101. Winkler, A.; Santo, N.; Ortenzi, M.A.; Bolzoni, E.; Bacchetta, R.; Tremolada, P. Does mechanical stress cause microplastic release from plastic water bottles? *Water Res.* **2019**, *166*, 115082. [CrossRef]
102. Cherian, A.G.; Liu, Z.; McKie, M.J.; Almuhtaram, H.; Andrews, R.C. Microplastic Removal from Drinking Water Using Point-of-Use Devices. *Polymers* **2023**, *15*, 1331. [CrossRef]
103. Mukherjee, A.G.; Wanjari, U.R.; Bradu, P.; Patil, M.; Biswas, A.; Murali, R.; Renu, K.; Dey, A.; Vellingiri, B.; Raja, G.; et al. Elimination of microplastics from the aquatic milieu: A dream to achieve. *Chemosphere* **2022**, *303*, 135232. [CrossRef] [PubMed]
104. Barbier, J.-S.; Dris, R.; Lecarpentier, C.; Raymond, V.; Delabre, K.; Thibert, S.; Tassin, B.; Gasperi, J. Microplastic occurrence after conventional and nanofiltration processes at drinking water treatment plants: Preliminary results. *Front. Water* **2022**, *4*, 1–14. [CrossRef]
105. Ding, H.J.; Zhang, J.; He, H.; Zhu, Y.; Dionysiou, D.D.; Liu, Z.; Zhao, C. Do membrane filtration systems in drinking water treatment plants release nano/microplastics? *Sci. Total Environ.* **2021**, *755*, 142658. [CrossRef] [PubMed]
106. Chu, X.X.; Zheng, B.; Li, Z.X.; Cai, C.; Peng, Z.; Zhao, P.; Tian, Y.M. Occurrence and distribution of microplastics in water supply systems: In water and pipe scales. *Sci. Total Environ.* **2022**, *803*, 150004. [CrossRef]
107. Devi, M.K.; Karmegam, N.; Manikandan, S.; Subbaiya, R.; Song, H.; Kwon, E.E.; Sarkar, B.; Bolan, N.; Kim, W.; Rinklebe, J.; et al. Removal of nanoplastics in water treatment processes: A review. *Sci. Total Environ.* **2022**, *845*, 157168. [CrossRef]
108. Uheida, A.; Mejia, H.G.; Abdel-Rehim, M.; Hamd, W.; Dutta, J. Visible light photocatalytic degradation of polypropylene microplastics in a continuous water flow system. *J. Hazard. Mater.* **2021**, *406*, 124299. [CrossRef] [PubMed]
109. Abomohra, A.; Hanelt, D. Recent Advances in Micro-/Nanoplastic (MNPs) Removal by Microalgae and Possible Integrated Routes of Energy Recovery. *Microorganisms* **2022**, *10*, 2400. [CrossRef]
110. Cella, C.; La Spina, R.; Mehn, D.; Fumagalli, F.; Ceccone, G.; Valsesia, A.; Gilliland, D. Detecting Micro- and Nanoplastics Released from Food Packaging: Challenges and Analytical Strategies. *Polymers* **2022**, *14*, 1238. [CrossRef] [PubMed]
111. Ivleva, N.P. Chemical Analysis of Microplastics and Nanoplastics: Challenges, Advanced Methods, and Perspectives. *Chem. Rev.* **2021**, *121*, 11886–11936. [CrossRef]
112. Liu, Q.R.; Chen, Y.L.; Chen, Z.; Yang, F.W.; Xie, Y.F.; Yao, W.R. Current status of microplastics and nanoplastics removal methods: Summary, comparison and prospect. *Sci. Total Environ.* **2022**, *851*, 157991. [CrossRef]
113. Lee, J.; Chae, K.J. A systematic protocol of microplastics analysis from their identification to quantification in water environment: A comprehensive review. *J. Hazard. Mater.* **2021**, *403*, 124049. [CrossRef] [PubMed]
114. Koelmans, A.A.; Nor, N.H.M.; Hermsen, E.; Kooi, M.; Mintenig, S.M.; De France, J. Microplastics in freshwaters and drinking water: Critical review and assessment of data quality. *Water Res.* **2019**, *155*, 410–422. [CrossRef] [PubMed]
115. Mintenig, S.M.; Loder, M.G.J.; Primpke, S.; Gerdts, G. Low numbers of microplastics detected in drinking water from ground water sources. *Sci. Total Environ.* **2019**, *648*, 631–635. [CrossRef]
116. Siegel, H.; Fischer, F.; Lenz, R.; Fischer, D.; Jekel, M.; Labrenz, M. Identification and quantification of microplastic particles in drinking water treatment sludge as an integrative approach to determine microplastic abundance in a freshwater river. *Environ. Pollut.* **2021**, *286*, 117524. [CrossRef] [PubMed]
117. Tong, H.Y.; Jiang, Q.Y.; Hu, X.S.; Zhong, X.C. Occurrence and identification of microplastics in tap water from China. *Chemosphere* **2020**, *252*, 126493. [CrossRef] [PubMed]
118. Bauerlein, P.S.; Hofman-Caris, R.; Pieke, E.N.; Laak, T.L.T. Fate of microplastics in the drinking water production. *Water Res.* **2022**, *221*, 118790. [CrossRef]
119. Taghipour, H.; Ghayebzadeh, M.; Ganji, F.; Mousavi, S.; Azizi, N. Tracking microplastics contamination in drinking water in Zahedan, Iran: From source to consumption taps. *Sci. Total Environ.* **2023**, *872*, 162121. [CrossRef]
120. Kosuth, M.; Mason, S.A.; Wattenberg, E.V. Anthropogenic contamination of tap water, beer, and sea salt. *PLoS ONE* **2018**, *13*, e0194970. [CrossRef] [PubMed]
121. Li, Y.; Wang, Z.Q.; Guan, B.H. Separation and identification of nanoplastics in tap water. *Environ. Res.* **2022**, *204*, 112134. [CrossRef]
122. Huang, Y.H.; Wong, K.K.; Li, W.; Zhao, H.R.; Wang, T.M.; Stanescu, S.; Boult, S.; van Dongen, B.; Mativenga, P.; Li, L. Characteristics of nano-plastics in bottled drinking water. *J. Hazard. Mater.* **2022**, *424*, 127404. [CrossRef] [PubMed]
123. Winkler, A.; Fumagalli, F.; Cella, C.; Gilliland, D.; Tremolada, P.; Valsesia, A. Detection and formation mechanisms of secondary nanoplastic released from drinking water bottles. *Water Res.* **2022**, *222*, 118848. [CrossRef] [PubMed]
124. Weisser, J.; Beer, I.; Hufnagl, B.; Hofmann, T.; Lohninger, H.; Ivleva, N.P.; Glas, K. From the Well to the Bottle: Identifying Sources of Microplastics in Mineral Water. *Water* **2021**, *13*, 841. [CrossRef]
125. Ossmann, B.E.; Sarau, G.; Holtmannspotter, H.; Pischetsrieder, M.; Christiansen, S.H.; Dicke, W. Small-sized microplastics and pigmented particles in bottled mineral water. *Water Res.* **2018**, *141*, 307–316. [CrossRef] [PubMed]
126. Song, K.; Ding, R.R.; Sun, C.Y.; Yao, L.G.; Zhang, W.C. Microparticles and microplastics released from daily use of plastic feeding and water bottles and plastic injectors: Potential risks to infants and children in China. *Environ. Sci. Pollut. Res.* **2021**, *28*, 59813–59820. [CrossRef] [PubMed]
127. Perez-Guevara, F.; Roy, P.D.; Elizalde-Martinez, I.; Kutralam-Muniasamy, G.; Shruti, V.C. Human exposure to microplastics from urban decentralized pay-to-fetch drinking-water refill kiosks. *Sci. Total Environ.* **2022**, *848*, 157722. [CrossRef] [PubMed]
128. Funck, M.; Yildirim, A.; Nickel, C.; Schram, J.; Schmidt, T.C.; Tuerk, J. Identification of microplastics in wastewater after cascade filtration using Pyrolysis-GC-MS. *Methodsx* **2020**, *7*, 100778. [CrossRef] [PubMed]

129. Anuar, S.T.; Altarawnah, R.S.; Ali, A.A.M.; Lee, B.Q.; Khalik, W.; Yusof, K.; Ibrahim, Y.S. Utilizing Pyrolysis-Gas Chromatography/Mass Spectrometry for Monitoring and Analytical Characterization of Microplastics in Polychaete Worms. *Polymers* **2022**, *14*, 3054. [CrossRef] [PubMed]
130. Lin, P.Y.; Wu, I.H.; Tsai, C.Y.; Kirankumar, R.; Hsieh, S.C. Detecting the release of plastic particles in packaged drinking water under simulated light irradiation using surface-enhanced Raman spectroscopy. *Anal. Chim. Acta* **2022**, *1198*, 339516. [CrossRef]
131. Gao, Q.; Wang, Y.Q.; Ji, Y.X.; Zhao, X.Z.; Zhang, P.P.; Chen, L.X. Tracking of realistic nanoplastics in complicated matrices by iridium element labeling and inductively coupled plasma mass spectroscopy. *J. Hazard. Mater.* **2022**, *424*, 127628. [CrossRef] [PubMed]
132. Kundu, A.; Shetti, N.P.; Basu, S.; Reddy, K.R.; Nadagouda, M.N.; Aminabhavi, T.M. Identification and removal of micro- and nano-plastics: Efficient and cost-effective methods. *Chem. Eng. J.* **2021**, *421*, 129816. [CrossRef]
133. Materic, D.; Holzinger, R.; Niemann, H. Nanoplastics and ultrafine microplastic in the Dutch Wadden Sea—The hidden plastics debris? *Sci. Total Environ.* **2022**, *846*, 157371. [CrossRef]
134. Corti, A.; Vinciguerra, V.; Iannilli, V.; Pietrelli, L.; Manariti, A.; Bianchi, S.; Petri, A.; Cifelli, M.; Domenici, V.; Castelvetro, V. Thorough Multianalytical Characterization and Quantification of Micro- and Nanoplastics from Bracciano Lake's Sediments. *Sustainability* **2020**, *12*, 878. [CrossRef]
135. Chialanza, M.R.; Samarra, S.F.; Parada, A.P. Modeling microplastic with polyethylene (PE) spherical particles: A differential scanning calorimetry approach for quantification. *Environ. Sci. Pollut. Res.* **2022**, *29*, 2311–2324. [CrossRef]
136. Jakubowicz, I.; Enebro, J.; Yarahmadi, N. Challenges in the search for nanoplastics in the environment-A critical review from the polymer science perspective. *Polym. Test.* **2021**, *93*, 106953. [CrossRef]
137. Yuan, Z.H.; Nag, R.; Cummins, E. Human health concerns regarding microplastics in the aquatic environment- From marine to food systems. *Sci. Total Environ.* **2022**, *823*, 153730. [CrossRef] [PubMed]
138. Chang, L.; Jiang, S.; Luo, J.; Zhang, J.F.; Liu, X.H.; Lee, C.Y.; Zhang, W. Nanowell-enhanced Raman spectroscopy enables the visualization and quantification of nanoplastics in the environment. *Environ. Sci.-Nano* **2022**, *9*, 542–553. [CrossRef]
139. Villacorta, A.; Rubio, L.; Alaraby, M.; Lopez-Mesas, M.; Fuentes-Cebrian, V.; Moriones, O.H.; Marcos, R.; Hernandez, A. A new source of representative secondary PET nanoplastics. Obtention, characterization, and hazard evaluation. *J. Hazard. Mater.* **2022**, *439*, 129593. [CrossRef] [PubMed]
140. Sobhani, Z.; Lei, Y.J.; Tang, Y.H.; Wu, L.W.; Zhang, X.; Naidu, R.; Megharaj, M.; Fang, C. Microplastics generated when opening plastic packaging. *Sci. Rep.* **2020**, *10*, 4841. [CrossRef] [PubMed]
141. Fadare, O.O.; Wan, B.; Guo, L.H.; Zhao, L.X. Microplastics from consumer plastic food containers: Are we consuming it? *Chemosphere* **2020**, *253*, 126787. [CrossRef]
142. Asamoah, B.O.; Roussey, M.; Peiponen, K.E. On optical sensing of surface roughness of flat and curved microplastics in water. *Chemosphere* **2020**, *254*, 126789. [CrossRef]
143. Yang, X.; Man, Y.B.; Wong, M.H.; Owen, R.B.; Chow, K.L. Environmental health impacts of microplastics exposure on structural organization levels in the human body. *Sci. Total Environ.* **2022**, *825*, 154025. [CrossRef]
144. Rahman, A.; Sarkar, A.; Yadav, O.P.; Achari, G.; Slobodnik, J. Potential human health risks due to environmental exposure to nano-and microplastics and knowledge gaps: A scoping review. *Sci. Total Environ.* **2021**, *757*, 143872. [CrossRef] [PubMed]
145. Fiore, M.; Conti, G.O.; Caltabiano, R.; Buffone, A.; Zuccarello, P.; Cormaci, L.; Cannizzaro, M.A.; Ferrante, M. Role of Emerging Environmental Risk Factors in Thyroid Cancer: A Brief Review. *Int. J. Environ. Res. Public Health* **2019**, *16*, 1185. [CrossRef]
146. Tisler, S.; Christensen, J.H. Non-target screening for the identification of migrating compounds from reusable plastic bottles into drinking water. *J. Hazard. Mater.* **2022**, *429*, 128331. [CrossRef]
147. Karbalaei, S.; Hanachi, P.; Walker, T.R.; Cole, M. Occurrence, sources, human health impacts and mitigation of microplastic pollution. *Environ. Sci. Pollut. Res.* **2018**, *25*, 36046–36063. [CrossRef] [PubMed]
148. Li, X.; Ying, G.G.; Su, H.C.; Yang, X.B.; Wang, L. Simultaneous determination and assessment of 4-nonylphenol, bisphenol A and triclosan in tap water, bottled water and baby bottles. *Environ. Int.* **2010**, *36*, 557–562. [CrossRef] [PubMed]
149. Rani, M.; Shim, W.J.; Han, G.M.; Jang, M.; Al-Odaini, N.A.; Song, Y.K.; Hong, S.H. Qualitative Analysis of Additives in Plastic Marine Debris and Its New Products. *Arch. Environ. Contam. Toxicol.* **2015**, *69*, 352–366. [CrossRef]
150. Cooper, J.E.; Kendig, E.L.; Belcher, S.M. Assessment of bisphenol A released from reusable plastic, aluminium and stainless steel water bottles. *Chemosphere* **2011**, *85*, 943–947. [CrossRef] [PubMed]
151. Ginter-Kramarczyk, D.; Zembrzuska, J.; Kruszelnicka, I.; Zajac-Woznialis, A.; Cislak, M. Influence of Temperature on the Quantity of Bisphenol A in Bottled Drinking Water. *Int. J. Environ. Res. Public Health* **2022**, *19*, 5710. [CrossRef]
152. Galloway, T.S. Micro-and Nano-plastics and Human Health. *Mar. Anthropog. Litter* **2015**, *13*, 343–366. [CrossRef]
153. Guart, A.; Wagner, M.; Mezquida, A.; Lacorte, S.; Oehlmann, J.; Borrell, A. Migration of plasticisers from Tritan (TM) and polycarbonate bottles and toxicological evaluation. *Food Chem.* **2013**, *141*, 373–380. [CrossRef] [PubMed]
154. Srivastava, R.K.; Godara, S. Use of polycarbonate plastic products and human health. *Int. J. Basic Clin. Pharm.* **2017**, *2*, 12–17. [CrossRef]
155. Le, H.H.; Carlson, E.M.; Chua, J.P.; Belcher, S.M. Bisphenol A is released from polycarbonate drinking bottles and mimics the neurotoxic actions of estrogen in developing cerebellar neurons. *Toxicol. Lett.* **2008**, *176*, 149–156. [CrossRef] [PubMed]
156. Moriyama, K.; Tagami, T.; Akamizu, T.; Usui, T.; Saijo, M.; Kanamoto, N.; Hataya, Y.; Shimatsu, A.; Kuzuya, H.; Nakao, K. Thyroid hormone action is disrupted by bisphenol A as an antagonist. *J. Clin. Endocrinol. Metab.* **2002**, *87*, 5185–5190. [CrossRef]

157. Cipelli, R.; Harries, L.; Okuda, K.; Yoshihara, S.; Melzer, D.; Galloway, T. Bisphenol A modulates the metabolic regulator oestrogen-related receptor-alpha in T-cells. *Reproduction* **2014**, *147*, 419–426. [CrossRef]
158. Melzer, D.; Osborne, N.J.; Henley, W.E.; Cipelli, R.; Young, A.; Money, C.; McCormack, P.; Luben, R.; Khaw, K.T.; Wareham, N.J.; et al. Urinary Bisphenol A Concentration and Risk of Future Coronary Artery Disease in Apparently Healthy Men and Women. *Circulation* **2012**, *125*, 1482–1490. [CrossRef]
159. Budhiraja, V.; Urh, A.; Horvat, P.; Krzan, A. Synergistic Adsorption of Organic Pollutants on Weathered Polyethylene Microplastics. *Polymers* **2022**, *14*, 2674. [CrossRef]
160. Chailurkit, L.O.; Srijaruskul, K.; Ongphiphadhanakul, B. Bisphenol A in Canned Carbonated Drinks and Plastic-Bottled Water from Supermarkets. *Expo. Health* **2017**, *9*, 243–248. [CrossRef]
161. Gomez, C.; Gallart-Ayala, H. Metabolomics: A tool to characterize the effect of phthalates and bisphenol A. *Environ. Rev.* **2018**, *26*, 351–357. [CrossRef]
162. Li, H.; Li, C.M.; An, L.H.; Deng, C.; Su, H.; Wang, L.F.; Jiang, Z.J.; Zhou, J.; Wang, J.; Zhang, C.H.; et al. Phthalate esters in bottled drinking water and their human exposure in Beijing, China. *Food Addit. Contam. Part B-Surveill.* **2019**, *12*, 1–9. [CrossRef]
163. Cheng, Z.; Nie, X.P.; Wang, H.S.; Wong, M.H. Risk assessments of human exposure to bioaccessible phthalate esters through market fish consumption. *Environ. Int.* **2013**, *57–58*, 75–80. [CrossRef]
164. Mihucz, V.G.; Zaray, G. Occurrence of antimony and phthalate esters in polyethylene terephthalate bottled drinking water. *Appl. Spectrosc. Rev.* **2016**, *51*, 163–189. [CrossRef]
165. Brennecke, D.; Ferreira, E.C.; Costa, T.M.M.; Appel, D.; da Gama, B.A.P.; Lenz, M. Ingested microplastics (>100 μm) are translocated to organs of the tropical fiddler crab Uca rapax. *Mar. Pollut. Bull.* **2015**, *96*, 491–495. [CrossRef]
166. Tavelli, R.; Callens, M.; Grootaert, C.; Abdallah, M.F.; Rajkovic, A. Foodborne pathogens in the plastisphere: Can microplastics in the food chain threaten microbial food safety? *Trends Food Sci. Technol.* **2022**, *129*, 1–10. [CrossRef]
167. Mortensen, N.P.; Johnson, L.M.; Grieger, K.D.; Ambroso, J.L.; Fennell, T.R. Biological interactions between nanomaterials and placental development and function following oral exposure. *Reprod. Toxicol.* **2019**, *90*, 150–165. [CrossRef]
168. Canesi, L.; Ciacci, C.; Bergami, E.; Monopoli, M.P.; Dawson, K.A.; Papa, S.; Canonico, B.; Corsi, I. Evidence for immunomodulation and apoptotic processes induced by cationic polystyrene nanoparticles in the hemocytes of the marine bivalve Mytilus. *Mar. Environ. Res.* **2015**, *111*, 34–40. [CrossRef] [PubMed]
169. Wright, S.L.; Kelly, F.J. Plastic and Human Health: A Micro Issue? *Environ. Sci. Technol.* **2017**, *51*, 6634–6647. [CrossRef]
170. Campanale, C.; Massarelli, C.; Savino, I.; Locaputo, V.; Uricchio, V.F. A Detailed Review Study on Potential Effects of Microplastics and Additives of Concern on Human Health. *Int. J. Environ. Res. Public Health* **2020**, *17*, 1212. [CrossRef]
171. Prata, J.C.; da Costa, J.P.; Lopes, I.; Duarte, A.C.; Rocha-Santos, T. Environmental exposure to microplastics: An overview on possible human health effects. *Sci. Total Environ.* **2020**, *702*, 134455. [CrossRef]
172. Ji, Y.; Wang, C.; Wang, Y.; Fu, L.; Man, M.; Chen, L. Realistic polyethylene terephthalate nanoplastics and the size- and surface coating-dependent toxicological impacts on zebrafish embryos. *Environ. Sci.-Nano* **2020**, *7*, 2313–2324. [CrossRef]
173. Monnery, B.D.; Wright, M.; Cavill, R.; Hoogenboom, R.; Shaunak, S.; Steinke, J.H.G.; Thanou, M. Cytotoxicity of polycations: Relationship of molecular weight and the hydrolytic theory of the mechanism of toxicity. *Int. J. Pharm.* **2017**, *521*, 249–258. [CrossRef] [PubMed]
174. Xia, T.; Kovochich, M.; Liong, M.; Zink, J.I.; Nel, A.E. Cationic polystyrene nanosphere toxicity depends on cell-specific endocytic and mitochondrial injury pathways. *ACS Nano* **2008**, *2*, 85–96. [CrossRef] [PubMed]
175. Choi, D.; Hwang, J.; Bang, J.; Han, S.; Kim, T.; Oh, Y.; Hwang, Y.; Choi, J.; Hong, J. In vitro toxicity from a physical perspective of polyethylene microplastics based on statistical curvature change analysis. *Sci. Total Environ.* **2021**, *752*, 142242. [CrossRef] [PubMed]
176. Magri, D.; Veronesi, M.; Sanchez-Moreno, P.; Tolardo, V.; Bandiera, T.; Pompa, P.P.; Athanassiou, A.; Fragouli, D. PET nanoplastics interactions with water contaminants and their impact on human cells. *Environ. Pollut.* **2021**, *271*, 116262. [CrossRef]
177. Magri, D.; Sanchez-Moreno, P.; Caputo, G.; Gatto, F.; Veronesi, M.; Bardi, G.; Catelani, T.; Guarnieri, D.; Athanassiou, A.; Pompa, P.P.; et al. Laser Ablation as a Versatile Tool to Mimic Polyethylene Terephthalate Nanoplastic Pollutants: Characterization and Toxicology Assessment. *ACS Nano* **2018**, *12*, 7690–7700. [CrossRef]
178. Banerjee, A.; Billey, L.O.; McGarvey, A.M.; Shelver, W.L. Effects of polystyrene micro/nanoplastics on liver cells based on particle size, surface functionalization, concentration and exposure period. *Sci. Total Environ.* **2022**, *836*, 155621. [CrossRef]
179. Park, I.; Yang, W.S.; Lim, D.K. Current Status of Organic Matters in Bottled Drinking Water in Korea. *ACS EST Water* **2022**, *2*, 738–748. [CrossRef]
180. Smith, M.; Love, D.C.; Rochman, C.M.; Neff, R.A. Microplastics in Seafood and the Implications for Human Health. *Curr. Environ. Health Rep.* **2018**, *5*, 375–386. [CrossRef]
181. Pentelovitch, N.; Nagel, J.K. Understanding the Use of Bio-Inspired Design Tools by Industry Professionals. *Biomimetics* **2022**, *7*, 63. [CrossRef] [PubMed]

182. Schmaltz, E.; Melvin, E.C.; Diana, Z.; Gunady, E.F.; Rittschof, D.; Somarelli, J.A.; Virdin, J.; Dunphy-Daly, M.M. Plastic pollution solutions: Emerging technologies to prevent and collect marine plastic pollution. *Environ. Int.* **2020**, *144*, 106067. [CrossRef] [PubMed]
183. Kiran, B.R.; Kopperi, H.; Mohan, S.V. Micro/nano-plastics occurrence, identification, risk analysis and mitigation: Challenges and perspectives. *Rev. Environ. Sci. Bio-Technol.* **2022**, *21*, 169–203. [CrossRef]

polymers

MDPI

Article

Study and Characterization of Regenerated Hard Foam Prepared by Polyol Hydrolysis of Waste Polyurethane

Xiaohua Gu [1,2,3,*], Xiaoyao Wang [2], Xinyu Guo [4], Siwen Liu [5], Qi Li [6] and Yan Liu [1,*]

1 School of Energy and Building Environment of Guilin University of Aerospace Technology, Guilin 541004, China
2 School of Material Science and Engineering of Qiqihar University, Qiqihar 161006, China
3 State Key Laboratory for Modification of Chemical Fibers and Polymer Materials, College of Materials Science and Engineering, Donghua University, Shanghai 201620, China
4 College of Chemistry and Chemical Engineering of Qiqihar University, Qiqihar 161006, China
5 College of Innovative Material & Energy, Hubei University, Wuhan 430062, China
6 Fishery College, Zhejiang Ocean University, Zhoushan 316022, China
* Correspondence: gxh218@163.com (X.G.); 2019009@guat.edu.cn (Y.L.); Tel.: +86-18121138868 (X.G.); +86-18078343019 (Y.L.)

Abstract: In this paper, four different kinds of diols were used for the alcoholysis of waste thermoplastic polyurethane elastomers. The recycled polyether polyols were used to prepare regenerated thermosetting polyurethane rigid foam through one-step foaming. We used four different kinds of alcoholysis agents, according to different proportions of the complex, and we combined them with an alkali metal catalyst (KOH) to trigger the catalytic cleavage of the carbamate bonds in the waste polyurethane elastomers. The effects of the different types and different chain lengths of the alcoholysis agents on the degradation of the waste polyurethane elastomers and the preparation of regenerated polyurethane rigid foam were studied. Based on the viscosity, GPC, FT-IR, foaming time and compression strength, water absorption, TG, apparent density, and thermal conductivity of the recycled polyurethane foam, eight groups of optimal components were selected and discussed. The results showed that the viscosity of the recovered biodegradable materials was between 485 and 1200 mPa·s. The hard foam of the regenerated polyurethane was prepared using biodegradable materials instead of commercially available polyether polyols, and its compressive strength was between 0.131 and 0.176 MPa. The water absorption rate ranged from 0.7265 to 1.9923%. The apparent density of the foam was between 0.0303 and 0.0403 kg/m^3. The thermal conductivity ranged from 0.0151 to 0.0202 W/(m·K). A large number of experimental results showed that the degradation of the waste polyurethane elastomers by the alcoholysis agents was successful. The thermoplastic polyurethane elastomers can not only be reconstructed, but they can also be degraded by alcoholysis to produce regenerated polyurethane rigid foam.

Keywords: waste polyurethane elastomer; degradation and recovery; glycolysis; diols; regenerated polyurethane foam

check for updates

Citation: Gu, X.; Wang, X.; Guo, X.; Liu, S.; Li, Q.; Liu, Y. Study and Characterization of Regenerated Hard Foam Prepared by Polyol Hydrolysis of Waste Polyurethane. *Polymers* **2023**, *15*, 1445. https://doi.org/10.3390/polym15061445

Academic Editors: Cristina Cazan and Mihaela Cosnita

Received: 29 December 2022
Revised: 13 February 2023
Accepted: 19 February 2023
Published: 14 March 2023

1. Introduction

Plastic is an indispensable material in our lives [1,2]. As a part of plastics, polyurethane (PU) has been rapidly developed in household goods, the construction industry, daily necessities, the transportation industry, household appliances, and other fields since its emergence [3–5]. PU can be used to prepare soft polyurethane foam, hard polyurethane foam, polyurethane elastomers, coatings, adhesives, and other products [6] which are deeply loved by people. Thermoplastic polyurethane elastomers, with their excellent properties and wide application, have become important thermoplastic elastomer materials whose molecules are largely linear, with no or little chemical cross-linking [7,8]. There are many physical cross-links formed by hydrogen bonds between linear polyurethane molecular chains, which strengthen the material's form and endow it with many excellent properties, such as a high modulus, high strength, and excellent wear resistance, chemical resistance, hydrolysis resistance, high and low temperature resistance, and mold resistance [9–14]. With the wide application of polyurethane elastomers in various industries, the global production and consumption of polyurethane have been increasing year over year at an amazing speed [15]. At the same time, under such tremendous consumption, polyurethane elastomer waste has caused serious environmental pollution and resource waste. With the implementation of the global concept of environmental protection and sustainable development, the recycling and sustainable utilization of polyurethane waste has become a global issue [16].

For the recycling and sustainable development of polyurethane waste, landfill and incineration were primarily used in the beginning of its development, but the pollution of land caused by landfill is immeasurable. Although incineration is better than a landfill as it does not occupy land resources, the carbon dioxide and other gases generated by incineration cause a serious burden on the atmospheric protective layer [17,18], and people's awareness of environmental protection and legal restrictions relating to polyurethane waste have caused its disposal via landfills and incineration to be banned. Due to its three-dimensional network structure [19], in recent years, various recycling methods for polyurethane waste have been studied, including the mechanical method, pyrolysis method, microbial method, and chemical recovery method [20,21]. Among the many recycling methods, chemical recycling can effectively break the carbamate bond in the chain segment of polyurethane waste so as to obtain valuable small molecular monomers for sustainable use [22,23]. Therefore, this method has been recognized by many scholars.

Among the many chemical recovery methods, alcoholysis has been favored for advantages such as green environmental protection, low energy consumption, less impurities in the recycled materials, and directional bond-breaking by catalysts [24,25]. In this paper, different types of small molecule diols were used for the combination ratio to study the degradation and recycling of used polyurethane elastomers. Alcoholysis agents with different ratios successfully degraded the used polyurethane elastomers into small molecule alcohols, and recycled polyurethane rigid foam was successfully prepared by using the small molecule alcohols obtained from the degradation and recovery of the elastomers. The effects of different types of alcoholysis agents on the degradation of polyurethane elastomers were studied, and they have provided certain reference significance for the selection of alcohols and the combination of two components in the degradation of polyurethane elastomers.

2. Materials and Methods

2.1. Materials and Reagents

The reagents used in this experiment are shown in Table 1.

Table 1. Raw materials of the paper experiment.

Reagent Name	The Purity	Factory of Production
Waste polyurethane elastomer	Industrial waste	Shanghai Hecheng Polymer Material Co., Ltd.
Ethylene glycol (EG)	AR	Tianjin Chemical Reagent Factory 1
1,2-Propanediol (PG)	AR	Tianjin Chemical Reagent Factory 1
Butane-1,4-diol (BDO)	AR	Tianjin Kaitong Chemical Reagent Co., Ltd.
Diethylene glycol (DEG)	AR	China Pharmaceutical Guangzhou Chemical Reagent Company
Potassium hydroxide (KOH)	AR	Tianjin Tianli Chemical Reagent Co., Ltd.
Silicone oil stabilizer	CP	Guangzhou Feirui Chemical Co., LRD
TEA	AR	Shanghai Demao Chemical Co., Ltd.
Dibutyltin dilaurate	AR	Shanghai Jieer Technology Co., Ltd.
Foaming agent	CP	Shenzhen Huachang Chemical Co., Ltd.
Polyether 4110	CP	Shandong Lianhaoyao New Material Co., Ltd.
Polyaryl polymethylene isocyanate (PAPI)	CP	Wuhan Fude Chemical Co., Ltd.

Note: Deionized water was used in all of the experiments.

The equipment used in this experiment is shown in Table 2.

Table 2. Experimental equipment used in the paper.

Name of Instrument	Model	Factory of Production
Electronic analytical balance	JA3003C	Sartorius Scientific Instruments (Beijing) Co., Ltd.
Cantilever constant speed power electric mixer	TJ-1200W	Changzhou Huaao Instrument Manufacturing Co., Ltd.
Spherical reactor (1 L)	ZNHW-200	Shanghai Leighton Industrial Co., Ltd
Digital blast drying oven	WX881	Wujiang Weixin Electric Heating Equipment Co., Ltd.
Digital viscometer	NDJ-5	Shanghai Pingxuan Scientific Instrument Co., Ltd.
Disposable plastic cup	350 mL	Topu Daily Chemicals (China) Co., Ltd.
Constant temperature heating sleeve	FDSG-420	Wuxi Huachang Chemical Co., Ltd.

2.2. Preparation of the Titanium Nanosystem Catalysts

In this experiment, four different alcoholysis agents were selected: butanediol, propylene glycol, monodiethylene glycol, and ethylene glycol. Through the previous cumulative work in the laboratory, the four alcoholysis agents were paired to form a two-component alcoholysis agent, and the alcoholysis reaction was carried out on the waste polyurethane elastomer. The specific ratios of the alcoholysis agents were: I = EG:PG (A1–A5); II = EG:BDO (B1–B5); III = PG:BDO (C1–C5); IV = DEG:BDO (D1–D5); V = DEG:EG (E1–E5), and VI = DEG:PG (F1–F5). The mass ratios of the alcoholysis agents were 40:40, 30:50, 50:30, 60:20, and 20:60. Based on the above experimental design, the degradation steps for the waste polyurethane elastomers were as follows (the process is also shown in Figure 1):

1. The waste polyurethane elastomer was cut and processed to 5–10 mm small sections.
2. We mixed 80 g small molecular alcohol and catalyst (KOH) and added them into a 1000 mL spherical reactor. We then placed the reactor in a heating jacket and stirred to 90 °C until the catalyst was completely dissolved.
3. We added 80 g of chopped waste polyurethane elastomer into the reactor and heated it up to 180 °C. At that time, we accelerated the stirring speed and waited for the polyurethane elastomer to dissolve, gently stirring for 0.5–1 h.
4. We stopped the reaction and cooled the mixture to room temperature before pouring it into a disposable plastic cup.
5. The hydroxyl value and viscosity of the sample were measured, and the foaming experiment was carried out after the value had reached the standard.
6. We took a certain amount of alcoholysis products and added the catalyst, foaming agent, and surfactant, stirring the mixture evenly. Then, we added polyphenyl polymethylene polyisocyanate (PAPI) at a ratio of 1:1 and immediately stirred with a high-speed mixer for 10 s until the mixture was uniform, and we waited for the foam to slowly form.
7. The sample preparation and testing could then be carried out on the regenerated polyurethane hard foam, which was to be prepared after curing and cross-linking for 24 h.

2.3. Performance Test and Structural Characterization of the Polyurethane Rigid Foam

Viscosity analysis: An NDJ-5S digital viscometer was used for the viscosity testing. A number of degradation products were taken and placed in the container and a suitable rotor was selected, as well as a suitable rotational speed, and the viscosity testing was carried out at 25 °C.

Compression strength test: With reference to the GB/T8813-2008 test standard, we selected a sample size of 50 mm × 50 mm × 50 mm and we used an EFS-24RE universal testing machine for the compression strength test.

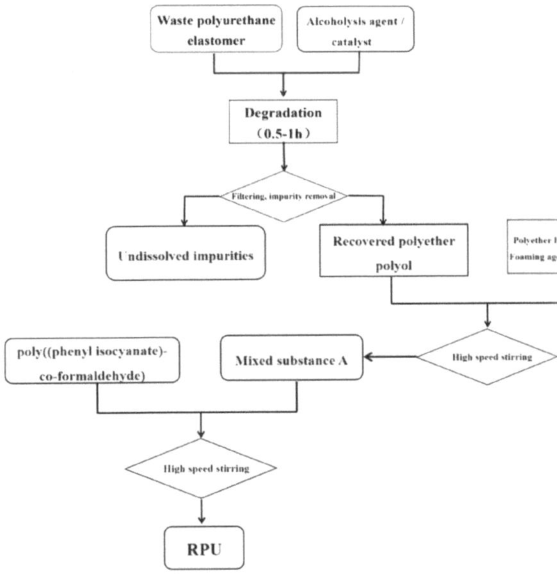

Figure 1. Experimental flow chart.

Water absorption test: With reference to the GB/T8810-1988 test standard, the sample size was 50 mm × 10 mm × 10 mm, and we used distilled water to determine the water absorption, weighing the mass of the sample before and after immersion, and then calculated its water absorption.

Hydroxyl value determination: With reference to the GB/T12008.3-2009 standard, we took an appropriate amount of oligomeric regenerated polyether polyol in 10 mL conical flasks and we used ester anhydride method—pyridine for the hydroxyl value determination.

Thermal conductivity analysis: With reference to the QB/T3806-1999 standard, the sample size was 200 mm × 200 mm × 20 mm, and we used a FEHC-S thermal conductivity tester acquired from Changzhou Hua'ao Instrument Manufacturing Co., Changzhou, China.

SEM analysis: We cut the regenerated polyurethane rigid foam specimens into appropriate thin slices (we did not squeeze them), and then we used SEM to observe the microstructures of the bubble pores of the foam (magnified 20 times) and identify the clear, complete, and uniform areas of the bubble pore structures for observation. We recorded the observation results.

FT-IR test: A GR-285 IR produced by Dalian Precision Scientific Instruments Co., Ltd., Dalian, China was used to analyze the structures of the foam samples. The samples were made using the KBr pressing method, and the test wavelength range was 500–4000 cm^{-1}.

TG analysis: TG analysis was performed by using TG under a nitrogen atmosphere at a rate of 30 °C/min, from room temperature (25 °C) to 500 °C. The gas flow rate was selected to be 50 mL/min, the carrier gas was air, and alumina was used as a reference object.

Determination of molecular weight distribution coefficient: The molecular weight distribution of the regenerated polyether polyol was determined by gel permeation chromatography (GPC). The measurements were performed using a thermal scientific chromatograph equipped with an isocratic DionexUltra3000 pump and a RefrtoMax521 refractive index detector. The separations were performed in four Phenomenex Phenogel GPC columns with a separation temperature of 30 °C, a particle size of 5 μm, and porosities of 105, 103, 102, and 50, respectively, located in an ultimate thermostatic column at 3000 °C. The mobile phase was tetrahydrofuran (THF) at a flow rate of 1 mL·min^{-1}. The samples were first dissolved with N, N—Dimethylformamide (DMF), followed by 1.6 wt.% THF, and then filtered through a nylon filter with a pore size of 2 mm and prepared for use.

3. Results and Discussion

3.1. Degradation Mechanism of the Waste Polyurethane Elastomers

Under the action of the two-component alcohols and catalysts, the carbamate in the chain segment of the polyurethane elastomer was broken and replaced by small molecular alcohol chains to generate polyether polyols [26,27]. The main reaction is:

$$R_1\text{-NHCOO-}R_2 + \text{HO-}R_3\text{-OH} \rightarrow R_1\text{-NHCOO-}R_3\text{-OH} + R_2\text{-OH} \quad (1)$$

$$R_1\text{-NHCOO-}R_3\text{-OH} \rightarrow R_1\text{-NH-}R_3\text{-OH} + CO_2 \quad (2)$$

Under the condition of a high temperature reaction, there will be many groups involved in the reaction, and many side reactions will occur. The primary side reactions are the urea group fracture in the alcoholysis agent, which generates small molecules of amines and small molecules of alcohols [28]. The specific reactions are as follows:

$$R_1\text{-NHCONH-}R_2 + \text{HO-}R_3\text{-OH} \rightarrow R_1\text{-NHCOO-}R_3\text{-OH} + R_2\text{-}NH_2 \quad (3)$$

Taking DEG and BDO alcoholysis agents as an example, under the catalysis of an alkaline catalyst (KOH) at a high temperature in the reaction reactor, the used polyurethane elastomer and the alcoholysis agent broke and were replaced by the alcoholysis agent, thus generating polyether polyols [29]. The specific reaction mechanism is shown in the following (Figure 2):

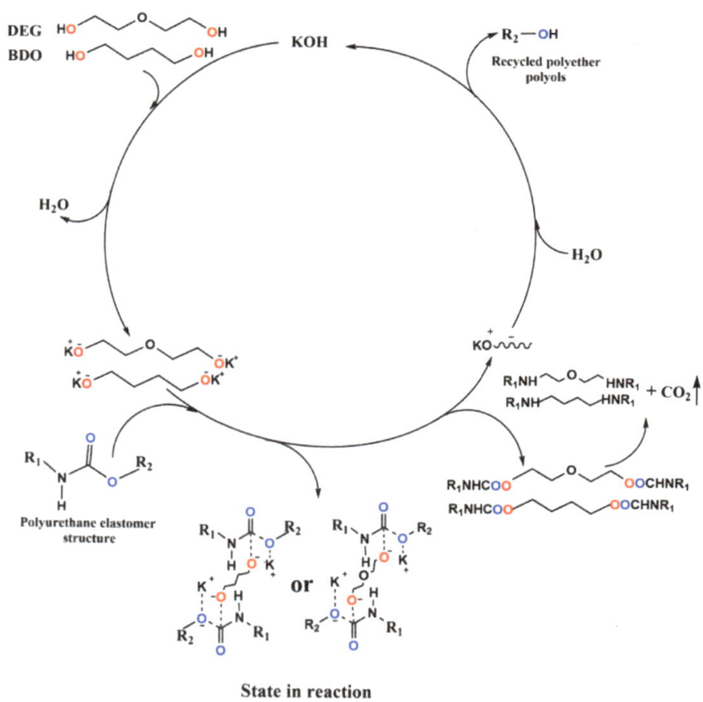

Figure 2. Catalytic degradation mechanism of the used polyurethane elastomer.

3.2. Small Molecular Alcohol Is Degraded and Foamed

The purpose of this study was to explore the route of industrialization such that different types of two-component alcohols could be compared so as to select the optimal ratio of alcoholysis agents based on the viscosity of the degradation material, the compression strength, the thermal conductivity of the prepared regenerated polyurethane rigid foam, and the properties of the milky white time and gel time in the foam. According to the experimental data, the data corresponding to the eight groups of optimal components are shown in Tables 3 and 4.

Table 3. Statistical table of the best component degradation materials of the eight groups of alcoholysis agents.

Number	Proportion	Viscosity (mPa·s)	Proportion of Catalyst	Wire Drawing Time (s)	Debonding Time (s)
A5	50:30	605.1	-	60	25
B1	40:40	992.4	1:2	37	20
B4	20:60	692.5	2:2	60	70
B5	60:20	660.2	1:1	45	75
			1:2	34	64
C1	40:40	750.6	-	67	45
			1:1	50	46
			2:1	46	50
C4	60:20	555.0	-	67	29
			1:1	48	19
D1	40:40	1208.4	1:1	56	32
			2:2	21	22
E3	60:20	484.2	2:2	74	66

Table 4. Statistical table of the hard foam data of the eight groups of alcoholysis agents with the best components.

Number	Proportion	Density ($Kg \cdot m^{-3}$)	Water Absorption Rate (%)	Strength of Compression (MPa)	Coefficient of Thermal Conductivity ($W \cdot (m \cdot K)^{-1}$)
A5	50:30	0.0399	1.9923	0.131	0.0180
B1	40:40	0.0335	1.6748	0.137	0.0173
B4	20:60	0.0312	1.5339	0.132	0.0202
B5	60:20	0.0328	0.8853	0.146	0.0191
		0.0397	1.0561	0.138	0.0182
C1	40:40	0.0347	1.1618	0.133	0.0180
		0.0303	1.2980	0.139	0.0182
		0.0366	1.2802	0.143	0.0173
C4	60:20	0.0340	0.8976	0.143	0.0174
		0.0359	0.7265	0.162	0.0166
D1	40:40	0.0362	0.9368	0.136	0.0164
		0.0403	0.7630	0.176	0.0151
E3	60:20	0.0325	1.0157	0.141	0.0155

3.3. FT-IR Analysis of the Degradation Materials

Three groups (B5, C4, and D1) were selected from the eight optimal components. FT-IR was used to identify the degradation products in the glycolysis process, and the infrared spectroscopy was compared with commercially available polyether polyol 4110. The results are shown in Figure 3.

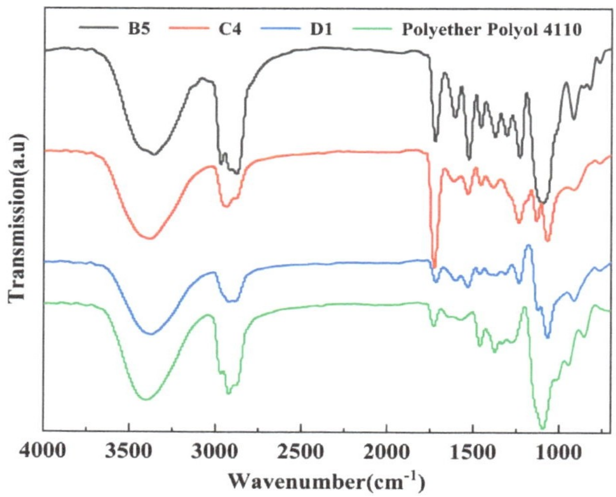

Figure 3. Infrared spectra of B5, C4, and D1 degradation products and polyether 4110.

Compared with commercially available polyether 4110, the degradation products of B5, C4, and D1 had stronger absorption bands within 3500–3300 cm^{-1}, which are the stretching vibration peaks of the alcohol hydroxyl group (-OH) [30]. There was a strong absorption band near 1732–1708 cm^{-1}, which is a benzene-type flood frequency peak [31]. At 1054 cm^{-1}, there was a clear strong absorption band, which was the polyether polyurethane ether group (-O-) absorption band, and this is the characteristic peak of polyether polyol [32,33], which is similar to the characteristic peak of commercially available polyether polyol. It can be concluded that the degradation product was the mixed product of polyether polyol and aromatic polyol. Therefore, the glycolysis product was expected to be a substitute for commercially available polyether 4110 for the preparation of the regenerated polyurethane foam.

3.4. Viscosity of Degradation Products

Polyether polyols are one of the most important raw materials for the synthesis of polyurethane foam, and their properties have attracted extensive attention. Their main properties are viscosity, hydroxyl value, acid value, etc. The viscosity of polyether polyols can be intuitively observed and is reflected in the differences in their fluidity. The synthesis of general polyurethane foam is optimized for polyether polyols with low viscosity and good fluidity. In this paper, the viscosity of all the degradation materials prepared by alcoholysis was less than 1700 mPa·s, while the viscosity of commercially available polyether 4110 is 4300–4500 mPa·s. Therefore, it could be seen that the degradation materials obtained by the alcoholysis degradation of waste polyurethane elastomers had excellent fluidity. The viscosity of the eight groups of degradation materials selected by the performance test of the degradation materials and the recycled polyurethane foam ranged from 400 to 1200 mPa·s. For the degradation materials recovered from the same waste polyurethane elastomer, the viscosity span was large because the alcoholysis agents used were different. Under the condition of the same catalyst, raw materials, and experimental conditions, the only element that may have affected the performance of the degradation materials was the difference in the type and proportion of alcoholysis agent used in the process of alcoholysis [34]. The length of the chain segment and the number and type of functional groups (-OH, -O-) in the alcoholysis agent used are all important indicators that will affect the final viscosity [35].

It can be clearly seen from the experimental data (Figure 4) that a longer chain segment and more functional groups (-OH, -O-) led to a greater level of viscosity in the degradation material obtained after alcoholysis. This is because the longer the chain segment of the small molecule alcohols was obtained after the carbamate in the chain segment of the waste polyurethane elastomer was broken and reacted with the alcoholysis agent at high temperature due to the longer the chain segment of the alcoholysis agent. Therefore, the viscosity of the resulting degradation product would also increase. The higher the number of functional groups of the alcoholysis agent, the denser the cross-linked network structure of the carbamate grafted onto the alcoholysis chain segment after fracture, and thus, the viscosity of the degradation product increased.

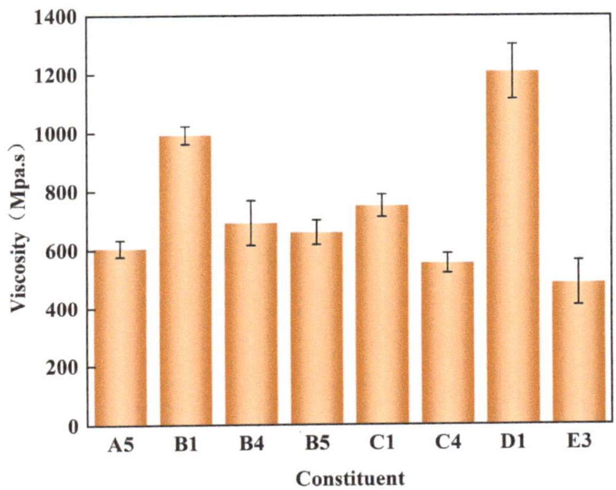

Figure 4. Determination of the viscosity of the degradation products of the eight groups of optimal components.

3.5. GPC Determination of the Degradation Products

In this experiment, under the catalysis of an alkaline metal catalyst (KOH), the carbamate bond of the waste polyurethane elastomer gradually broke at a high temperature and was replaced by alcoholysis agents, forming into small molecule polyether polyols. The side reaction at high temperatures generated amines, TDA, and other substances. Due to the poor solubility of the polyether polyols with the amines and TDA, the final degradation material was divided into two layers [36,37], where the upper layer was comprised of polyether polyols and the lower layer was comprised of amines, TDA, and other substances. In the preparation process for the polyurethane foam, the recovered polyether polyols (upper phase) played an important role, but this study intended to recycle waste polyurethane elastomers. Therefore, recycling as much as possible was what we needed to focus on. In this study, although the product was divided into upper and lower layers, in order to achieve higher recovery rates, we mixed the upper and lower layers evenly before foaming. Therefore, the GPC measured was also a mixture of the upper and lower phases.

As shown in Figure 5, GPC analysis was performed on commercially available polyether 4110 and on the recovered polyether polyol mixtures D1 and B5, respectively. The primary component of commercially available polyether 4110 is polyether polyol, and so it can be seen in the figure that D1, B5, and polyether 4110 had corresponding peaks in a, while the peaks in b appeared in degradation materials D1 and B5. These can be attributed to the amines, TDA, and other substances in the lower phase generated by the byproducts of the high-temperature degradation.

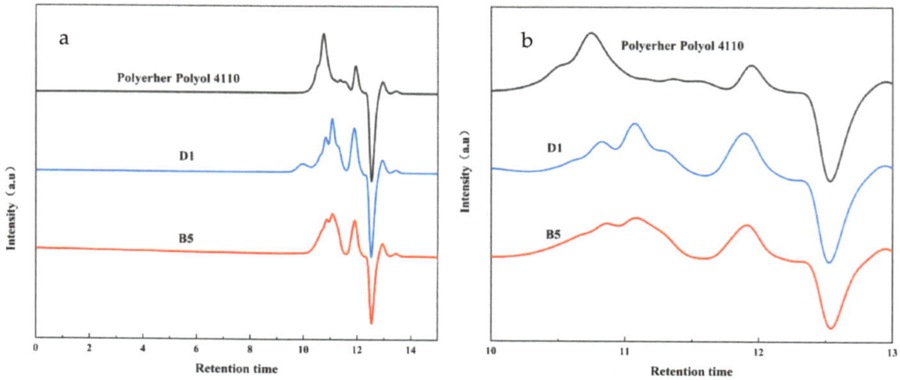

Figure 5. (**a**) Product retention time of polyether 4110 and degradation products D1 and B5 (**b**) Enlarged view of a specific location in (**a**).

3.6. Foaming Time of the Regenerated PU Rigid Foam

The preparation methods of polyurethane foam can be divided into one-step foaming and two-step foaming [38]. As the name implies, one-step foaming involves mixing polyether polyols, catalysts (dibutyltin dilaurate and triethanolamine) [39,40], foaming agents, surfactants, and black materials and stirring, and the reaction and foaming occur at the same time. The hydroxyl in the polyether polyol and the isocyanate in the black material can be used to create a chain extension reaction for two-step foaming, which is divided into the prepolymer method and the semi-prepolymer method, and the common point of the two is that all or part of the white material and the black material are mixed to form prepolymer before foaming, and then other additives are added, as is another part of the mixture of the black material and white material.

The method adopted in this paper was one-step foaming, and the catalyst was the most direct factor affecting foaming time. Generally, amine and tin compounds are used as catalysts for polyurethane foaming. According to the synergistic effect between the two catalysts, the milky white time and gel time in the foaming process of the degradation material obtained from the degradation of the polyurethane elastomer were roughly measured using the simple industrial empirical method, and the time range obtained is shown in Figure 6.

It can be seen from the figure that the total foaming time of the selected components was between 40–140 s. Generally speaking, within a certain range, with an increase in catalyst addition, the foaming time will be actively shortened [41]. However, the experimental results in this paper were obviously inconsistent with this conclusion because the polyurethane elastomer used in this paper was industrial waste. Although the waste had been cleaned, to a certain extent, before undergoing degradation, during the glycolysis process [42], the metals attached to or wrapped in the waste turned into metal ions in the alcoholysis agent at high temperatures. These metal ions also played a catalytic role in the foaming process. Thus, the final knot was skewed by the presence of metal ions. This was also the primary reason why the foaming time of the components with triethanolamine and dibutyltin dilaurate was shorter than that of the components with the catalyst.

In actual production, the uses of polyurethane foam are different, and the requirements for foaming time are also different. For polyurethane foam wrapped in directly buried pipe, it is hoped that the foaming time is relatively longer and the fluidity is better because of the pipe's length. For wall-spraying polyurethane foam, the foaming time is attempted to be as short as possible for faster forming.

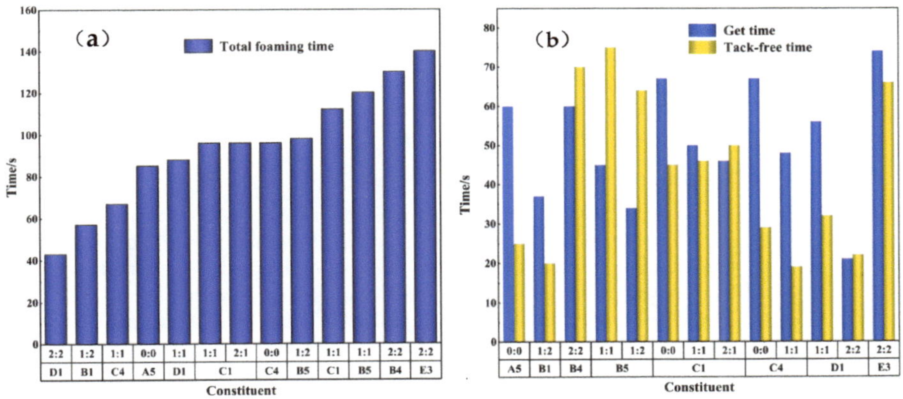

Figure 6. Comparison of the total foaming time (**a**) and the milky white time and gel time (**b**) in the foaming process of the regenerated polyurethane rigid foam prepared by the eight groups of optimal degradation products.

3.7. Water Absorption of the Regenerated PU Rigid Foam

Figure 7 shows the water absorption measurement data of the regenerated polyurethane rigid foam prepared by the one-step foaming of the selected eight groups of degradation products. In the foaming process for each group of degradation materials, the ratios of amine to tin varied (0:0, 1:1, 1:2, 2:1, 2:2) so as to prepare five groups of regenerated polyurethane foam. The prepared five groups of regenerated polyurethane foam were cut into pieces for sample preparation, and the water absorption of each group of foam was determined so as to select the regenerated polyurethane rigid foam that best met the standards.

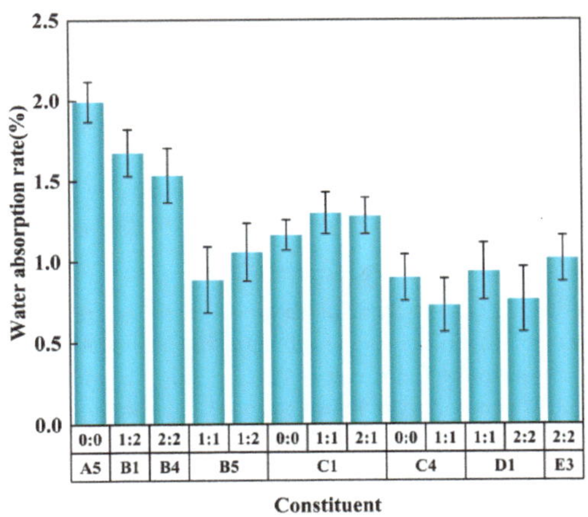

Figure 7. Determination of the water absorption of the regenerated polyurethane rigid foam prepared by the eight groups of optimal degradation products.

It can be clearly seen in Figure 7 that the water absorption rate of the optimal regenerated polyurethane foam was no more than 0.6%, and this can be used to judge the degree of density between the holes of the polyurethane foam. The most direct factor affecting density is the molecular weight of the polyether polyol and the proportion of isocyanate added. Generally speaking, reducing the molecular weight of polyether polyol or increasing the proportion of isocyanate can improve the density of foam and reduce foam water absorption. The lower the water absorption rate, the more complete the internal bubble structure of the regenerated polyurethane foam, the closer the fit between the bubble and the bubble, and the higher the uniformity of the bubble and the obturator rate [43]. This indicates that the prepared regenerated polyurethane foam had good air tightness and a good thermal insulation effect.

3.8. Determination of Density, Strength, and Thermal Conductivity of the Recycled PU Hard Foam

As an important performance reference index of polyurethane foam, the apparent density directly affects its compression strength and thermal conductivity. Figure 8 shows the density, compressive strength, and thermal conductivity of the regenerated polyurethane foam of the preferred component. As can be seen in the figure, the density of the regenerated polyurethane foam had a similar variation trend in its compressive strength, and the compressive strength changed with the density. The apparent density of the optimized regenerated polyurethane foam ranged from 30 kg/m^3 to 40 kg/m^3, which was consistent with the range of low-density polyurethane foam. The range of compressive strength was 0.13–0.18 MPa, the highest thermal conductivity was 0.0192 W (m·K) $^{-1}$, and the lowest thermal conductivity was 0.0132 W $(m{\cdot}K)^{-1}$.

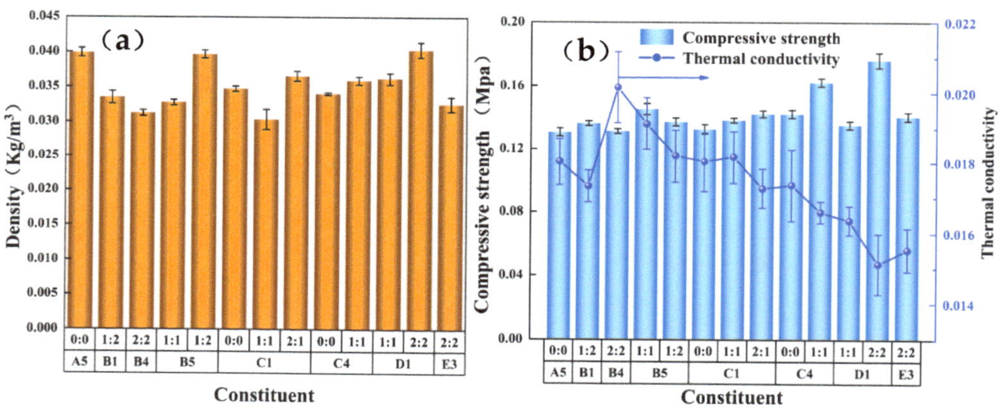

Figure 8. Determination of (**a**) foam density and (**b**) compression strength and thermal conductivity of the regenerated polyurethane rigid foam prepared by the eight groups of optimal degradation products.

3.9. SEM Analysis of the Regenerated PU Rigid Foam

The degradation material obtained from the degradation of the waste polyurethane elastomer by alcoholysis was foamed using the one-step method, and the prepared regenerated polyurethane foam was tested by SEM. The test results are shown in Figure 9.

Figure 9 compares the foams prepared by two types of commercially available polyether 4110 from the eight optimal products. A–C are the electron microscopic images of the foam prepared by commercially available polyether 4110 at different magnifications, and A is the electron microscopic image of the polyurethane foam prepared by commercially available polyether 4110 at different sizes (2 mm, 1 mm, and 500 μm). B is the electron microscopy of the regenerated polyurethane foam prepared by polyether polyols degraded by an alcoholysis agent (DEG: BDO) at different sizes (2 mm, 1 mm, and 500 μm). C is the electron microscopy of the regenerated polyurethane foam prepared by polyether polyols degraded by an alcoholysis agent (EG: BDO) at different sizes (2 mm, 1 mm, and 500 μm). It can be seen that the micromorphology of the regenerated polyurethane foam prepared by the alcoholysis method is similar to the foam prepared by the commercially available polyether 4110 in terms of the integrity of the bubble holes and the strength of the skeletons. Therefore, the alcoholysis method can successfully degrade the waste polyurethane elastomer and prepare regenerated polyurethane foam.

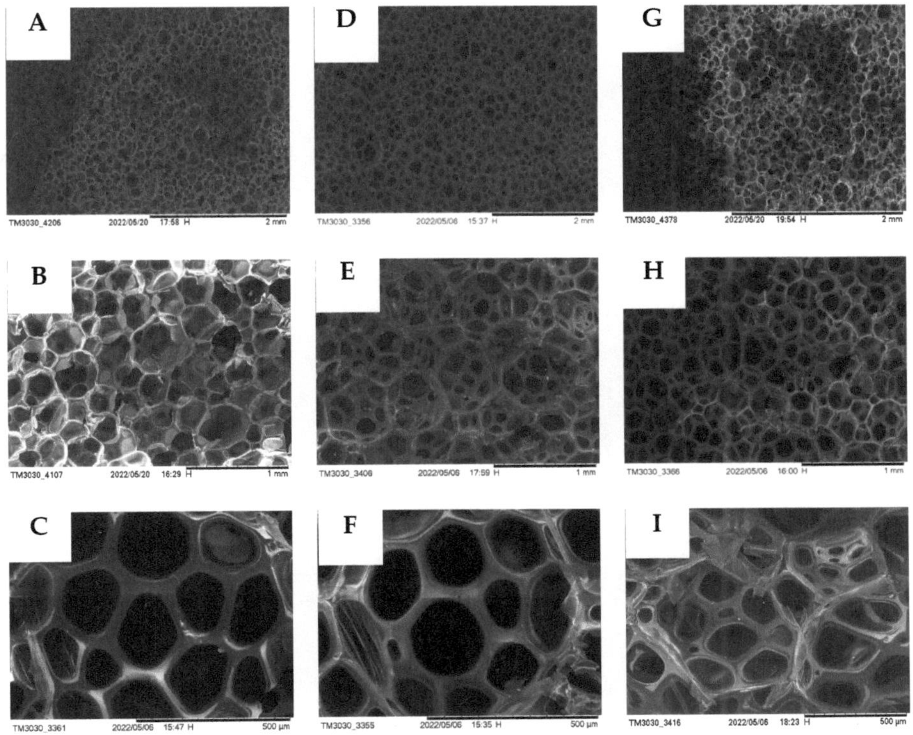

Figure 9. The SEM images of D1 and B5 of the optimal degradation products at three different sizes (2 mm, 1 mm, and 500 μm) were compared with those of polyether 4110. (**A–C**) show polyether 4110, (**D–F**) show D1, and (**G–I**) show B5.

Under SEM, the foam holes of the regenerated polyurethane foam showed a uniform "honeycomb" shape [44]. Under the SEM with a size of 2 mm, the bubble holes were tightly and evenly arranged in the visible field of vision, and it could be intuitively observed that the degradation material recovered from degradation could be used as a substitute for polyether polyol to foam and to prepare the regenerated polyurethane foam. From the figure at 500 μm, it can be clearly seen that the shape of the bubble presented regular pentagonal and hexagonal geometric shapes, the skeleton was strong, and the connections between the bubbles were tight, which indicated that the regenerated polyurethane foam prepared using the degradation material of the waste polyurethane elastomer recovered by the alcoholysis method had good sealing properties. The thick skeleton indicated that the foam had a strong compression performance. Under SEM, the test also showed that the bubble holes prepared using the degradation material displayed similar phenomena to the bubble holes prepared using polyether 4110. The tight connection between the bubble holes indicated that the foam had good air tightness and a good thermal insulation performance. Figure 10 shows physical pictures of the regenerated polyurethane foam prepared by one-step foaming in the experiment. It can be clearly seen in the pictures that the regenerated polyurethane foam freely foamed by one-step foaming had smooth surfaces, complete cross-linking, and no slag loss.

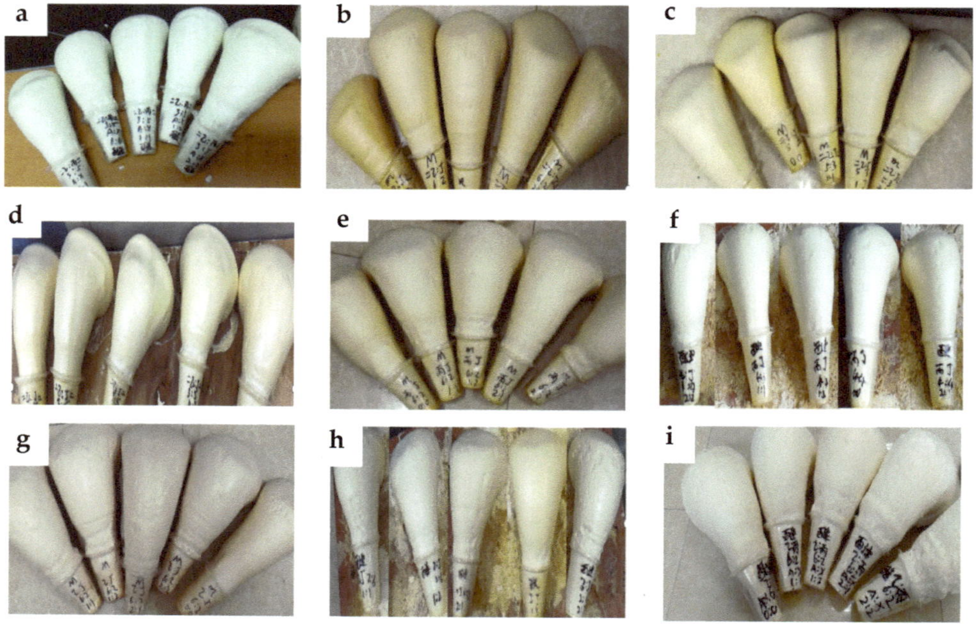

Figure 10. Actual picture of the polyurethane rigid foam partially prepared in the experiment ((**a–i**) are regenerated polyurethane rigid foam prepared by degradation products).

3.10. TG Analysis of the Regenerated PU Rigid Foam

Figure 11 shows the thermogravimetric analysis of the regenerated polyurethane rigid foam prepared using polyether 4110 and the regenerated polyurethane rigid foam prepared using the degradation materials D1 and B5 and obtained by recycling the waste polyurethane elastomer.

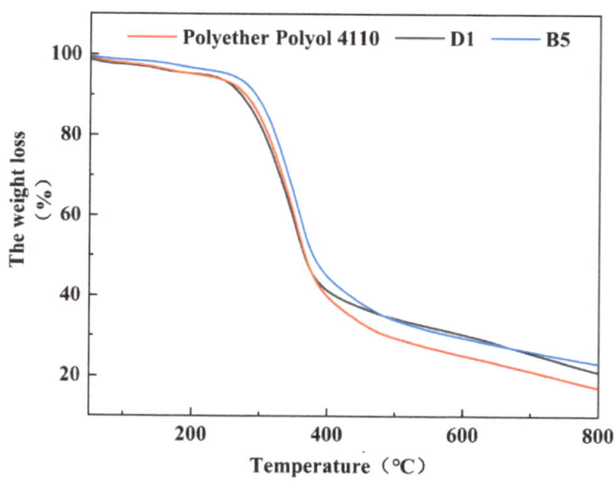

Figure 11. Thermogravimetric diagram of the hard foam of the regenerated polyurethane prepared using polyether 4110 and the D1 and B5 degradation materials.

As can be seen from the figure, the thermal weight loss of the two foams with different ratios could be divided into three stages. The first stage was 100–200 °C, and the volatilization of the free and bound water in the foam led to its weight loss. The second stage was 200–390 °C, during which the isocyanic acid in the polyurethane joint segment broke with the hard segment. At this stage, the polyether satin in the end of the polyurethane chain broke [45]. The mass loss temperature of the foam prepared by the two ratios was approximately 250 °C, and the decomposition stopped at approximately 780 °C. Finally, the fastest weight loss temperature of polyether 4110 and D1 was approximately 338 °C, while the fastest weight loss temperature of B5 was approximately 350 °C. It was obvious that the weight loss speeds of polyether 4110 and D1 were significantly faster than that of B5. This indicated that the chemical bond energy of the foam prepared by degradation material B5 was greater, and the human stability of the regenerated polyurethane foam prepared by the degradation material recovered from the waste polyurethane elastomer was similar, or even better, than that of polyether 4110. Therefore, it was shown that the thermal stability of the recycled polyurethane foam prepared by recycling and degradation was more stable, and the strain capacity of the polyurethane foam to environmental changes had improved in the actual use process.

4. Conclusions

In this paper, four kinds of diols (butanediol, propylene glycol, ethylene glycol, and monodiethylene glycol) were used to degrade waste polyurethane elastomers successfully, and the regenerated polyether polyol, which was similar to polyether polyol 4110 in its structure and properties, could be used as a raw material for polyurethane synthesis. Eight groups of optimal formulations were selected according to the viscosity of the degraded materials and the compression strength, density, water absorption, and thermal conductivity of the regenerated polyurethane rigid foam. The indexes of the products of the optimal formulations were all within the range of industry requirements, and the viscosity of the recovered degraded materials was between 485 and 1200 MPa·s. The regenerated polyurethane rigid foam was prepared by using the degradation materials rather than commercially available polyether polyols, and the foaming time of the regenerated polyurethane rigid foam was generally between 40 and 140 s. The compressive strength was between 0.131 and 0.176 MPa, and the water absorption rate ranged from 0.7265 to 1.9923%. The apparent density of the foam was between 0.0303 and 0.0403 kg/m^3. The thermal conductivity ranged from 0.0151 to 0.0202 W/(m·K). After synthesizing all the data, the optimal group was divided into D1: DEG:BDO = 40:40. The compressive strength, water absorption, thermal conductivity, and foam density of the recycled polyurethane foam prepared from this component were the best. From the above data, it can be seen that the recovery of waste polyurethane elastomers by the alcoholysis method and different proportions of alcoholysis agents was successful, and this has certain practical significance for the selection of alcohol types and proportions in later stages.

Author Contributions: Conceptualization, X.G. (Xiaohua Gu) and X.W.; Methodology, X.G. (Xinyu Guo) and Q.L.; Software, X.W., X.G. (Xinyu Guo) and Q.L.; Validation, S.L.; Formal analysis, X.G. (Xiaohua Gu); Investigation, X.W., S.L. and Q.L.; Resources, X.G. (Xiaohua Gu) and Y.L.; Data curation, X.W. and X.G. (Xinyu Guo); Writing—original draft, X.W.; Writing—review and editing, X.W. and X.G. (Xinyu Guo); Supervision, Y.L.; Project administration, X.G. (Xiaohua Gu). All authors have read and agreed to the published version of the manuscript.

Funding: This research was funded by Heilongjiang Provincial Department of Education Project (CLKFKT2021Z3, 145109301), China, and the key research and development guidance projects in Heilongjiang Province (GZ20210034).

Institutional Review Board Statement: Not applicable.

Informed Consent Statement: Not applicable.

Data Availability Statement: Not applicable.

Acknowledgments: The authors gratefully acknowledge the support from the Heilongjiang Provincial Department of Education Project (CLKFKT2021Z3, 145109301), China, and the key research and development guidance projects in Heilongjiang Province (GZ20210034).

Conflicts of Interest: The authors declare no conflict of interest.

References

1. Karamanlioglu, M.; Preziosi, R.; Robson, G.D. Abiotic and biotic environmental degradation of the bioplastic polymer poly(lactic acid): A review. *Polym. Degrad. Stab.* **2017**, *137*, 122–130. [CrossRef]
2. Evode, N.; Qamar, S.A.; Bilal, M.; Barceló, D.; Iqba, H.M.N. Plastic waste and its management strategies for environmental sustainability. *Case Stud. Chem. Environ. Eng.* **2021**, *4*, 100142. [CrossRef]
3. Santamaria-Echart, A.; Fernandes, I.; Barreiro, F.; Corcuera, M.A.; Eceiza, A. Advances in Waterborne Polyurethane and Polyurethane-Urea Dispersions and Their Eco-friendly Derivatives: A Review. *Polymers* **2021**, *13*, 409. [CrossRef]
4. Wang, Y.; Tian, H.; Bian, Q.; Xiang, A.; Ge, X.; Liu, Q. Effect of Trolamine and Dibutyltindilaurate on the Structure and Properties of Polyurethaneimide Foams. *Cell. Polym.* **2015**, *34*, 119–136. [CrossRef]
5. Hernandez, M.D.R.; Garcia, A.N.; Marcilla, A. Catalytic flash pyrolysis of HDPE in a fluidized bed reactor for recovery of fuel-like hydrocarbons. *J. Anal. Appl. Pyrolysis* **2007**, *78*, 272–281. [CrossRef]
6. Gang, H.; Lee, D.; Choi, K.-Y.; Kim, H.-N.; Ryu, H.; Lee, D.-S.; Kim, B.-G. Development of High Performance Polyurethane Elastomers Using Vanillin-Based Green Polyol Chain Extender Originating from Lignocellulosic Biomass. *ACS Sustain. Chem. Eng.* **2017**, *5*, 4582–4588. [CrossRef]
7. Zhang, H.B.; Li, W.; Kong, X.; Yang, X.J. Effects of synthesis technology route on mechanical properties and phases/microstructures of polyurethane elastomers. *Mater. Sci. Technol.* **2007**, *23*, 1256–1259. [CrossRef]
8. Xu, Z.; Wang, X.; Huang, H. Thermoplastic polyurethane–urea elastomers with superior mechanical and thermal properties prepared from alicyclic diisocyanate and diamine. *J. Appl. Polym. Sci.* **2020**, *137*, 49575. [CrossRef]
9. Jin, X.; Guo, N.; You, Z.; Tan, Y. Design and Performance of Polyurethane Elastomers Composed with Different Soft Segments. *Materials* **2020**, *13*, 4991. [CrossRef]
10. Król, P.; Król, B. Structures, properties and applications of the polyurethane ionomers. *J. Mater. Sci.* **2020**, *55*, 73–87. [CrossRef]
11. Elkhaoulani, A.; Arrakhiz, F.Z.; Benmoussa, K.; Bouhfid, R.; Qaiss, A. Mechanical and thermal properties of polymer composite based on natural fibers: Moroccan hemp fibers/polypropylene. *Mater. Des.* **2013**, *49*, 203–208. [CrossRef]
12. Fazli, A.; Rodrigue, D. Waste Rubber Recycling: A Review on the Evolution and Properties of Thermoplastic Elastomers. *Materials* **2020**, *13*, 782. [CrossRef] [PubMed]
13. Hepburn, C. *Polyurethane Elastomers*; Applied Science Publisher, Ltd.: London, UK, 2020.
14. Steube, M.; Johann, T.; Barent, R.D.; Müller, A.H.E.; Frey, H. Rational design of tapered multiblock copolymers for thermoplastic elastomers. *Prog. Polym. Sci.* **2022**, *124*, 101488. [CrossRef]
15. Feghali, E.; Tauk, L.; Ortiz, P.; Vanbroekhoven, K.; Eevers, W. Catalytic chemical recycling of biodegradable polyesters. *Polym. Degrad. Stab.* **2020**, *179*, 109241. [CrossRef]
16. Geyer, R.; Jambeck, J.R.; Law, K.L. Production, use, and fate of all plastics ever made. *Sci. Adv.* **2017**, *3*, e1700782. [CrossRef] [PubMed]
17. Foundation, M.C. *The New Plastics Economy: Rethinking the Future of Plastics*; Ellen MacArthur Foundation: Cowes, UK, 2015.
18. Hopewell, J.; Dvorak, R.; Kosior, E. Plastics recycling: Challenges and opportunities. *Philos. Trans. R. Soc. B Biol. Sci.* **2009**, *364*, 2115–2126. [CrossRef]
19. Mizera, K.; Sałasińska, K.; Ryszkowska, J.; Kurańska, M.; Kozera, R. Effect of the Addition of Biobased Polyols on the Thermal Stability and Flame Retardancy of Polyurethane and Poly(urea)urethane Elastomers. *Materials* **2021**, *14*, 1805. [CrossRef]
20. Zhang, G.; Yin, T.; Nian, G.; Suo, Z. Fatigue-resistant polyurethane elastomer composites. *Extrem. Mech. Lett.* **2021**, *48*, 101434. [CrossRef]
21. Watando, H.; Saya, S.; Fukaya, T.; Fujieda, S.; Yamamoto, M. Improving chemical recycling rate by reclaiming polyurethane elastomer from polyurethane foam. *Polym. Degrad. Stab.* **2006**, *91*, 3354–3359. [CrossRef]
22. Borda, J.; Pásztor, G.; Zsuga, M. Glycolysis of polyurethane foams and elastomers. *Polym. Degrad. Stab.* **2000**, *68*, 419–422. [CrossRef]
23. Wang, X.; Zhan, S.; Lu, Z.; Li, J.; Yang, X.; Qiao, Y.; Men, Y.; Sun, K.J. Healable, Recyclable, and Mechanically Tough Polyurethane Elastomers with Exceptional Damage Tolerance. *Adv. Mater.* **2020**, *32*, 2005759. [CrossRef] [PubMed]
24. Li, H.; Hou, X.; Chai, L.; Cui, X.; Wang, Y.; Deng, T. Efficient and green catalytic degradation of high crosslinked rigid PU foam and recovery value-added products via selective cleavage of C-O and C-N bonds. *Polym. Degrad. Stab.* **2020**, *181*, 109262. [CrossRef]
25. Taourit, S.; Gac, P.; Fayolle, B. Relationship between network structure and ultimate properties in polyurethane during a chain scission process. *Polym. Degrad. Stab.* **2022**, *201*, 109971. [CrossRef]
26. Akindoyo, J.O.; Beg, M.D.H.; Ghazali, S.; Islam, M.R.; Jeyaratnam, N.; Yuvaraj, A.R. Polyurethane types, synthesis and applications—A review. *RSC Adv.* **2016**, *6*, 114453–114482. [CrossRef]

27. Wu, C.-H.; Chang, C.-Y.; Li, J.-K. Glycolysis of rigid polyurethane from waste refrigerators. *Polym. Degrad. Stab.* **2002**, *75*, 413–421. [CrossRef]
28. Wu, C.-H.; Chang, C.-Y.; Cheng, C.-M.; Huang, H.-C. Glycolysis of waste flexible polyurethane foam. *Polym. Degrad. Stab.* **2003**, *80*, 103–111. [CrossRef]
29. Zhao, L.; Semetey, V. Recycling Polyurethanes through Transcarbamoylation. *ACS Omega* **2021**, *6*, 4175–4183. [CrossRef]
30. Rane, A.V.; Sabnis, A. Synthesis and Investigation of Aminolysates and obtained Poly Ester Amides. In Proceedings of the 3rd International Conference on Radiation Medicine (ICRM), Kottayam, India, 11–13 April 2014.
31. Xu, W.; Lu, B.; Hu, Y.; Song, L.; Nie, S. Synthesis and characterization of novel fluorinated polyurethane elastomers based on 2,2-bis[4-(4-amino-2-trifluoromehyloxyphenyl) phenyl]propane. *Polym. Adv. Technol.* **2012**, *23*, 877–883. [CrossRef]
32. Sang, S.; Li, Y.; Wang, K.; Tang, J. Application of blocked isocyanate in preparation of polyurethane(urea) elastomers. *J. Appl. Polym. Sci.* **2021**, *138*, 50582. [CrossRef]
33. Salgado, C.; Arrieta, M.P.; Sessini, V.; Peponi, L.; López, D.; Fernández-García, M. Functional properties of photo-cross link able biodegradable polyurethane nanocomposites. *Polym. Degrad. Stab.* **2020**, *178*, 109204. [CrossRef]
34. Vanbergen, T.; Verlent, I.; Geeter, J.D.; Haelterman, B.; Claes, L.; De Vos, D. Recycling of Flexible Polyurethane Foam by Split-Phase Alcoholysis: Identification of Additives and Alcoholyzing Agents to Reach Higher Efficiencies. *ChemSusChem* **2020**, *13*, 3835–3843. [CrossRef] [PubMed]
35. Gu, X.; Lyu, S.; Cheng, W.; Liu, S. Effect of different catalysts on recovery and reuse of waste polyurethane rigid foam. *Mater. Res. Express* **2021**, *8*, 035105. [CrossRef]
36. Simón, D.; Borreguero, A.M.; Lucas, A.D. Glycolysis of flexible polyurethane wastes containing polymeric polyols—ScienceDirect. *Polym. Degrad. Stab.* **2014**, *109*, 115–121. [CrossRef]
37. Wang, X.; Chen, H.; Chen, C.; Li, H. Chemical degradation of thermoplastic polyurethane for recycling polyether polyol. *Fibers Polym.* **2011**, *12*, 857. [CrossRef]
38. Blattmann, H.R.; Lauth, M.; Mülhaupt, R. Flexible and Bio-Based Nonisocyanate Polyurethane (NIPU) Foams. *Macromol. Mater. Eng.* **2016**, *301*, 944–952. [CrossRef]
39. Strachota, A.; Strachotová, B.; Špírková, M. Comparison of Environmentally Friendly, Selective Polyurethane Catalysts. *Mater. Manuf. Process.* **2008**, *23*, 566–570. [CrossRef]
40. Mondal, P.; Khakhar, D.V. Hydraulic resistance of rigid polyurethane foams. II. Effect of variation of surfactant, water, and nucleating agent concentrations on foam structure and properties. *J. Appl. Polym. Sci.* **2010**, *93*, 2830–2837. [CrossRef]
41. Shabani, A.; Fathi, A.; Erlwein, S.; Altstädt, V. Thermoplastic polyurethane foams: From autoclave batch foaming to bead foam extrusion. *J. Cell. Plast.* **2020**, *57*, 391–411. [CrossRef]
42. Molero, C.; Lucas, A.D.; Rodríguez, J.F. Recovery of polyols from flexible polyurethane foam by "split-phase" glycolysis: Glycol influence. *Polym. Degrad. Stab.* **2006**, *91*, 221–228. [CrossRef]
43. Thirumal, M.; Khastgir, D.; Singha, N.K.; Manjunath, B.S.; Naik, Y.P. Effect of foam density on the properties of water blown rigid polyurethane foam. *J. Appl. Polym. Sci.* **2008**, *108*, 1810–1817. [CrossRef]
44. Nelson, M.C. *The Relationship of Cell Morphology, Density, and Mechanical Properties in a Rigid Polyurethane Foam*; University of Nevada: Las Vegas, NV, USA, 2003.
45. Yang, S.; Liu, X.; Tang, G.; Long, H.; Wang, B.; Zhang, H.; Ji, Y.; Yang, Y. Fire retarded polyurethane foam composites based on steel slag/ammonium polyphosphate system: A novel strategy for utilization of metallurgical solid waste. *Polym. Adv. Technol.* **2022**, *33*, 452–463. [CrossRef]

polymers

Article

One-Pot Tandem Alcoholysis-Hydrogenation of Polylactic Acid to 1,2-Propanediol

Jialin Xu [1,2], Kuo Zhou [3], Linlin Qin [1], Zaiming Tan [1], Shijing Huang [1], Peigao Duan [2] and Shimin Kang [1,*]

1 Engineering Research Center of None-Food Biomass Efficient Pyrolysis and Utilization Technology of Guangdong Higher Education Institutes, Guangdong Provincial Key Laboratory of Distributed Energy Systems, Dongguan University of Technology, Dongguan 523808, China
2 School of Chemical Engineering and Technology, Xi'an Jiaotong University, Xi'an 710049, China
3 Department of Chemistry, Lishui University, Lishui 323000, China
* Correspondence: kangshimin@dgut.edu.cn

Abstract: The chemical recycling of end-of-life polylactic acid (PLA) plays roles in mitigating environmental pressure and developing circular economy. In this regard, one-pot tandem alcoholysis and hydrogenation of PLA was carried out to produce 1,2-propanediol, which is a bulk chemical in polymer chemistry. In more detail, the commercially available Raney Co was employed as the catalyst, and transformation was conducted in ethanol, which acted as nucleophilic reagent and solvent. Single-factor analysis and Box–Behnken design were used to optimize the reaction conditions. Under the optimized condition, three kinds of PLA materials were subjected to degradation. Additionally, 74.8 ± 5.5%, 76.5 ± 6.2%, and 71.4 ± 5.7% of 1,2-propanediol was yielded from PLA powder, particle, and straws, respectively, which provided a recycle protocol to convert polylactic acid waste into value-added chemicals.

Keywords: polylactic acid; alcoholysis; hydrogenation; 1,2-propanediol; PLA straws

check for updates

Citation: Xu, J.; Zhou, K.; Qin, L.; Tan, Z.; Huang, S.; Duan, P.; Kang, S. One-Pot Tandem Alcoholysis-Hydrogenation of Polylactic Acid to 1,2-Propanediol. *Polymers* **2023**, *15*, 413. https://doi.org/10.3390/polym15020413

Academic Editors: Cristina Cazan and Mihaela Cosnita

Received: 26 December 2022
Revised: 7 January 2023
Accepted: 9 January 2023
Published: 12 January 2023

1. Introduction

Polylactic acid (PLA) is an environmental-friendly alternative to petroleum-derived plastic due to biodegradability and biocompatibility [1], which can be used in the following sectors: packaging, agriculture, textiles, clinical sector, and others [2]. In 2015, the global production of PLA was around 200 kt [3], and it is expected to reach 560 kt in 2025 [4], with an increase of 180% over the past decade. Thus, its tremendous growth in single-use products has led to serious social problems, e.g., economical inefficiency, excess wastes, and associated environmental concerns [5]. Although PLA is biodegradable, it takes decades to completely degrade in landfill [6], and 60–100 days are also needed under elevated temperature (50 ± 5 °C) in composting conditions [7]. These two routine degradation methods are not recommended choices since no material is recovered. In this regard, the chemical end-of-life scenario of short-life PLA is essential to determine its sustainability and circular economy, where monomer regeneration occurs, or value-added chemicals are produced. Chemical degradation of PLA generally includes hydrolysis, pyrolysis, alcoholysis, and ammonolysis [8]. A serious of products were derived from the above recycling options, such as lactic acid [9,10], lactide [11,12], lactate ester [13,14], biofuel [15,16], and others [17]. Hydrolysis occurs through random scission of the ester bond around 200 °C and the monomer of lactic acid is formed, which can further convert into new PLA and a variety of chemicals [10]. Pyrolysis takes place via intramolecular transesterification of PLA and results in lactide over the range of 200–500 °C [11,12]. Alcoholysis allows the formation of lactate esters, known as green solvents, and they can be transformed into lactide, resulting in circular economy [13,14]. Ammonolysis depolymerizes PLA into alanine, which also expands the application of PLA wastes [17].

In addition, PLA was found as a potential feedstock to produce 1,2-propanediol (1,2-PDO) as well [18,19], a bulk commodity chemical, which is extensively used as antifreeze, monomer for production of unsaturated polyester resins and biodegradable plastic, and additive in drug, food, and cosmetics [20]. Commonly used 1,2-PDO is obtained from propylene, which is primarily combined with ethylbenzene and chlorine to form propylene oxide, and then yield 1,2-PDO via hydration [21]. Regarding the depletion of fossil fuel and the environmental concern, there has been increased demand for bio-based processes using renewable feedstocks and waste streams, such as glycerol [22,23], glucose [24], cellulose [25], lactic acid [26,27], etc. (Table S1). Additionally, selective hydrogenolysis of glycerol is the most common renewable route for producing commercial 1,2-PDO, as glycerol is a low valued byproduct of biodiesel [28]. However, excess H_2 and high energy input are necessary in the process, and byproducts (e.g., the acetol intermediate, ethylene glycol, propanol) are derived via further hydrogenolysis of glycerol [29]. Therefore, it provides a strong motivation to operationalize a new route for green and efficient production of 1,2-propanediol.

In recent years, reductive depolymerization of PLA has emerged as an efficient alternative methodology to recover 1,2-PDO, employing ruthenium- [30,31], copper- [18], or zinc-based [32] catalysts, and H_2 or silane as the reducing agent (Table S1). These recycling concepts from PLA to 1,2-PDO can not only greatly diminish the environmental concerns of PLA wastes, but also offer an attractive and renewable approach to 1,2-PDO. Nevertheless, one of the disadvantages of reported methods is the use of toxic, noble metal-based, and homogeneous catalysts. Moreover, expensive and not eco-friendly reagents should be avoided, such as toluene [32], 1,4-dioxane [31], THF, and anisole [19] (Table S1). In this regard, for sustainable industrial application of depolymerization PLA, it is inevitable to develop a mild reaction system with green solvents and reusable heterogeneous catalysts. Herein, we developed a one-pot tandem alcoholysis–hydrogenation route to transform polylactic acid into 1,2-PDO, in which ethanol was used as green solvent [33], and Raney Co, an efficient heterogeneous catalyst for hydrogenation of biomass [34], was employed for the catalysis reaction.

2. Materials and Methods

2.1. Materials

PLA powder (100 mesh) was purchased from Huachuang Plastic, Dongguan, China. PLA particle (Mw ~80,000) was purchased from Shanghai Macklin Biochemical Technology Co., Ltd. (Shanghai, China). PLA straws were obtained by Ningbo Sizhuo Plastic Industry Co., Ltd., Ningbo, China, and were shredded into particles with size < 5 mm before use. Standard substances (1,2-propanediol, ethyl lactate, and ethyl propionate, purity $\geq$ 99.5%), Raney Co (50 μm, dispersed in water), and internal standard substance (dodecane, purity of 99.5%) were obtained from Shanghai Aladdin Industrial Development Co., Ltd. (Shanghai, China). Other chemicals (analytical purity) were purchased from Sinopharm Reagent Co., Ltd. (Beijing, China).

2.2. Typical Alcoholysis and Hydrogenation Process

Typically, 1 g PLA powder, 10 mL ethanol, 0.1 g Raney Co, and 0.1 g dodecane were charged into a 50 mL Hastelloy batch autoclave. The vessel was sealed, and the mixture was flushed with H_2 at least 5 times to remove air. Afterward, the reactor was purged with hydrogen to pre-set pressure (e.g., 3 MPa). The reactor was heated, and the reaction time was marked once the desired temperature was reached. After reaction, products were sampled and subjected to analysis. Moreover, methanolization of PLA straws without catalyst and hydrogen was carried out to determine the content of PLA.

2.3. Analysis

The reactants identification was carried out by Shimadzu QP 2010 Plus gas chromatography-mass spectrometry (GC-MS) with an Rtx-5MS column (30.0 m × 0.25 mm

× 0.25 μm). The oven temperature was held at 60 °C for 5 min, and ramped to 260 °C for another 5 min at 12 °C/min. The injector and detector temperature were set at 280 °C and 285 °C, respectively.

The content of 1,2-propanediol (mol/L) was determined by a gas chromatograph (GC, Shimazu GC-2014C, Shimazu, Kyoto, Japan) equipped with a flame ionization detector (FID) and WondaCap FFAP capillary column (30.0 m × 0.25 mm × 0.25 μm). The initial column temperature was 60 °C holding for 5 min, and ramped to 260 °C at a rate of 12 °C/min. Both the injector and the detector temperature were set at 280 °C. The quantification was analyzed by the standard curves of each standard substance, using dodecane as the internal standard. The yield (%) of 1,2-PDO in the reaction bulk was calculated as Equation (1).

$$\text{Yield (\%)} = \frac{m_i}{m_0} \times 100\% \tag{1}$$

where mi (g) is the mass quantity of 1,2-PDO and m_0 (g) refers to the mass of the PLA material.

The conversion (%) of PLA feedstock was calculated as Equation (2).

$$\text{Conversion (\%)} = \frac{(m_0 - m_r)}{m_0} \times 100\% \tag{2}$$

where m_r (g) is the mass quantity of residues and m_0 (g) refers to the mass of the PLA material.

2.4. Digestion PLA in Ethanol

A total of 1 g PLA powder and 10 mL ethanol were mixed in a 15-mL stoppered test tube, and then subjected to a digestion apparatus. The mixture was heated to a set temperature (100–240 °C), and the dissolution status of PLA was recorded at different times.

2.5. Calcination of PLA Straws

A total of 1 g of PLA straw pieces was submitted to an air furnace and heated to 600 °C at 10 °C/min, and then kept for another 3 h. After roasting, the residual inorganic ash was collected and weighed.

2.6. Fourier Transform Infrared (FTIR) Spectroscopy Analysis

The infrared spectra of PLA materials and its residues were recorded using a Spectrum Two LiTA Spectrometer (PerkinElmer, Waltham, MA, USA) with $LiTaO_3$ Detector in the range of 4000 to 400 cm^{-1} with the resolution of 1 cm^{-1}. The samples were grounded, mixed with KBr, and pressed into pellets for analysis.

2.7. Box-Benhken Optimization Design

Response surface methodology (RSM) was used to optimize experimental factors during the depolymerization process of PLA to recover 1,2-propanediol and investigate the correlation between the response and parameters. In this study, Box–Behnken design (BBD) was adopted as fewer experiment groups are needed to build a model equation than central composite design (CCD) [35]. A three-factor and three-level BBD was employed, requiring 17 runs (Table S2) to optimize the independent variables, reaction time (A, h), temperature (B, °C), and ethanol/PLA ratio (C, mL/g), using a Design Expert 11.0.4 software (Stat-Ease, Minneapolis, MN, USA).

3. Results and Discussion

3.1. Analysis of Alcoholysis and Hydrogenation Products

During the tandem alcoholysis–hydrogenation process of PLA, ethanol was employed both as nucleophile and solvent. Nucleophile attacks the carbonyl group in PLA macro-molecule, leading to its alcoholytic depolymerisation with the formation of lactate esters [14]. Afterwards, residual ethanol can be used as reaction medium, and ethyl lactate as bio-based solution for further dissolving PLA [33,36]. Figure 1a shows the distribution of PLA powder-derived products, in which 1,2-propanediol was the primary product, along with ethyl lactate and negligible ethyl propionate as byproducts. In this reaction, toxic or ambiguity solvents were avoided, such as 1,4-dioxane [31], THF [30], and anisole [19], despite their solubility to PLA obviously stimulating the formation of 1,2-PDO. Meanwhile, methanol was also applied in this reaction system as it is a better nucleophile [37,38], and methyl lactate was the dominant product (Figure S1). Thence, ethanol is more optional in our alcoholysis–hydrogenation process owing to its safety and recyclability. Additionally, ethanol could be reused although it was involved in the reaction, and residual ethyl lactate could be further hydrogenated into 1,2-PDO, which catered to circular economy and eco-production.

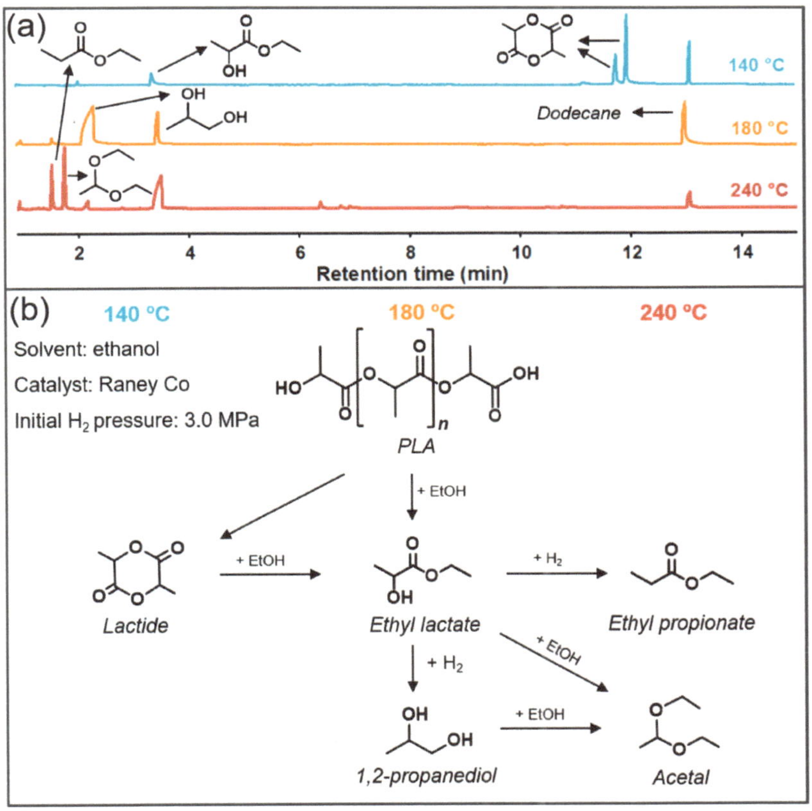

Figure 1. GC-MS analysis (**a**) and proposed mechanism (**b**) for tandem alcoholysis–hydrogenation of PLA. Reaction conditions: 1 g PLA, 10 mL ethanol, 0.1 g Raney Co, 3 MPa H_2, 140–240 °C, and reacting for 12 h.

Here, Raney Co was necessary in alcoholysis and hydrogenation of PLA to produce 1,2-PDO, and the proposed pathway is exhibited in Figure 1b. Under proper temperature (180 °C), ethyl lactate was evidenced to be the important intermediate during the degradation process, as it was the unique product of ethanolization for PLA without catalyst (Figure S2). Subsequently, ethyl lactate was transformed into 1,2-propanediol via hydrogenation in the presence of Raney Co and H_2. In addition to Co-derived catalysts [39,40], Cu- [41,42] and Ru-based [27,43] catalysts have also been studied in catalyzing the hydrogenation of ethyl lactate to 1,2-PDO, but the low activity of Cu and the exorbitant price of Ru makes non-noble Co-derived catalysts more preferable in industrial application. As Figure 1a shows, inadequate alcoholysis and other side reactions occurred under low temperature and high temperature conditions, respectively. In the 140 °C run, PLA tended to degrade into dimer, different configurations of lactide, rather than alcoholytic depolymerisation into ethyl lactate. As feedstock, lactide would yield 1,2-PDO via tandem alcoholysis and hydrogenation process under suitable conditions. While at elevated temperature, ethyl propionate was more likely to be formed through the hydrogenation of -OH in ethyl lactate. Moreover, it was notable that a certain amount of acetal was available in degradation products, which might be the coupling product of decarboxylative acceptorless dehydrogenation of ethyl lactate [44].

3.2. *Effect of Single Factor*

To scan the effect of different parameters on the yield of 1,2-propanediol from the alcoholysis and hydrogenation of PLA, single-factor tests were carried out. Due to the insolubility of PLA powder in ethanol, the degradation temperature was scanned by a digestion test of PLA in ethanol (Figure 2). When the temperature was below 120 °C, the PLA powder was almost deposited in the bottom of tube and the solvent was kept as transparent. Further increasing the digestion temperature from 140 to 160 °C, the PLA powder was gradually dissolved and dispersed in ethanol, and total digestion was accomplished at 180 °C. Considering the reaction adequacy, a temperature range of 140–240 °C was investigated towards the effect on PLA conversion and 1,2-PDO yield in Figure 3a. As the data showed, temperature violently affected the conversion of raw material and the composition of products. Under relatively low temperature (140–160 °C), total conversion of PLA could not be achieved, and PLA was more easily to form lactide rather than ethyl lactate, through back-biting depolymerization or radical process [45,46] (Figure 1a). However, elevated temperature (220–240 °C) favored to form more byproducts, such as ethyl propionate and acetal, which resulted in barely any formation of 1,2-propanediol.

Sufficient reaction time was necessary to completely degrade PLA and promote the formation of 1,2-propanediol. Figure 3b demonstrates that 6 h was required for 100% conversion of PLA, and the longer reaction time still was needed to gain higher yield of 1,2-PDO. However, further increasing the reaction time (>18 h) even led to a slight decrease in the yield of target product. Meanwhile, adequate ethanol dosage (≥2 mL/g PLA) was also indispensable for full conversion, and larger amount of ethanol (10–30 mL /g PLA) was necessary to yield 1,2-PDO (Figure 3c). Considering cost and energy consumption, the range of 160–200 °C, 6–18 h, and 10–30 mL ethanol/g PLA was selected for further optimization.

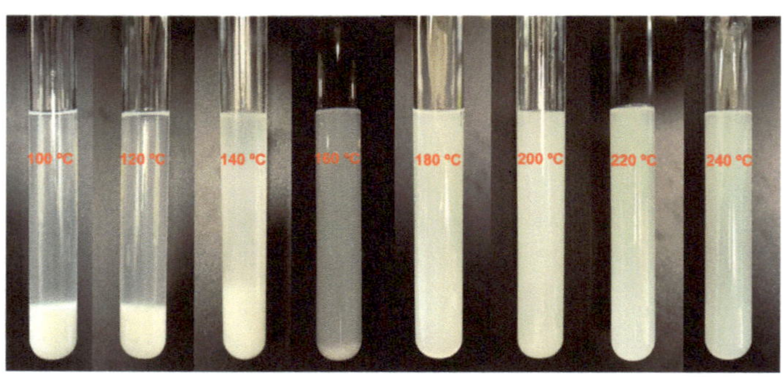

Figure 2. Digestion of PLA powder in ethanol. Reaction conditions: 1 g PLA powder, 10 mL ethanol, andheating to 100, 120, 140, 160, 180, and 200 °C for 10 min.

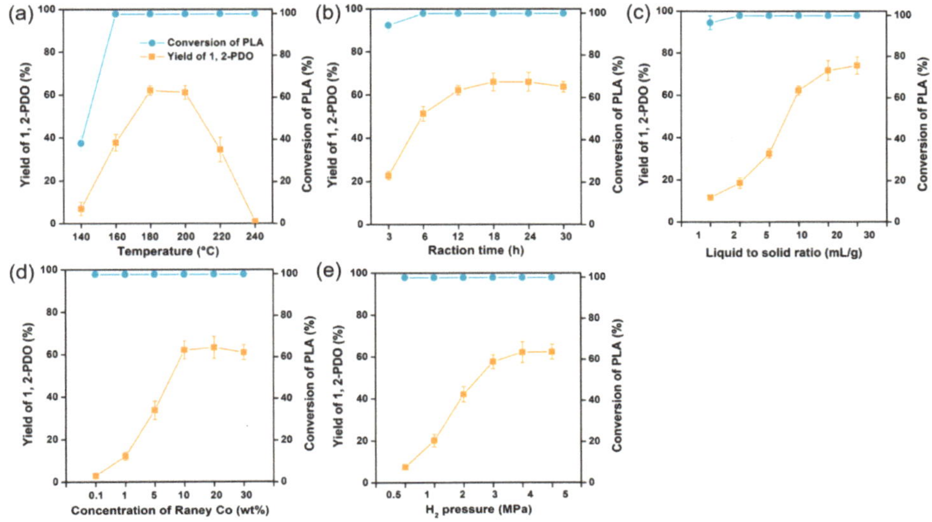

Figure 3. Effect of different parameters: (**a**) temperature, (**b**) reaction time, (**c**) liquid to solid ratio, (**d**) concentration of Raney Co, and (**e**) H_2 pressure on the yield of 1,2-PDO and conversion of PLA.

In addition, another two parameters were investigated to select a suitable value of Raney Co dosage and H_2 pressure (Figure 3d,e). Figure S2 indicated that Raney Co and hydrogen were integral for the formation of 1,2-PDO, and ethyl lactate was the main product in the absence of Raney Co or H_2 or both. The yield of 1,2-PDO increased sharply along with the increase in catalyst dosage and H_2 pressure at initial stage, and then tended to invariability. Thus, 10 wt% Raney Co and 3 MPa H_2 were used in the following tests.

3.3. Optimization of Reaction Procedure

A three-factor and three-level BBD consisting of 17 experimental runs was used to optimize the tandem alcoholysis–hydrogenation process of PLA. Table S2 summarizes the experimental yield of 1,2-propanediol and predicted values by model, which was fitted as a quadratic polynomial equation shown as Equation (3):

$$\text{Yield (1,2-propanediol, \%)} = 69.3 + 5.1\text{ A} + 13.7\text{ B} + 5.3\text{ C} + 0.6\text{ AB} + 2.1\text{ AC} - 0.2\text{ BC} - 4.3\text{ A}^2 - 18.0\text{ B}^2 - 4.0\text{ C}^2 \quad (3)$$

where A, B, and C are the coded value of reaction time, temperature, and ethanol/PLA ratio, respectively. The sign of coefficients implied how the parameters influence the response. Therein, positive coefficients represent related factors have synergistic effect, but otherwise factors have antagonist effect towards response [45].

Statistical testing of regression model was checked by F-test, and the analysis of variance (ANOVA) for the fitted quadratic polynomial equation is shown in Table S3. Additionally, the model was significant as the F-value was 56.7, while the lack of fit F-value of 2.6 indicated no statistical significance relative to the pure error as the *p*-value is higher than 0.05, which is desirable as the model exhibits good fitting to the relative response [47]. Moreover, the validation of the polynomial regression model was confirmed by the comparative plot between experimental and predicted values in Figure S3a, in which the coefficient (R^2 = 0.99) indicated that the fitted model was well in agreement with actual results [48]. In addition, the normal plot residuals in Figure S3b displayed the normal probability and studentized residuals lie reasonably in a straight line, implying the significance of the fitted model, and confirming that the assumption of the analysis was satisfied [49].

The 3D response surface plots and 2D contour plots in Figure 4, i.e., the graphical and visualized representations of the polynomial regression model, could help to better understand the individual and interactions of the variables during the process of alcoholysis and hydrogenation of PLA. Additionally, the optimal reaction conditions were optimized by the 3D response surface with Box–Behnken design as follows: 187 °C of reaction temperature, 15.6 h of reaction time, and 27 mL ethanol/g PLA, which contributed to the predicted maximum yield of 1,2-PDO as 76.1%.

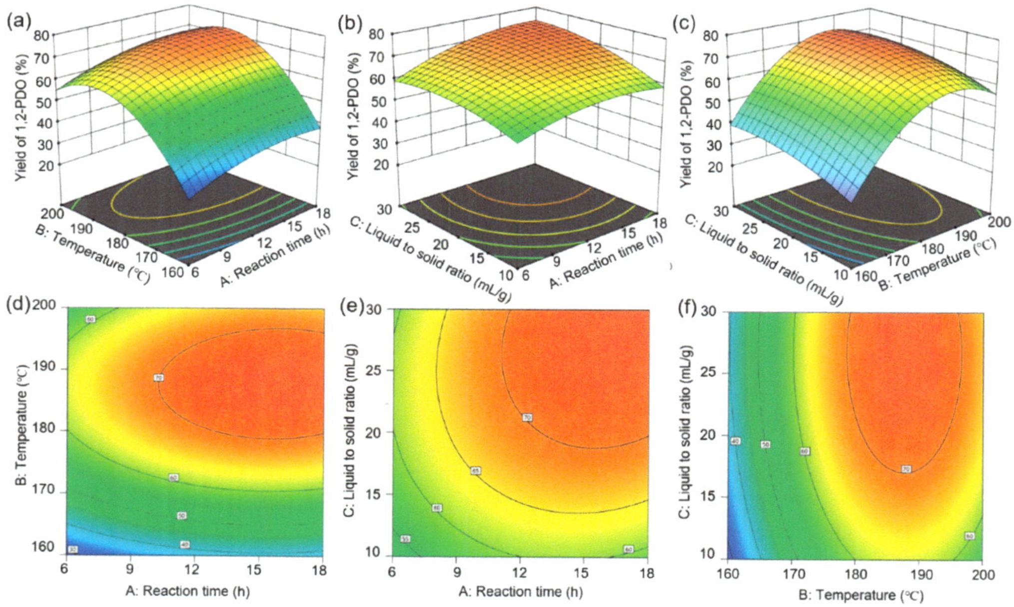

Figure 4. Response surface (**a–c**) and contour plots (**d–f**) for yield of 1,2-PDO.

3.4. Verification for Optimal Model

As discussed above, the model of tandem alcoholysis–hydrogenation of PLA to produce 1,2-PDO was established, and the predicted optimal conditions were optimized. To verify the reliability of the model, three PLA materials (PLA powder, PLA particle, and commercially available PLA-straws) were submitted to the reaction under the optimal conditions in Figure 5. Both PLA powder and particle were totally depolymerized, and 74.8 ± 5.5% and 76.5 ± 6.2% of 1,2-PDO were obtained, respectively, which indicated the applicability of the model as the difference between actual and predicted value was not significant ($p > 0.05$).

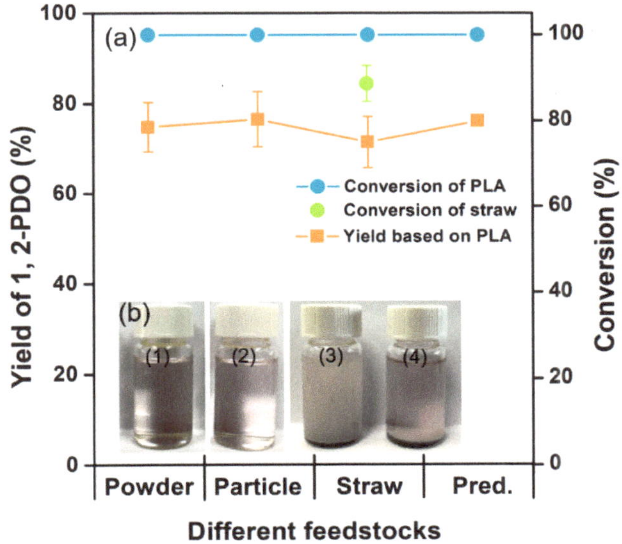

Figure 5. Verification experiments for optimal model. (**a**) Yield of 1,2-propanediol and conversion derived from different feedstocks vs. predicted (Pred.) values from the model. The recovery of 1,2-PDO was based on the mass of feedstock. (**b**) Depolymerized samples of PLA powder (1), particle (2), dispersed (3), and sedimented insoluble particles (4) of PLA-straws. Reaction conditions: 1 g feedstock, 3 MPa H_2, 0.1 g Raney Co, 27 mL ethanol, 187 °C, and reacting for 15.6 h.

Under the same conditions, total conversion of PLA and 71.4 ± 5.7% yield of 1,2-PDO were achieved based on the mass of PLA in the straws, and the conversion of PLA straws was 88.9 ± 4.2%. As the FTIR spectra of neat PLA powder, reaction residues and roasting residues of PLA straws were compared in Figure S4, the PLA ingredient was totally converted as the specific peaks of PLA were absent in the spectra of residues. However, insoluble particles remained in the final samples (Figure 5b), which might be some inorganic additives (such as talcum [50], diatomite [51]) in PLA straws as there were 10.4% residues after roasting PLA straws at 600 °C. Moreover, some additional C4 derivatives were formed in PLA straws degradation reaction, such as butanediol, diethyl butanedioate, and ethyl hydroxybutyl butanedioate (Figure S5), which was supposed to be the organic agents (e.g., co-polymers: poly(butylene succinate) [52], poly(butylene adipate-co-terephthalate) [53], and biomass: lignin [54] and coffee ground [55]). These additives in PLA straws are used to modify its physical limitations, such as softening temperature, brittleness, slow crystallization, etc. [56]. Consequently, the model is also suitable for the depolymerization of PLA straws and other commercial products if the PLA content could be determined.

3.5. Reusability Test of Catalyst

Under the optimized condition, Raney Co was recycled and reused ten times, and the results are shown in Figure 6. Overall, total conversion of PLA powder was achieved for all runs, and the catalytic activity for alcoholysis–hydrogenation of PLA was maintained at least 7 times with only a 4.9% decrease in the yield of 1,2-propanediol. Continuing reusing the catalyst gradually led to an obvious decrease in catalytic properties, as there was a 29.1% reduction of 1,2-PDO yield in the tenth run. Thus, further effects should be focused on improving the selectivity and stability of Co-based catalysts.

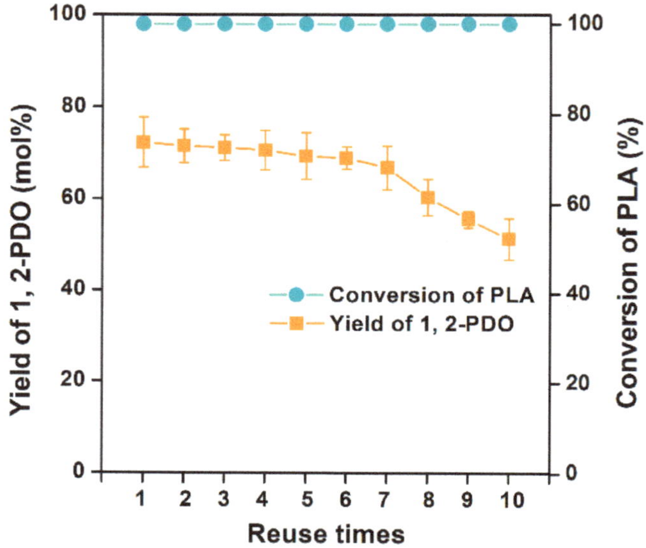

Figure 6. Reusability of Raney Co for alcoholysis and hydrogenation of PLA. Reaction conditions: 1 g PLA, 3 MPa H_2, 0.1 g Raney Co, 27 mL ethanol, 187 °C, and reacting for 15.6 h.

4. Conclusions

In this work, a one-pot tandem alcoholysis and hydrogenation of PLA was successfully conducted to produce 1,2-PDO using Raney Co catalyst, and ethanol was used both as reagent for esterification as well as solvent for subsequent hydrogenation. The reaction conditions were optimized by single-factor analysis and response surface optimization as 15.9 h reaction time, 187 °C temperature, and 27 mL ethanol/g PLA, and 1,2-PDO yield of 74.8 ± 5.5%, 76.5 ± 6.2%, and 71.4 ± 5.7% were derived from powder, particle, and straws of PLA, respectively. The Raney Co catalyst was efficient on a gram scale, and can be reused in at least seven degradation cycles, suggesting the possibility of industrial application of this method. Finally, this work demonstrated that it is possible to recycle PLA wastes using inexpensive and commercially available catalyst, and eco-friendly and multi-functional solvent, contributing to reduce and digest the end-of-life PLA residues in the environment, which offers a sustainable alternative to the depolymerization of PLA wastes.

Supplementary Materials: The following supporting information can be downloaded at: https://www.mdpi.com/article/10.3390/polym15020413/s1. Figure S1. GC-MS analysis of methanolization and hydrogenation products from PLA powder and methyl lactate; Figure S2. GC-MS analysis of controlled experiments; Figure S3. Validation of fitted model; Figure S4. FTIR spectra of PLA and its residues; Figure S5. GC-MS analysis of different PLA products-derived degradation liquids; Table S1. Bio-based processes examples towards 1,2-PDO production [18,19,24,27,30–32,43,57–66]; Table S2. Experimental design and observed response; Table S3. ANOVA results of the quadratic model.

Author Contributions: Conceptualization, J.X. and S.K.; methodology, J.X.; validation, J.X., L.Q., S.H., and Z.T.; formal analysis, J.X. and S.K.; resources, S.K.; investigation, L.Q., Z.T., and S.H.; writing—original draft preparation, J.X.; writing—review and editing, K.Z., P.D., and S.K.; supervision, K.Z., P.D., and S.K.; project administration, S.K.; funding acquisition, K.Z. and S.K. All authors have read and agreed to the published version of the manuscript.

Funding: This work was supported by the funding of Guangdong Provincial Key Laboratory of Distributed Energy Systems (2020B1212060075, S.K.), the Characteristic Innovation Project of Guangdong Provincial Department of Education (2021KTSCX133, S.K.), Guangdong Basic and Applied Basic

Research Foundation (2022A1515010578, S.K.; 2020A1515110127, K.Z.), and the College Students' Climbing Plan of Guangdong Province (PDJH2022A0508, Z.T.; PDJH2023A0507, L.Q.).

Institutional Review Board Statement: Not applicable.

Data Availability Statement: The data that support the findings of this study are available on request from the corresponding author.

Conflicts of Interest: The authors declare no conflict of interest.

References

1. Lunt, J. Large-scale production, properties and commercial applications of polylactic acid polymers. *Polym. Degrad. Stab.* **1998**, *59*, 145–152. [CrossRef]
2. Maga, D.; Hiebel, M.; Thonemann, N. Life cycle assessment of recycling options for polylactic acid. *Resour. Conserv. Recy.* **2019**, *149*, 86–96. [CrossRef]
3. Lorenzo, M.L.D.; Androsch, R. *Applications of Poly(Lactic Acid) in Commodities and Specialties*, 1st ed.; Springer: Berlin/Heidelberg, Germany, 2018; pp. 1–16.
4. Bioplastics Market Development Update 2019. Available online: https://docs.european-bioplastics.org/conference/Report_Bioplastics_Market_Data_2020_short_version.pdf (accessed on 22 February 2021).
5. Hong, M.; Chen, Y.X. Chemically recyclable polymers: A circular economy approach to sustainability. *Green Chem.* **2017**, *19*, 3692–3706. [CrossRef]
6. Rossi, V.; Cleeve-Edwards, N.; Lundquist, L.; Schenker, U.; Dubois, C.; Humbert, S.; Jolliet, O. Life cycle assessment of end-of-life options for two biodegradable packaging materials: Sound application of the European waste hierarchy. *J. Clean. Prod.* **2015**, *86*, 132–145. [CrossRef]
7. Laycock, B.; Nikolic, M.; Colwell, J.M.; Gauthier, E.; Halley, P.; Bottle, S.; George, G. Lifetime prediction of biodegradable polymers. *Prog. Polym. Sci.* **2017**, *71*, 144–189. [CrossRef]
8. Sun, C.; Wei, S.; Tan, H.; Huang, Y.; Zhang, Y. Progress in upcycling polylactic acid waste as an alternative carbon source: A review. *Chem. Eng. J.* **2022**, *446*, 136881. [CrossRef]
9. Siddiqui, M.N.; Kolokotsiou, L.; Vouvoudi, E.; Redhwi, H.H.; Al-Arfaj, A.A.; Achilias, D.S. Depolymerization of PLA by phase transfer catalysed alkaline hydrolysis in a microwave reactor. *J. Polym. Environ.* **2020**, *28*, 1664–1672. [CrossRef]
10. Cristina, A.M.; Sara, F.; Fausto, G.; Vincenzo, P.; Rocchina, S.; Claudio, V. Degradation of post-consumer PLA: Hydrolysis of polymeric matrix and oligomers stabilization in aqueous phase. *J. Polym. Environ.* **2018**, *26*, 4396–4404. [CrossRef]
11. Sun, C.; Li, C.; Tan, H.; Zhang, Y. Synergistic effects of wood fiber and polylactic acid during co-pyrolysis using TG-FTIR-MS and Py-GC/MS. *Energy Convers. Manage.* **2019**, *202*, 112212. [CrossRef]
12. Garg, M.; White, S.R.; Sottos, N.R. Rapid degradation of poly(lactic acid) with organometallic catalysts. *ACS Appl. Mater. Interface* **2019**, *11*, 46226–46232. [CrossRef]
13. McKeown, P.; Kamran, M.; Davidson, M.G.; Jones, M.D.; Román-Ramírez, L.A.; Wood, J. Organocatalysis for versatile polymer degradation. *Green Chem.* **2020**, *22*, 3721–3726. [CrossRef]
14. Majgaonkar, P.; Hanich, R.; Malz, F.; Brüll, R. Chemical recycling of post-consumer pla waste for sustainable production of ethyl lactate. *Chem. Eng. J.* **2021**, *423*, 129952. [CrossRef]
15. Xu, J.; Zhou, K.; Fu, J.; Tan, Z.; Qin, L.; Duan, P.; Xu, Y.; Kang, S. Near-zero-waste hydrogenolysis of poly(lactic acid) to biofuel. *Fuel* **2023**, *334*, 126609. [CrossRef]
16. Saeaung, K.; Phusunti, N.; Phetwarotai, W.; Assabumrungrat, S.; Cheirsilp, B. Catalytic pyrolysis of petroleum-based and biodegradable plastic waste to obtain high-value chemicals. *Waste Manage.* **2021**, *127*, 101–111. [CrossRef] [PubMed]
17. Tian, S.; Jiao, Y.; Gao, Z.; Xu, Y.; Fu, L.; Fu, H.; Zhou, W.; Hu, C.; Liu, G.; Wang, M.; et al. Catalytic amination of polylactic acid to alanine. *J. Am. Chem. Soc.* **2021**, *143*, 16358–16363. [CrossRef]
18. Shuklov, I.A.; Dubrovina, N.V.; Schulze, J.; Tietz, W.; Kuhlein, K.; Borner, A. Propane-1,2-diols from dilactides, oligolactides, or poly-L-lactic acid (PLLA): From plastic waste to chiral bulk chemicals. *Chem. Eur. J.* **2014**, *20*, 957–960. [CrossRef]
19. Krall, E.M.; Klein, T.W.; Andersen, R.J.; Nett, A.J.; Glasgow, R.W.; Reader, D.S.; Dauphinais, B.C.; Mc Ilrath, S.P.; Fischer, A.A.; Carney, M.J.; et al. Controlled hydrogenative depolymerization of polyesters and polycarbonates catalyzed by ruthenium(ii) PNN pincer complexes. *Chem. Commun.* **2014**, *50*, 4884–4887. [CrossRef]
20. Vivek, N.; Hazeena, S.H.; Alphy, M.P.; Kumar, V.; Magdouli, S.; Sindhu, R.; Pandey, A.; Binod, P. Recent advances in microbial biosynthesis of C3 - C5 diols: Genetics and process engineering approaches. *Bioresour. Technol.* **2021**, *322*, 124527. [CrossRef]
21. Kirk, R.E. *Encyclopedia of Chemical Technology*; John Wiley & Sons Inc.: Hoboken, NJ, USA, 2005; pp. 157–170.
22. Tabassum, N.; Pothu, R.; Pattnaik, A.; Boddula, R.; Balla, P.; Gundeboyina, R.; Challa, P.; Rajesh, R.; Perugopu, V.; Mameda, N.; et al. Heterogeneous catalysts for conversion of biodiesel-waste glycerol into high-added-value chemicals. *Catalysts* **2022**, *12*, 767. [CrossRef]
23. Sun, D.; Yamada, Y.; Sato, S.; Ueda, W. Glycerol hydrogenolysis into useful C3 chemicals. *Appl. Catal. B Environ.* **2016**, *193*, 75–92. [CrossRef]

24. Xiao, Z.; Jin, S.; Sha, G.; Williams, C.T.; Liang, C. Two-step conversion of biomass-derived glucose with high concentration over Cu–Cr catalysts. *Ind. Eng. Chem. Res.* **2014**, *53*, 8735–8743. [CrossRef]
25. Dastidar, R.G.; Galebach, P.H.; Lanci, M.P.; Wang, C.; Du, Y.; Huber, G.W. Elucidation of reaction network and kinetics between cellulose-derived 1,2-propanediol and methanol for one-pot biofuel production. *Green Chem.* **2022**, *24*, 350–364. [CrossRef]
26. Liu, K.; Huang, X.; Pidko, E.A.; Hensen, E.J.M. Hydrogenation of lactic acid to 1,2-propanediol over Ru-based catalysts. *ChemCatChem* **2018**, *10*, 810–817. [CrossRef] [PubMed]
27. Iqbal, S.; Kondrat, S.A.; Jones, D.R.; Schoenmakers, D.; Edwards, J.K.; Lu, L.; Yeo, B.R.; Wells, P.P.; Gibson, E.K.; Morgan, D.J.; et al. Ruthenium nanoparticles supported on carbon—an active catalyst for the hydrogenation of lactic acid to 1,2-propanediol. *ACS Catal.* **2015**, *5*, 5047–5059. [CrossRef]
28. Zhao, H.; Zheng, L.; Li, X.; Chen, P.; Hou, Z. Hydrogenolysis of glycerol to 1,2-propanediol over Cu-based catalysts: A short review. *Catal. Today* **2020**, *355*, 84–95. [CrossRef]
29. Marchesan, A.N.; Oncken, M.P.; Maciel Filho, R.; Wolf Maciel, M.R. A roadmap for renewable C2–C3 glycols production: A process engineering approach. *Green Chem.* **2019**, *21*, 5168–5194. [CrossRef]
30. Kindler, T.O.; Alberti, C.; Fedorenko, E.; Santangelo, N.; Enthaler, S. Ruthenium-catalyzed hydrogenative degradation of end-of-life poly(lactide) to produce 1,2-propanediol as platform chemical. *ChemistryOpen* **2020**, *9*, 401–404. [CrossRef]
31. Stefan, W.; Jasmine, I.; Klankermayer, J. Molecular catalyst systems as key enablers for tailored polyesters and polycarbonate recycling concepts. *Sci. Adv.* **2018**, *4*, eaat9669.
32. Fernandes, A.C. Reductive depolymerization of plastic waste catalyzed by $Zn(OAc)_2 \cdot 2H_2O$. *ChemSusChem* **2021**, *14*, 4228–4233. [CrossRef]
33. Calvo-Flores, F.G.; Monteagudo-Arrebola, M.J.; Dobado, J.A.; Isac-García, J. Green and bio-based solvents. *Top. Curr. Chem.* **2018**, *376*, 18. [CrossRef]
34. Huang, L.; Zhu, Y.; Zheng, H.; Du, M.; Li, Y. Vapor-phase hydrogenolysis of biomass-derived lactate to 1,2-propanediol over supported metal catalysts. *Appl. Catal. A Gen.* **2008**, *349*, 204–211. [CrossRef]
35. Ferreira, S.L.; Bruns, R.E.; Ferreira, H.S.; Matos, G.D.; David, J.M.; Brandao, G.C.; da Silva, E.G.; Portugal, L.A.; dos Reis, P.S.; Souza, A.S.; et al. Box-Behnken design: An alternative for the optimization of analytical methods. *Anal. Chim. Acta* **2007**, *597*, 179–186. [CrossRef] [PubMed]
36. Pereira, C.S.M.; Silva, V.M.T.M.; Rodrigues, A.E. Ethyl lactate as a solvent: Properties, applications and production processes—A review. *Green Chem.* **2011**, *13*, 2658–2671. [CrossRef]
37. Carné Sánchez, A.; Collinson, S.R. The selective recycling of mixed plastic waste of polylactic acid and polyethylene terephthalate by control of process conditions. *Eur. Polym. J.* **2011**, *47*, 1970–1976. [CrossRef]
38. Piemonte, V.; Sabatini, S.; Gironi, F. Chemical recycling of PLA: A great opportunity towards the sustainable development? *J. Polym. Environ.* **2013**, *21*, 640–647. [CrossRef]
39. Xue, J.; Cui, F.; Huang, Z.; Zuo, J.; Chen, J.; Xia, C. Effect of metal additives on structure and properties of a Co/SiO_2 hydrogenation catalyst. *Chin. J. Catal.* **2012**, *33*, 1642–1649. [CrossRef]
40. Zhang, X.; Wang, H.; Liu, X.; Han, J.; Zhu, X.; Ge, Q. Effect of calcination and metal loading on the characteristics of Co/NaY catalyst for liquid-phase hydrogenation of ethyl lactate to 1,2-propanediol. *Micropor. Mesopor. Mat.* **2016**, *233*, 184–193. [CrossRef]
41. Zhang, S.; Huo, Z.; Ren, D.; Luo, J.; Fu, J.; Li, L.; Jin, F. Catalytic conversion of ethyl lactate to 1,2-propanediol over CuO. *Chin. J. Chem. Eng.* **2016**, *24*, 126–131. [CrossRef]
42. Kasinathan, P.; Yoon, J.W.; Hwang, D.W.; Lee, U.H.; Hwang, J.S.; Hwang, Y.K.; Chang, J.S. Vapor-phase hydrogenation of ethyl lactate over copper–silica nanocomposites. *Appl. Catal. A Gen.* **2013**, *451*, 236–242. [CrossRef]
43. Luo, G.; Yan, S.; Qiao, M.; Zhuang, J.; Fan, K. Effect of tin on Ru-B/γ-Al_2O_3 catalyst for the hydrogenation of ethyl lactate to 1,2-propanediol. *Appl. Catal. A Gen.* **2004**, *275*, 95–102. [CrossRef]
44. Liu, K.; Litke, A.; Su, Y.; Van Campenhout, B.G.; Pidko, E.A.; Hensen, E.J. Photocatalytic decarboxylation of lactic acid by Pt/TiO_2. *Chem. Commun.* **2016**, *52*, 11634–11637. [CrossRef]
45. Dong, H.; Esser-Kahn, A.P.; Thakre, P.R.; Patrick, J.F.; Sottos, N.R.; White, S.R.; Moore, J.S. Chemical treatment of poly(lactic acid) fibers to enhance the rate of thermal depolymerization. *ACS Appl. Mater. Interface* **2012**, *4*, 503–509. [CrossRef] [PubMed]
46. Feng, L.; Feng, S.; Bian, X.; Li, G.; Chen, X. Pyrolysis mechanism of poly(lactic acid) for giving lactide under the catalysis of tin. *Polym. Degrad. Stab.* **2018**, *157*, 212–223. [CrossRef]
47. Yuan, W.; Cheng, J.; Huang, H.; Xiong, S.; Gao, J.; Zhang, J.; Feng, S. Optimization of cadmium biosorption by *Shewanella putrefaciens* using a Box-Behnken design. *Ecotoxicol. Environ. Saf.* **2019**, *175*, 138–147. [CrossRef] [PubMed]
48. Honary, S.; Ebrahimi, P.; Hadianamrei, R. Optimization of particle size and encapsulation efficiency of vancomycin nanoparticles by response surface methodology. *Pharm. Dev. Technol.* **2014**, *19*, 987–998. [CrossRef] [PubMed]
49. Deshmukh, R.K.; Naik, J.B. The impact of preparation parameters on sustained release aceclofenac microspheres: A design of experiments. *Adv. Powder Technol.* **2015**, *26*, 244–252. [CrossRef]
50. Somsunan, R.; Mainoiy, N. Isothermal and non-isothermal crystallization kinetics of PLA/PBS blends with talc as nucleating agent. *J. Therm. Anal. Calorim.* **2019**, *139*, 1941–1948. [CrossRef]
51. Cacciotti, I.; Mori, S.; Cherubini, V.; Nanni, F. Eco-sustainable systems based on poly(lactic acid), diatomite and coffee grounds extract for food packaging. *Int. J. Biol. Macromol.* **2018**, *112*, 567–575. [CrossRef]

52. Nazrin, A.; Sapuan, S.M.; Zuhri, M.Y.M.; Ilyas, R.A.; Syafiq, R.; Sherwani, S.F.K. Nanocellulose reinforced thermoplastic starch (TPS), polylactic acid (PLA), and polybutylene succinate (PBS) for food packaging applications. *Front. Chem.* **2020**, *8*, 213. [CrossRef]
53. Moustafa, H.; El Kissi, N.; Abou-Kandil, A.I.; Abdel-Aziz, M.S.; Dufresne, A. PLA/PBAT bionanocomposites with antimicrobial natural rosin for green packaging. *ACS Appl. Mater. Interface* **2017**, *9*, 20132–20141. [CrossRef]
54. Yang, W.; Weng, Y.; Puglia, D.; Qi, G.; Dong, W.; Kenny, J.M.; Ma, P. Poly(lactic acid)/lignin films with enhanced toughness and anti-oxidation performance for active food packaging. *Int. J. Biol. Macromol.* **2020**, *144*, 102–110. [CrossRef] [PubMed]
55. Terroba-Delicado, E.; Fiori, S.; Gomez-Caturla, J.; Montanes, N.; Sanchez-Nacher, L.; Torres-Giner, S. Valorization of liquor waste derived spent coffee grains for the development of injection-molded polylactide pieces of interest as disposable food packaging and serving materials. *Foods* **2022**, *11*, 1162. [CrossRef] [PubMed]
56. Kovalcik, A.; Pérez-Camargo, R.A.; Fürst, C.; Kucharczyk, P.; Müller, A.J. Nucleating efficiency and thermal stability of industrial non-purified lignins and ultrafine talc in poly(lactic acid) (PLA). *Polym. Degrad. Stab.* **2017**, *142*, 244–254. [CrossRef]
57. Pandhare, N.; Pudi, S.M.; Mondal, S.; Pareta, K.; Kumar, M.; Biswas, P. Development of Kinetic Model for Hydrogenolysis of Glycerol over Cu/MgO Catalyst in a Slurry Reactor. *Ind. Eng. Chem. Res.* **2017**, *57*, 101–110. [CrossRef]
58. Kim, N.D.; Park, J.R.; Park, D.S.; Kwak, B.K.; Yi, J. Promoter effect of Pd in $CuCr_2O_4$ catalysts on the hydrogenolysis of glycerol to 1,2-propanediol. *Green Chem.* **2012**, *14*, 2638–2646. [CrossRef]
59. Xia, S.; Yuan, Z.; Wang, L.; Chen, P.; Hou, Z. Catalytic production of 1,2-propanediol from glycerol in bio-ethanol solvent. *Bioresour. Technol.* **2012**, *104*, 814–817. [CrossRef] [PubMed]
60. Yuan, Z.; Wang, L.; Wang, J.; Xia, S.; Chen, P.; Hou, Z.; Zheng, X. Hydrogenolysis of glycerol over homogenously dispersed copper on solid base catalysts. *Appl. Catal. B Environ.* **2011**, *101*, 431–440. [CrossRef]
61. Ardila, A.N.; Sánchez-Castillo, M.A.; Zepeda, T.A.; Villa, A.L.; Fuentes, G.A. Glycerol hydrodeoxygenation to 1,2-propanediol catalyzed by CuPd/TiO_2-Na. *Appl. Catal. B Environ.* **2017**, *219*, 658–671. [CrossRef]
62. Wang, X.; Meng, L.; Wu, F.; Jiang, Y.; Wang, L.; Mu, X. Efficient conversion of microcrystalline cellulose to 1,2-alkanediols over supported Ni catalysts. *Green Chem.* **2012**, *14*, 758–765. [CrossRef]
63. Cortright, R.; Sanchez-Castillo, M.; Dumesic, J. Conversion of biomass to 1,2-propanediol by selective catalytic hydrogenation of lactic acid over silica-supported copper. *Appl. Catal. B Environ.* **2002**, *39*, 353–359. [CrossRef]
64. Ma, X.; Sun, D.; Zhao, F.; Du, C. Liquid phase hydrogenation of biomass-derived ethyl lactate to propane-1,2-diol over a highly active CoB amorphous catalyst. *Catal. Commun.* **2015**, *60*, 124–128. [CrossRef]
65. Feng, J.; Xiong, W.; Jia, Y.; Wang, J.; Liu, D.; Qin, R. Hydrogenation of ethyl lactate over ruthenium catalysts in an additive-free catalytic system. *React. Kinet. Catal. Lett.* **2011**, *104*, 89–97. [CrossRef]
66. Balaraman, E.; Fogler, E.; Milstein, D. Efficient hydrogenation of biomass-derived cyclic di-esters to 1,2-diols. *Chem. Commun.* **2011**, *48*, 1111–1113. [CrossRef] [PubMed]

polymers

MDPI

Review

Sustainable Reuse of Waste Tire Textile Fibers (WTTF) as Reinforcements

Ali Fazli and Denis Rodrigue *

Department of Chemical Engineering, Université Laval, Quebec, QC G1V 0A6, Canada
* Correspondence: denis.rodrigue@gch.ulaval.ca; Tel.: +1-418-656-29031

Abstract: Waste tire textile fibers (WTTF), as a by-product (10–15% by weight of tires) of end-of-life tires (ELT) mechanical recycling (grinding), are classified as hazardous wastes and traditionally burnt (thermal recycling) or buried (landfilling), leading to several environmental and ecological issues. Thus, WTTF still represent an important challenge in today's material recycling streams. It is vital to provide practical and economical solutions to convert WTTF into a source of inexpensive and valuable raw materials. In recent years, tire textile fibers have attracted significant attention to be used as a promising substitute to the commonly used natural/synthetic reinforcement fibers in geotechnical engineering applications, construction/civil structures, insulation materials, and polymer composites. However, the results available in the literature are limited, and practical aspects such as fiber contamination (~65% rubber particles) remain unsolved, limiting WTTF as an inexpensive reinforcement. This study provides a comprehensive review on WTTF treatments to separate rubber and impurities and discusses potential applications in expansive soils, cement and concrete, asphalt mixtures, rubber aerogels and polymer composites.

Keywords: waste tire textile fibers; soil composite; reinforced concrete; asphalt mixtures; rubber aerogels; polymer blend

Citation: Fazli, A.; Rodrigue, D. Sustainable Reuse of Waste Tire Textile Fibers (WTTF) as Reinforcements. *Polymers* **2022**, *14*, 3933. https://doi.org/10.3390/polym14193933

Academic Editors: Cristina Cazan and Mihaela Cosnita

Received: 23 August 2022
Accepted: 15 September 2022
Published: 20 September 2022

1. Introduction

As a consequence of a significant increase in the number of vehicles, more than 1.2 billion tires are manufactured annually, generating a large quantity of scrap tires ending up in landfills over the years and causing significant damage to the environment [1]. The management of used tires from its inception to its final disposal has become a major concern around the globe [2,3]. It is of great importance to develop safe and efficient disposal techniques for these waste streams, which has been an important challenge to the environment and ecosystems [4,5]. Based on a report of the United States Tire Manufacturers Association (USTMA), about 4.2 million tons (249 million scrap tires) were generated in the US in 2017, while the Canadian Association of Tire Recycling Agencies (CATRA) reported the generation of about 386 thousand tons of scrap tires in 2016 [6,7]. Similarly, data from the European Tire and Rubber Manufacturers Association (ETRA) showed that 3.6 million tons of scrap tires were generated in Europe in 2013, while 51 million waste tires were discarded in Australia between 2013 and 2014 [8,9]. All these numbers point towards the generation of high amounts of waste tires to be managed. Table 1 presents a brief overview of a tire composition in which each component in a tire formulation contributes to a specific characteristic, leading to a long lifetime, and achieves dimensional stability with high performances under severe conditions (high temperatures, radiations, mechanical stresses and chemical reagents) [10,11].

Vulcanized tire rubber as an elastic, insoluble and infusible thermoset material containing about 1.52–1.64% sulfur requires several decades to degrade, naturally causing significant health and environmental problems if not recycled or discarded properly [12,13].

The main applications for end-of-life tires (ELT) disposal include reusing (5–23%), landfilling (20–30%), energy recovery and pyrolysis (25–60%), as well as blending with other materials to produce composites (3–15%) [14].

Table 1. Typical tire composition (wt.%) [10].

Type of Tire	Rubber/Elastomers	Carbon Black	Metal	Textile	Zinc Oxide	Others
Passenger Car	47	21.5	16.5	5.5	1	8.5
Lorry	45	22	23	3	2	5
Off Road	47	22	12	10	2	7

Figure 1 shows that ELT can be processed into rubber granulates, such as rubber chips, rubber crumbs and rubber powder, as well as other components from tire reinforcement: steel and textile fibers [15]. Besides rubber as the main component of used tires, steel wires and textile fibers are two by-products derived from the treatment of ELT, but both are generally discarded as undervalued resources. The separated steel wires can simply be used in tunnel linings, hydraulic structures, bridge decks, pavements and slope stabilization [16]. Furthermore, waste tire textile fibers (WTTF) are generated in significant amounts (10–15% by weight of waste tires) and are classified as special wastes (EWC code 19.12.08) which have to be carefully disposed of or incinerated [10].

Figure 1. The different materials extracted from end-of-life tires such as (1) textile fiber, (2) metal cord, (3) large size fraction of rubber crumb (1–5 mm), (4) mean size fraction (1–2 mm), and (5) small size fraction (0.1–1 mm) [15].

ELT can be reused by stripping off the tire tread and applying a new one via cold or hot processes, which is called retreading with the purpose of increasing the lifetime of used tires. Although tire retreading is an ecofriendly and waste-free method of ELT disposal, the low quality and safety concerns at high speeds restrain this approach for passenger car tires [2,13]. Burying ELT in landfills causes important environmental issues since not only the materials occupy large spaces, but they are the origin of self-initiated fires during hot seasons, as well as storing rain water, creating a breeding habitat for rodents and insects [17,18]. About 49% of waste tires are incinerated to produce energy since tire rubber benefits from a high calorific value (32.6 MJ/kg), competing with common fuel to produce steam, electrical energy, pulp, paper, lime and steel [2]. However, because of incomplete/inefficient combustion, tire combustion leads to the emission of air pollutants, such as carbon oxides (CO_x), nitrogen oxides (NO_x), sulfur oxides (SO_x) and polycyclic aromatic hydrocarbons (PAH) [19]. Accordingly, energy generation through tire incineration was banned in most developed countries following the Paris agreement to prevent air pollution. However, pyrolysis can be considered as a practical strategy to transform ELT as an available source of hydrocarbons into gases, carbon blacks, liquid fuels, pyrolytic oils and pyrolysis char through catalytic or non-catalytic reactions above 400 °C in an oxygen-free environment [12,20]. Figure 2 presents a wide range of possible value-added products from the pyrolysis of ELT [21]. However, the main limitations to use pyrolysis at an industrial scale are high costs (purchase, installation and operation), pyrolysis conditions (high temperature with low pressure), and the generation of toxic hydrogen sulfide (gas) as the main by-product [22].

Figure 2. Classification of the different waste tire pyrolysis products with possible applications [21].

To avoid high environmental pollution associated with ELT landfilling and combustion, it is imperative to develop alternative modes of waste tires disposal and value-added potential applications. Composites production involves turning tire residues into valuable raw materials through compounding (mixing) with different components. As an upcycling strategy, scrap tires are valorized at the end of their service life to develop low costs and sustainable composites [3,11,23]. For instance, tire rubber crumbs have been used as additives or aggregates in automotive parts [24], concretes [25], asphalt mixtures [26,27], pavements [28,29], soil structures [30], etc.

Different grinding processes have been developed such as cryogenic, ambient, wet and water jet, converting used tires into triturated rubber particles of various sizes and shapes. The rubber granulates, as the main portion of ELT, can be exploited as aggregates in composite manufacturing with positive effects in terms of cost, weight, mechanical performance, durability and environmental sustainability [31,32]. Figure 3 shows that ELT grinding consists of three main steps. Initially, ground tire rubber (GTR) is produced using grinder blades and knife elements for tire downsizing into 7–10 cm particles followed by the removal of the metallic fraction. Next, crunching is performed to obtain tire rubber granulates of smaller sizes (around 2 cm). Finally, pulverization and separation of the tire residues takes place to produce granulated or pulverized rubber fractions with specific particle sizes (smaller than 1 mm) using pneumatic separators (sieving) and electromagnets for the separation of textile fibers and metal components from the rubbers, respectively.

About 15–50% by mass of uncleaned fibers is generated from this process as a mixture of rubber–fiber, which still needs to be fully valorized [10].

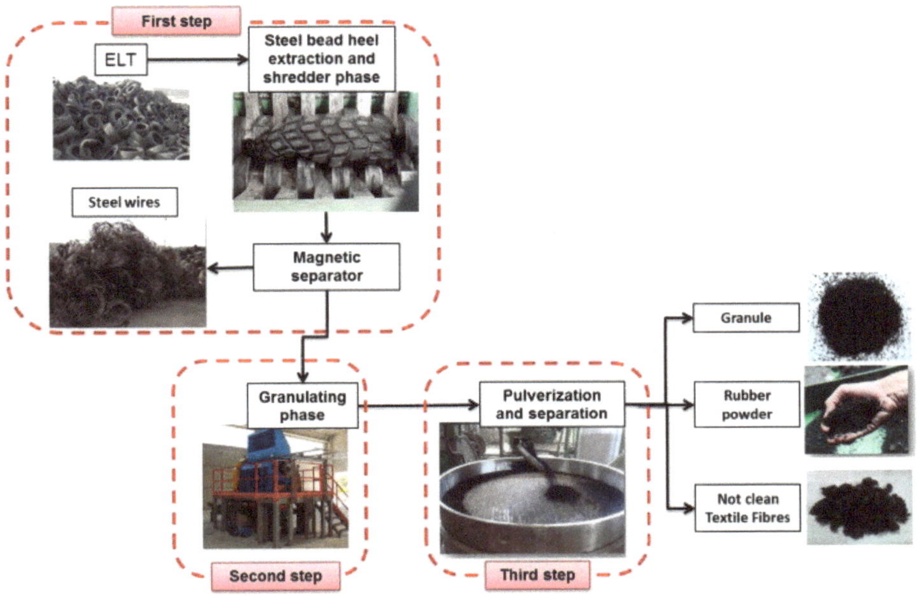

Figure 3. Different ELT disposal processes [10].

Most of the current treatment processes for reusing, recovering, and recycling of ELT are aiming at recovering steel and rubber, while waste management strategies are not implemented for WTTF. This apparent lack of interest is associated with the fact that WTTF represent an obstacle for widespread ELT recycling because of dust accumulating in the workplace, resulting in health problems for operators [33,34]. Moreover, the fibers obtained from the processing of used vehicle tires are contaminated with rubber particles accumulating electrostatic charges, making it difficult for melt processing (inhomogeneous mixing and feeding problems in extrusion or injection), as well as high volume and low specific weight (approximately 140 kg/m^3), making their transport/handling very difficult and expensive. In addition, rubber particles attached to WTTF contain varying amounts of sulfur, nitrogen, oxygen and hydrogen leading to the emission of toxic gases during combustion [35,36]. As an example, around 5–8 tons of WTTF are buried daily in Iran, indicating the role of tire fiber residues on polluting the soil, underground waters and wildfire, in addition to the high amount of valuable urban lands being filled with this waste [37]. One of the most efficient ways of tackling this challenge is to apply efficient cleaning operations to separate the fine crumb rubber and steel particles from the textile fibers to generate added-value applications [10,38,39].

Recently, alternative strategies for WTTF are gradually being developed based on important environmental and financial incentives instead of landfilling or burning. Among the products acquired in the recycling process of waste tires, WTTF are introduced as promising reinforcements in geotechnical engineering practices [37,40,41], building materials [38,42], and polymer blends [43,44]. According to the literature, Nylon, especially Nylon 6.6, and polyester are the main polymers found in WTTF [10,42,43,45]. WTTF can be incorporated as a reinforcement into different matrices with a processing temperature lower than the melting temperature of Nylon 6.6 (259 °C) [46] and polyester (253 °C) [43] as the main components of WTTF. Conversely, they can be used in "colder" materials such as soil, cement or bituminous conglomerates [10]. For instance, the application of WTTF as reinforcing agents in expansive soil for impermeable liners and covers in landfill applications results in improved geotechnical characteristics such as split tensile strength, swelling–consolidation, volumetric shrinkage and desiccation cracking tests [47].

Since ELT can be of various sources, such as truck tires, off-the-road tires, car tires, tractor tires, etc., the WTTF obtained after grinding have different lengths, diameters and mechanical properties with typical values and characteristics as presented in Table 2. As reported in Table 3, energy dispersive X-ray (EDAX) spectroscopy confirms the predominance of carbon and oxygen in the chemical composition of these fibers, while small amounts of Na, Zn, S, Al and Si are also detected due to the different additives used in vulcanized rubber formulations [40,48].

Table 2. Physical and mechanical characteristics of WTTF [40].

Property	Unit	Testing Method	Values	
			General	Most Frequent
Fiber type	–	–	Yarned	–
Equivalent diameter	mm	ASTM D885M-10A e1 (2014)	0.030–1.50	0.80
Length	mm	ASTM D885M-10A e1 (2014)	0–70	20–40
Tensile strength	MPa	ASTM D885M-10A e1 (2014)	300–2000	600
Twist S (Folded)	T/10 cm	ASTM D885M-10A e1 (2014)	30–50	39
Elongation at break	%	ASTM D885M-10A e1 (2014)	18–25	22
Elastic modulus	GPa	ASTM D885M-10A e1 (2014)	2–7.5	2.7
Hot air shrinkage (at 177 °C × 2 min × 143 g)	%	ASTM D5591-04 (2016)	3–5	4.5
Linear density	Denier	ASTM D885M-10A e1 (2014)	840–1890	1260
Melting point	°C	ASTM D885M-10A e1 (2014)	250–260	256
Water absorption	%	ASTM D885M-10A e1 (2014)	5–13	9.5

Table 3. Chemical composition (EDAX result) of WTTF [40].

Element	Symbols	Content (%)
Carbon	C	62.3
Oxygen	O	23.9
Sodium	Na	4.75
Zinc	Zn	2.91
Sulfur	S	2.19
Aluminum	Al	1.83
Silicon	Si	1.38
Magnesium	Mg	0.72

Some reviews are available on the subject of tire recycling into thermoplastic elastomer (TPE), rubber compounding, building/construction materials, and tire-derived fuels. For instance, the review by Ramarad et al. [2] covers the recycling of the rubber fraction of ELT with a focus on the addition of GTR into polymeric matrices. Fazli and Rodrigue [13] reviewed in detail the curing, rheological, mechanical, aging, thermal, dynamic mechanical and swelling properties of rubber compounds reinforced with recycled tire rubber (rubber review). Bockstal et al. [49] provided a comprehensive review about physical and chemical devulcanization processes used to recycle rubber granules as reclaimed rubber (RR), as well as the properties of rubber compounds filled with RR. Wahid Ferdous et al. [17] reviewed the recent progress using tire rubber residues as replacements of aggregates, binders and fibers in concrete formulations. Compared to a large body of studies on tire rubber waste management, only a few studies were devoted to study the reuse opportunities and potential applications of WTTF in composite preparation and the effect of different fiber treatments on the final product properties. Nevertheless, these treatments are not always very clear and need better understanding to fully exploit the fiber properties into high-value applications.

The current study aims to show that using WTTF in composite materials is an important move towards a circular economy that can minimize virgin materials consumption and develop recyclable fiber-reinforced composites having similar properties to virgin materials at lower costs. In this context, a distinctive effort is made here to summarize all the works dealing with the recycling of treated or untreated WTTF. As a step towards a more sustainable development, a special focus on the possibility of reusing WTTF as reinforcement in soil composites, cementitious materials, asphalt mixture, rubber aerogels and polymer blends is presented and discussed in terms of the advantages and limitations for each application.

2. Soil Composite

The creation and expansion of tension cracks and fissures are prevalent phenomena in soil structures such as dams, hydraulic barriers, slopes, runway subgrades, highways and railway embankments [40,50]. In addition, the soil structures used for the construction of impermeable layers (liners and covers in landfills) to prevent from leachate and pollutant migration into the soil and ground water are susceptible to volume changes and desiccation cracking [50]. This is why soil stabilization/reinforcement techniques are being developed to increase the shear and compressive strengths, bearing capacity and permeability of the structures [51]. Chemical techniques are being used for expansive soil reinforcement and stabilization through application of fly ash [52], lime [53], geo-polymers [54], cement [55], and ground granulated blast furnace slag [56]. Furthermore, mechanical techniques have been proposed as a more sustainable solution for expansive soil reinforcement based on randomly distributed synthetic fibers such as polypropylene (PP) [57], polyester [58], polyvinyl alcohol (PVA) [59], waste carpet [60], glass [61], and carbon [62]), as well as natural (biobased) fibers such as kenaf [63], corn husk [64], hemp [65], bagasse [66], sisal [67], and jute [68].

Tire-derived textile tire fibers are among the main reinforcing agents gaining more attention in geotechnical engineering applications with a special focus on soil structures due to their efficient and economical nature [30,37,41,69]. In one of the earliest attempts towards a new area for WTTF reuse in soil reinforcement, Abbaspour et al. [37] reported that up to 10% was possible to use as an effective soil reinforcement strategy for two types of soils based on clay and sand. As shown in Figure 4, the failure pattern of fiber-reinforced soils in unconfined compressive strength (UCS) tests reveals a brittle failure of the unreinforced specimen due to the presence of a continuous shear plane (vertical crack) from top to bottom (Figure 4a) [37]. Furthermore, incorporation of up to 4% WTTF generated multiple undeveloped cracks and bulging deformation as indicative of the fiber effect to prevent the expansion of the original shear plane leading to higher specimen strength [37].

Expansive soils are mainly used as impermeable liners or covers which are prone to volume changes and suffer from a lack of high load bearing capacity. Therefore, these soils must be modified and improved via various methods prior to construction [40,50]. As a step towards the sustainable development of fiber-reinforced soil structures, Abbaspour et al. [47] and Narani et al. [40] introduced WTTF (2%) as prospective soil reinforcements in sandy soils and expansive clays, resulting in higher load bearing capacity (up to 46.5%), as well as swelling reduction (44%). The overall thickness of the superstructure in flexible pavement/railway structures can be minimized due to improved load-bearing capacity and decreased permanent deformation under repeated loadings of the WTTF-reinforced subgrade soil [70]. Narani et al. [40] studied the effect of WTTF as reinforcing agents on the swelling–consolidation and volumetric shrinkage of the expansive soil. As reported in Figure 5, the volumetric shrinkage strain of the reinforced specimens with different fiber contents (f_c) of 2%, 3% and 4% decreased by 4.5%, 10.8% and 12.3%, respectively, compared to that of an unreinforced sample in agreement with the results of Olgun [71] who also reported lower volume changes in clay soils stabilized with polypropylene (PP) fiber additions (0.25%, 0.50%, 0.75% and 1.0%). These results confirm that replacing a fraction of expansive solids (soil grains) with WTTF as a non-expansive material with negligible water absorption (highly hydrophobic) leads to lower volume change during drying.

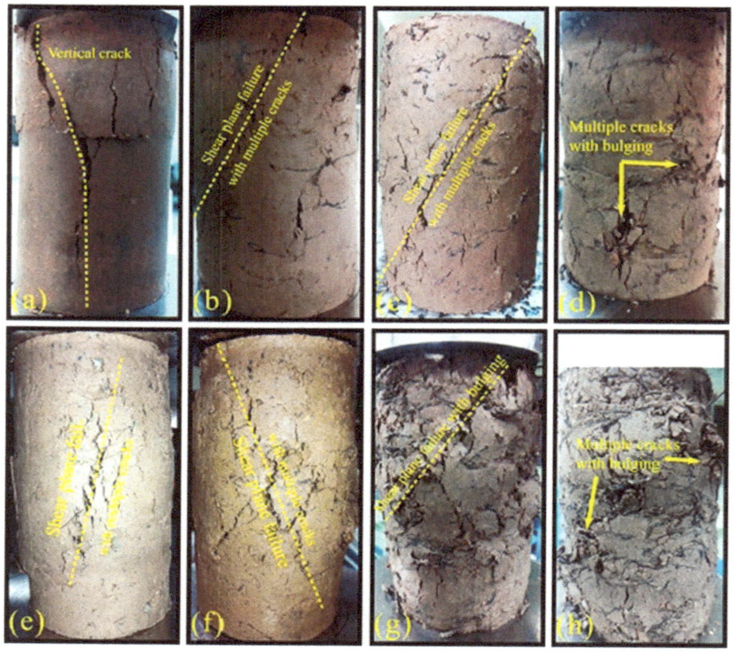

Figure 4. Typical failure patterns under UCS tests: (**a**) CF0, (**b**) CF0.5, (**c**) CF1, (**d**) CF4, (**e**) SF0, (**f**) SF0.5, (**g**) SF3, (**h**) SF4 (clayey and sandy composites are coded as CFχ and SFχ, respectively, where χ represents the weight ratio of WTTF to dry soil) [37].

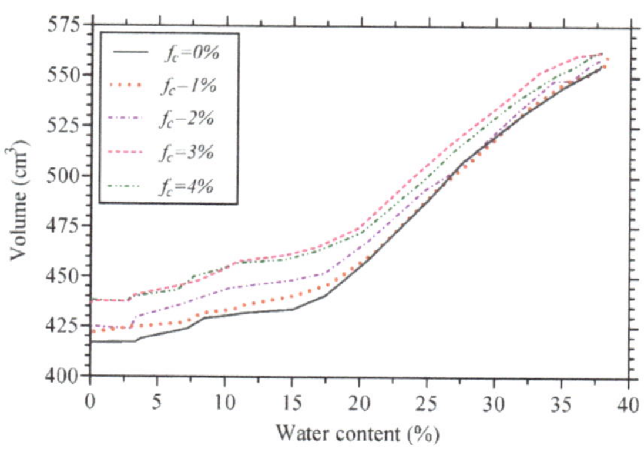

Figure 5. Volumetric shrinkage deformation as a function of water content for different WTTF contents [40].

Very few investigations compared the performance/efficiency limitation of WTTF with other fiber options (glass fibers). Valipour et al. [70] investigated the effect of WTTF and glass fibers (0.5, 1.0 and 1.5%) having varying lengths (5 and 10 mm) on the strength/deformation properties of clays. They observed an increasing shear strength of the fiber-reinforced soils in terms of higher internal friction angle (ϕ_p) and cohesion intercept (c). The maximum shear strength for the composite soils was obtained at 1.0% glass fibers and 0.5% WTTF because of the reinforcing effects of each fiber and their interaction. As schematically illustrated in Figure 6, after cracking or shear banding/displacement to the fiber-reinforced soil composite, the fibers may be distorted and pulled out from the soil matrix. If the reinforcing fibers can sustain the applied forces before being pulled out from the soil, the stresses within the fibers as a function of the strain level keep the soil element together, leading to increased shear strength. The bonding stress (τ_B) between the fiber and the soil determines the pullout resistance of fibers depending on the adhesion (chemical) and interlocking (mechanical and physical) forces between the fiber and soil, as well as interfacial friction [70,72].

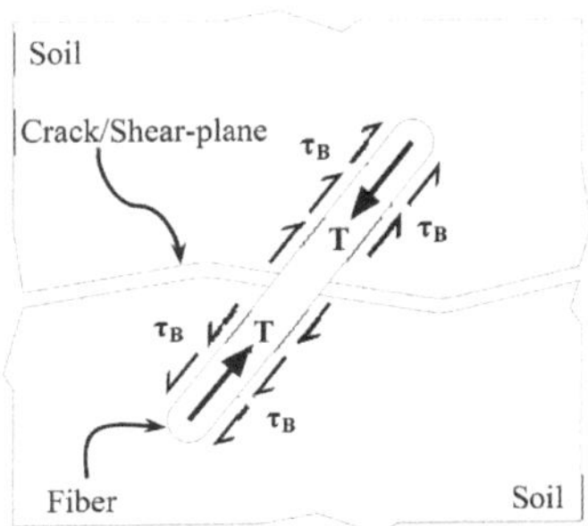

Figure 6. Schematic illustration of fiber–soil interaction (not to scale) [70].

When dealing with different fiber-reinforced composites, it is of great importance to understand how the fibers type, diameter and length can affect the properties and processing of the resulting reinforced samples. For example, an increase in fiber length results in higher probability of fiber agglomeration, which increases the required time of mixing [73]. For example, Tabakouei et al. [73] reinforced sandy soil with three different lengths of fibers including WTTF, date palm (DP) and PP fibers with diameters of 71, 100 and 151 mm, respectively. In WTTF-reinforced soil, contrary to the role of fiber diameter, an increase in fiber length contributed to higher ductility and toughness combined with a drop of the elastic modulus due to the reduced stiffness of longer fibers compared to shorter ones [73].

Very few investigations were carried out on the effect of fiber reinforcement on the performance of subgrade soils. Recently, WTTF-reinforced soil structures were used as subgrade layers of pavements having high function in the distribution of traffic loads [74–76]. The incorporation of WTTF into a soil structure may contribute to increasing the resilient modulus (M_R), which is the material recovery ratio after a series of deviatoric stresses applied to a specimen and expressed as:

$$M_R = \frac{\sigma}{\varepsilon_r} \quad (1)$$

where σ is the applied stress, and ε_r is the recoverable axial strain. Another possibility is to use the damping ratio, which is the energy dissipated by a vibrating structure caused by dynamic or cyclic loading and is defined as:

$$D = \frac{C}{C_c} \quad (2)$$

where C is the damping coefficient, and C_c is the critical damping related to the stiffness (k) and the mass (m) via:

$$C_C = 2\sqrt{km} \quad (3)$$

In general, the overall performance of such pavements was influenced by interactions between the materials in road layers being complex and rarely studied [76]. Abbaspour et al. [76] investigated the response of soil subgrades as a function of WTTF concentrations by studying the static and dynamic behavior of sandy soil. They also proposed an innovative model for the resilient modulus prediction in fiber-reinforced soils. Suitable interaction and interlocking between WTTF and soil grains was ensured due to simultaneous existence of friction and cohesion to hold the fibers and soil grains together during slippage, leading to increased shear strength. According to their results, adding between 1 and 4% WTTF led to increased shear strength of the fiber-reinforced specimen, reaching up to 17.5% improvement at a 2% fiber content compared to unreinforced soil [76].

3. Reinforced Concrete

In recent years, waste tire-filled concretes have been extensively investigated to develop high strength and toughened reinforced concretes. Although the introduction of tire rubber crumbs into cementitious material was reported to improve the resistance to cracking, acid and sulphate attack, chloride ion penetration and sound absorption ability, noticeable decreases in drying shrinkage resistance, as well as compressive and tensile strengths of rubberized concretes were reported. These loses were associated with weak adhesion between rubber particles and the cementitious matrix, as well as the formation of more porous concrete matrices due to rubber particles' air adhesion/entrapment and hydrophobicity [25,77]. Angelin et al. [78] observed high porosity (void content up to 40%) and significant density reduction (27%) of rubberized mortar due to void spaces entrapped in the cement matrix associated with the presence of rubber aggregates (30%). Contrary to the ductile failure mode with high deformation resistance of rubberized mortar, the compressive and flexural strength of mortars filled with 5% rubber crumbs was reported to be 85% and 96% of the reference specimen, respectively [79]. It is well stablished that the addition of micro- and macro-synthetic fibers changes the brittle behavior of concrete from brittle to plastic or quasi-plastic, resulting in significant improvement of tensile strength, flexural strength, toughness, dynamic load resistance, energy-absorbing capacity and cracking resistance of concrete [80].

Textile fibers derived from ELT can also act as substitutes for polymeric/synthetic fibers such as carbon [81], PVA [82], PP [83], polyethylene (PE) [84], and steel [81] in cement-based materials. These options received increasing attraction to resolve some of the environmental problems caused by waste tires, as well as improving the performance and durability of cement-based materials [38,39,85,86]. Long-term durability and performance of cement composites can be deteriorated by plastic shrinkage cracking caused by self-desiccation and external drying, especially for applications having large surface areas such as slabs on grade, thin surface repairs, patching, tunnel linings, etc. [87]. To overcome this problem, fiber reinforcement of concrete may hinder plastic shrinkage cracking since the fibers can induce bridging forces across the cracks, preventing their growth [88]. Banthia and Onuaguluchi [42] observed that the presence of scrap tire fibers (STF) reduced plastic shrinkage cracking of mortar mixtures with small-sized and segmented multiple cracks compared to large transverse cracks as in reference specimens. They claimed that the presence of tire-derived fibers (0.1–0.4%) and polyethylene terephthalate (PET) fibers (0.1–0.3%) both led to lower maximum crack width of fiber-reinforced mixtures compared to a neat (unreinforced) mortar. Figure 7 shows that lower total crack area was observed for STF and PET fiber-reinforced mortar mixtures by 74–97.5% and 96–99.4%, respectively. This improvement was associated with increased matrix ductility and crack bridging caused by the fibers preventing further crack initiation and propagation, as well as lower capillary pressure in the matrix [42]. Given similar findings in different reports, it was expected to observe higher efficiency of tire fibers in reducing plastic shrinkage cracking compared to that of PET fibers. However, the presence of rubber particles into the tire fibers makes the material fluffier, resulting in an entangled mass of fibers with increased fiber spacing. This state of the material leads to dispersion problems (inhomogeneous distribution) and inaccurate fiber content (fluctuating mixing ratio). All these factors contributed to less effective crack mitigation of the fiber-reinforced specimens compared to specimens containing PET [42]. Similarly, Serdar et al. [89] concluded that compared to neat concrete, the addition of PP fibers and WTTF produced a smaller crack width reflected by the value of restrained shrinkage and higher stresses that the composites can withstand. However, higher WTTF content was required (5 kg/m^3) compared to PP fiber (1 kg/m^3) to reach the same properties.

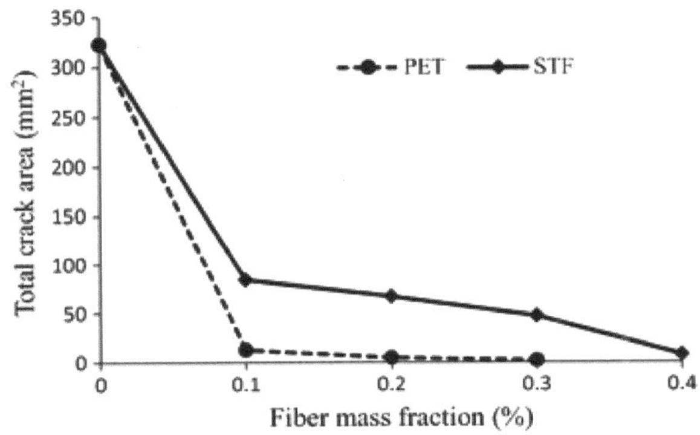

Figure 7. Maximum crack width and total crack area in specimen for STF and virgin PET fibers [42].

Baričević et al. [39] evaluated the effect of ELT fibers cleaning on the properties of fiber-reinforced concrete. For this purpose, a small-scale cleaning system was designed based on fiber residue shaking and gravitational forces, as rubber particles fell to the bottom of the device. Figure 8 shows that the first step includes a cleaning device consisting of four sieves (openings of 0.25, 0.71, 2 and 4 mm) placed on a shaking table at a frequency of 1.5 Hz, while small rubber balls (top sieve) and compressed air (8 bars) were used to increase the fiber cleaning efficiency. Partially cleaned fibers as a product of the first step were used for further rubber–fiber separation with a shaking frequency and time as in the first step, while a rotating airscrew was placed on top of the cleaning machine (second step) to spin out the contaminated fibers, while all the other impurities fell to the bottom of the sieves. A statistical analysis based on 117 samples showed that the obtained products from the second cleaning step were composed of 65% of rubber particles, 20% contaminated fibers with rubber, and 15% cleaned fibers with respect to the original (as-received) fibers (Figure 9) [39].

Addition of up to 10 kg/m^3 of uncleaned WTTF and up to 2 kg/m^3 of cleaned WTTF fibers showed positive effects on the stress distribution and resistance to scaling under freeze–thaw conditions. However, compared to the neat reference mixture, further introduction of cleaned fibers showed up to 5% compressive strength reduction, while higher contamination of uncleaned fibers contributed to more significant decreases (25%) because of higher air content entrapped and residual rubber (crosslinked particles) acting as stress concentration points [39].

To improve on the previous results, several options were proposed by various researchers to optimize the reinforcing efficiency of short fibers used in concrete/mortar through fiber hybridization (two or more types of fibers blended together). The purpose of fiber hybridization is to benefit from synergistic effects generated by each fiber with different lengths or moduli to enhance the overall properties of the resulting composites [81,84].

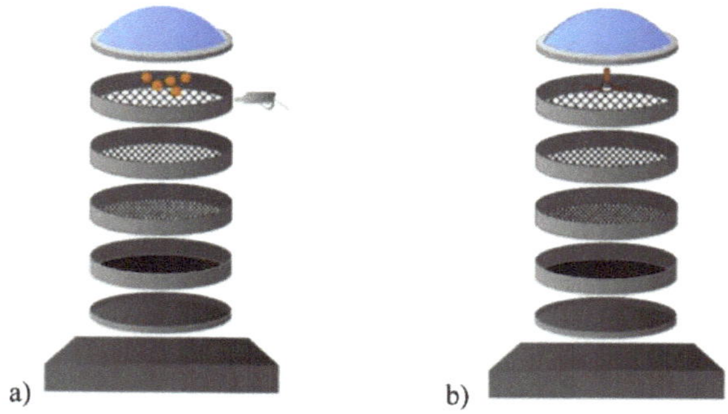

Figure 8. Schematic representation of a fiber cleaning procedure: (**a**) 1st step and (**b**) 2nd step [39].

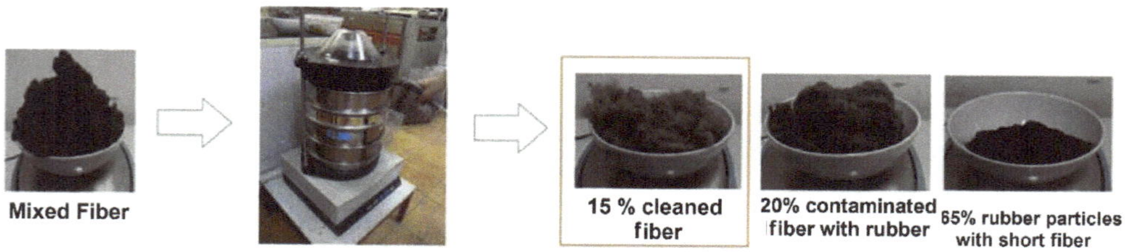

Figure 9. By-products obtained after cleaning mixed fiber [39].

Chen et al. [90] developed a sustainable cementitious composite reinforced by a combination of recycled tire steel fiber (RTS) (0.5, 1.0 and 1.5 vol%) and recycled tire textile fiber (RTF) (0.5 and 1.0 vol%). To prevent the negative effect of the large numbers of rubber particles attached to the fiber's surface, weakening the mechanical properties, a sieving method, as proposed by Chen et al. [38], was applied to clean the fibers from the rubber particles and impurities (43.2% of the total mass of as-received fibers). As the compressive strength of fiber-reinforced cementitious composite highly depends on the interfacial interaction between the reinforcing fibers and the matrix, strong fiber–matrix interaction leads to high load-carrying capacity. For example, Chen et al. [90] reported a 76.1% improvement of the 28-day compressive strength by increasing the RTS fiber content from 0.5 to 1.0 vol% for a fixed RTF content (0.5%). The results were associated with good fiber–matrix interactions within the composite [91]. However, increasing the RTF content from 0.5 to 1.0% at fixed RTS (1.0%) decreased the compressive strength by 23.1% due to excessive fiber content increasing the porosity in the interfacial transition zone (ITZ = the weakest part of concrete mixtures) and cementitious matrix leading to lower compressive strength [90].

The shear stress caused by the tensile loading transferred to the fiber–matrix interface can be sustained as a consequence of the fiber bridging ability, thus delaying crack growth/propagation [91]. As depicted in Figure 10a, the tensile strength of RTS fibers can resist the shear stress and hence delay the crack initiation and propagation, while the accumulation of tensile stress results in fiber break-up (Figure 10b). However, a hybridization of RTF and RTS (Figure 10c) was more effective to bridge the crack rather than the presence of RTS alone due to a synergistic effect between the combination of these fibers, which not only improved the composites' strength, but also their post-cracking behavior [90].

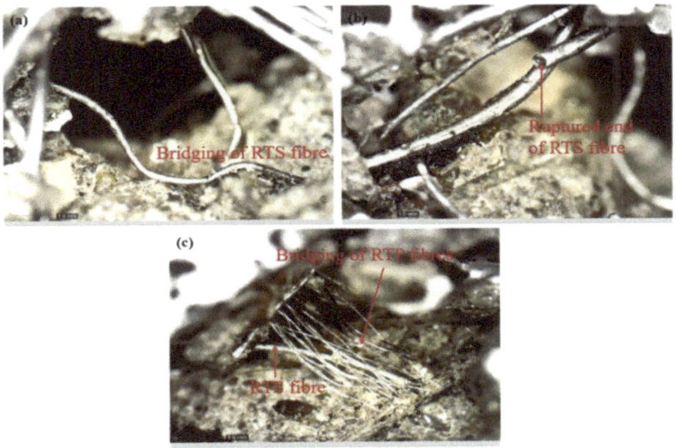

Figure 10. Images of the fiber–matrix interaction taken via optical microscopy [90].

In a similar work, Onuaguluchi et al. [36] investigated the use WTTF as a value-added reinforcement in cement composites through hybridization of such microfibers with macrofibers including PP, steel hook-end (HE) and micro/macro scrap tire steel fibers (SSF). Compared to the very low post-crack flexural residual strengths of plain mortar (0.10 MPa), adding WTTF showed negligible improvement (up to 0.20 MPa), while hybridization of 0.35% WTTF with 0.2–0.5% HE and PP fibers led to a noticeable increase in post-crack residual strengths ranging from 1.4 to 3.1 MPa due to the positive synergy between the fibers [36].

Fiber-reinforced concrete with superior crack-resistance and post-cracking performance has a wide range of potential applications including, but not limited to, pavements (airport and highway), bridge decks, tunnel linings and offshore platforms. However, concrete composites are generally subjected to repetitive cyclic loading during their entire service life that may lead to gradual degradation (mechanical fatigue). Thus, optimization of the fatigue behavior of these composites is mandatory [38,92]. Chen et al. [38] investigated for the first time the flexural fatigue behavior of fiber-reinforced concrete using hybridization of WTTF and PP fibers to account for two main factors: stress level and fiber content. They found that for all their reinforced composites, the fatigue life increased with decreasing stress level as reported in similar studies [38]. This behavior can be associated with the high number of initial microcracks and crack propagation induced by cyclic loading at higher stress levels facilitating failure as a result of faster cumulative cracks inside the concrete [38,93]. As expected, the introduction of WTTF not only restricted the formation and development of microcracks due to a fiber bridging effect, but also the rubber crumb attached on the fibers absorbing mechanical energy during fatigue, leading to longer fatigue life [88]. Compared to PP fibers, the distance between WTTF was lower due to their shorter length, increasing the bridging efficiency at a single microcrack within the matrix (more fibers are located inside each individual crack). Thus, WTTF-reinforced concrete showed the longest fatigue life under different failure probabilities [38].

4. Asphalt Mixture

The use of ELT to reinforce asphalt mixture is considered as a sustainable solution for paving material to minimize the cost and pollution with favorable engineering benefits [26]. Asphalt mixtures need to be resistant towards permanent deformation and cracking caused by a lack of foundation-bearing capacity and temperature cracking (climatic effects) [94]. For example, fatigue cracking of asphalt pavements is more prevalent in cold climate regions, while asphalt rutting more generally occurs in hot climatic regions [95]. The main advantages of rubberized asphalt mixtures are to delay reflection cracking, reduce traffic noise, minimize maintenance cost, and obtain hardened asphalt mixtures at high service temperatures since tire rubbers do not melt under heat and do not crack at low temperatures [26,28]. Recent studies have identified potential benefits of asphalt mixture reinforcement with recycled rubber derived from ELT grinding processes to produce tire rubber crumbs [27,96]. For instance, it was found that tire rubber crumbs contributed to lower temperature susceptibility and higher resistance towards moisture damage of reinforced asphalt compared with traditional mixtures [4]. Adding tire rubber particles produced better performances in permanent deformation resistance and rutting resistance compared with the unmodified binder and asphalt mixture [97]. Among the various modifiers for asphalt reinforcement, WTTF as a sub-product of ELT treatment have gradually attracted more attention due to their economic and environmental benefits instead of being disposed of in landfills or used in energetic valorization [98]. Waste tire-derived textile fibers have the potential to increase the toughness and fracture resistance of hot mix asphalt (HMA), as well as serve as stabilizers to hinder drain down of the asphalt binder (porous asphalt mixtures and stone mastic asphalt (SMA) mixtures) [94,95,99,100]. WTTF can be considered as a cheap alternative for common fibers such as PP, polyester, asbestos, cellulose, carbon, glass, nylon, coconut, basalt, steel and aramid to lower the use of virgin raw materials and develop asphalt mixtures with good resistance to aging, fatigue cracking, moisture damage, bleeding and reflection cracking [99]. Only few studies investigated the reuse WTTF in asphalt binder modification to induce superior performance of flexible road pavement layers subjected to traffic loading [94,95]. As an early step to introduce textile fibers derived from ELT in asphalt formulations, Putman and Amirkhanian [48] used cellulose, polyester, scrap tire and waste carpet fibers (0.3% wt. for all) to reduce the drain down of SMA mixtures. The results showed that WTTF not only prevented drain down (drain down limit at 10.1%), but also increased the toughness of the mixture without making any significant difference in permanent deformation of fiber-reinforced SMA mixtures. As shown in Figure 11, Landi et al. [99] observed drain down of porous asphalt mixtures composed of a fixed amount of bitumen (5% by mix weight) reinforced by natural (cellulose fibers) and synthetic fibers (tire textile fibers with Nylon 6.6). They placed 500 g of loose mixtures in glass beakers and kept them at 180 °C for 1 h followed by removing the mixtures and weighing the amount of bitumen drained to the bottom of the recipients. The beneficial effect of short fibers to prevent bitumen draining down in porous asphalt mixtures was more significant for WTTF compared to cellulose fibers. This observation might be ascribed to the presence of rubber aggregates stuck on the textile fibers not only reacting with the oily components of bitumen increasing its consistency, but also leading to a thicker bitumen film coating the aggregate particles (better compatibility between the phases present) [99].

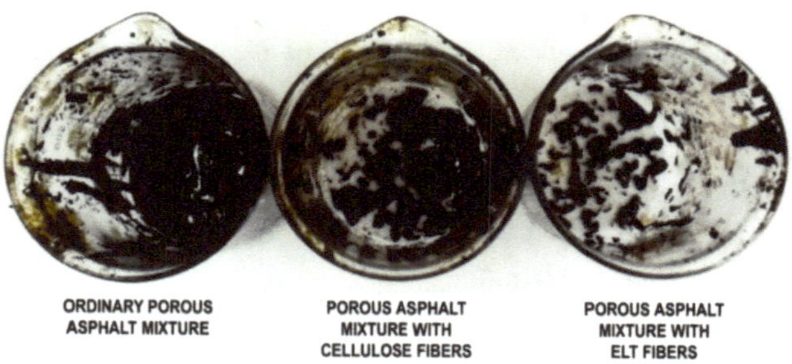

Figure 11. Effect of the fiber type in reducing bitumen drain down from porous asphalt mixtures [99].

Landi et al. [10] proposed an innovative approach to apply ELT fibers in asphalt focusing on cleaning the fibrous material and its subsequent reuse as a secondary raw material (compacted into pellets) in asphalt mixtures. In particular, they designed a new system for ELT treatment in which the whole discarded tires were transformed into ground rubber particles by grinding (downsizing) followed by the removal of the metallic fraction [10]. The fiber cleaning and compaction process to produce pellets were developed to reuse textile fibers (contaminated with 60% rubber) and facilitated their transport and dosage during the bitumen production [10]. The fiber cleaning step was carried out using a dry washing machine made up of a drum and reel with forward movement through a rotational motion, while fiber compaction was performed below the melting temperature of Nylon 6.6 (259 °C) as the main component of WTTF using another machine by applying a pressure vertically on a grid through the rotation of two cylinders around the main axis [46]. As shown in Figure 12, the use of a paraffin wax was necessary for pellets production to allow for higher fiber compaction (10 times increase in density). The resulting pellets were able to be loaded in mixing chambers via standard hoppers for the production of bituminous conglomerates without significant modification of the production cycle [10]. The fatigue resistance (number of cycles to failure) and the indirect tensile modulus of WTTF-reinforced bituminous were improved by 15% and 55% respectively, compared to cellulose fiber-reinforced samples (Table 4) [10].

Figure 12. Fiber pellets obtained with (**left**) and without (**right**) paraffin wax [10].

Table 4. Comparison between two different types of asphalt with cellulose fibers and ELT fibers [10].

Property	Asphalt with Cellulose Fibers	Asphalt with ELT Fibers
Bitumen (%)	5.4	5.4
Filler (%)	6	6
Indirect tensile module (MPa)	4482	5212
Indirect tensile strength (MPa)	1.37	1.38
Number of cycles to failure	3085	4785

In another work, Navarro et al. [29] used the steel belts (SB) and carcass ply (CP) of ELT (also containing textile fibers) as an anti-reflective cracking mat (ACM) within a pavement structure (Figure 13). It was found that the presence of rubber (elastic particles) in the ACM-CP system delayed the fatigue cracking process by absorbing a portion of the energy applied to the specimen, thus limiting the propagation of fatigue cracks in flexible pavement structures [29].

For fiber-reinforced asphalt, fine rubber particles sticking to tire textile fibers having resilient properties may result in low compaction and increased air voids, making the asphalt mixture susceptible to moisture/water/fluids at high filler content [101]. Bocci et al. [100] evaluated the effect of WTTF for binder layers to increase the mechanical performance of HMA. They claimed that WTTF decreased the HMA mix compatibility, and higher filler content was required to increase the volume of the bituminous mastic up to 5% and to decrease the air voids content to 2.4% (at a compaction energy of 180 gyrations) [100]. They also observed a substantial increase in the HMA resistance to fatigue due to the fibers' ability to oppose the micro-cracks evolution into macro-cracks by maintaining the micro-crack surfaces closer together [100]. Pais et al. [94] applied fibers as a by-product of used truck and car tires grinding in asphalt mixtures without any treatment. They reported that the tire-derived fiber residues were mainly composed of rubber (about 60%) including fine crumb rubber (33.2%), large crumb rubber (24.2%), string interlaced textile fibers (30.5%), and isolated interlaced textile fibers (12.1%). Although WTTF are susceptible to water absorption affecting the asphalt performance and appearance, the average water absorption of tire textile fibers was only 4.18%, which was half of the value obtained for polyacrylonitrile and polyester fibers [102].

Figure 13. Description of the tire parts used for the development of the ACM systems based on carcass ply (ACM-CP) and steel belts (ACM-SB) [29].

From an environment and economic point of view, improving HMA performances with elevated service life contributes to lower environmental issues of HMA-based pavements with reduced demand for virgin aggregates and bitumen (serving as adhesive to bind the HMA components), road infrastructures and energy consumption [26,27]. In this context, Landi et al. [99] presented a life cycle assessment (LCA) of hot HMA mixtures by evaluating the environmental performance of a standard mixture compared to cellulose-reinforced and WTTF-reinforced HMA mixtures. According to their results, WTTF-reinforced HMA may have longer service lives leading to high amounts of aggregates and bitumen (19.6 kg) being required during a complete life cycle perspective (a period of 30 years was used) compared to a standard HMA (38.3 kg bitumen) and cellulose fiber-reinforced HMA (24.3 kg of bitumen) [99]. Thus, switching from a standard HMA to a WTTF-reinforced HMA may result in a savings of about 405 MJ as a direct result of less bitumen being used and other raw materials being non-renewable petroleum-based resources [99].

To limit global warming and reduce greenhouse gas (GHG) emissions, the asphalt industries made recent efforts to develop new asphalt mixtures called warm-mix asphalt (WMA), which are suitable for lower construction temperatures without the sacrifice of properties [103]. As an example of limited research focusing on WMA reinforced with recycled fibers, Jin et al. [95] investigated the effect of both scrap tire rubber and ELT fiber on WMA modified with Sasobit (synthetic commercial wax). They observed higher crack resistance of tire rubber–fiber-modified WMA by 24.3% and 7.7% compared to conventional HMA and conventional asphalt mixture, respectively [95].

5. Rubber Aerogel

Successful application of WTTF in insulation products has attracted a great deal of interest as environmental and economic incentives to recycle waste tire fiber residues [10]. Due to industrialization and rapid urbanization, noise pollution caused by road traffic, construction sites, industries or social activities have raised concerns as important health issues. Prolonged exposure to noisy environments may cause stress, insomnia, cardiac dysfunction and high blood pressure [104]. For instance, the Ministry of Manpower (Singapore) reported in 2019 that noise pollution adversely affected the lives of about 75.49 thousand workers in the industry, having disability or other medical problems [105].

Sound insulators are categorized into resonant and porous insulation materials addressing the above-mentioned issue by absorbing the sound energy during a noise propagation process. Despite the limited efficiency of resonant materials caused by their narrow absorption frequency bands and their heavy weights, porous materials can be applied in a wide range of sound absorption frequencies with a wide range of structural designs [104,106]. However, the high thicknesses of sound absorption materials prevent their efficiency in viscous–thermal acoustic energy practices. This indicates the urgent need to develop high-value products such as aerogels [107]. Aerogels are highly porous (70–99%) materials with an ultralow thermal conductivity (0.012–0.050 W/m.K), an extremely low density (0.003–0.5 g/cm^3), large specific surface area (520–1590 m^2/g) and outstanding acoustic insulation properties. These materials are now being used for both noise and thermal insulations/filters (building insulation materials), as well as for oil absorption [107,108]. The development of rubber aerogels having very low density, high porosity and high compressibility not only has wide potential markets (heat and sound insulation, oil cleaning applications, etc.) and environmental benefits (minimize the risks of soil pollution and groundwater contamination by waste tire landfill, reducing the solvent exchange steps in the synthesis procedure by using water instead of toxic organic solvents), but can also address the waste generation issues of the tire industries. Rubber aerogels were first introduced by Thai et al. [109] as they prepared rubber aerogel crosslinkers through a freeze-drying method using WTTF, PVA and glutaraldehyde (GA) (crosslinking agent). The WTTF (30–50 μm diameter, 1–1.8 cm length) were soaked in acetone inducing a hydrolysis reaction, followed by swelling and sonication (80 °C for 20 min) in an acetone/PVA/GA/H_2O mixture for crosslinking. The cured mixture (85 °C for 3 h) was gelled in a refrigerator and then freeze-dried to obtain a rubber aerogel as shown in Figure 14b. Figure 14d shows that the presence of PVA led to the formation of hydrogen bonds and ester bonds between the carboxyl groups of the polymer chains in WTTF, mainly composed of Nylon 6.6, and the oxide group of PVA chains, while the addition of GA further reinforced the WTTF-PVA structure as a result of acetal bridges being formed between the oxide groups of the polymer fibers and PVA chains [109].

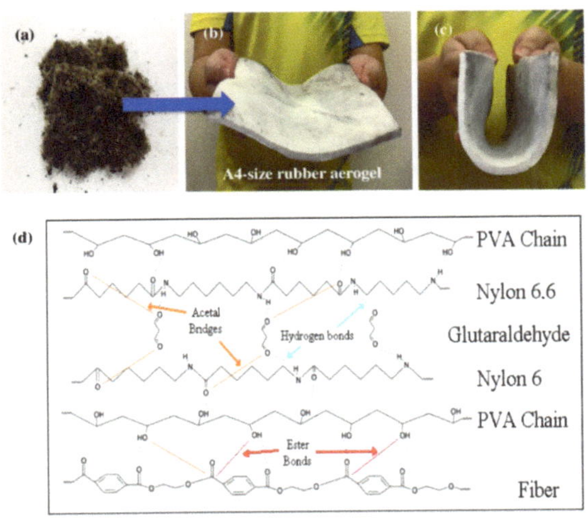

Figure 14. (**a**) Raw WTTF obtained from shredding the car tire waste, (**b**) A4-size rubber aerogel, (**c**) high flexibility of the rubber aerogel and (**d**) proposed formation mechanism between the WTTF and the crosslinkers inside the rubber aerogel [109].

Based on the acoustic insulation properties of rubber aerogels (Figure 15), the 11.2 mm thick aerogel containing 4.0 wt.% WTTF showed the best sound insulation (highest absorption coefficient). This observation can be related to the high fiber–fiber contact area and fiber entanglements at high WTTF loading (4.0 wt.% for RB 04 sample) contributing to more energy being lost (dissipated) by the propagating sound waves due to increasing surface friction and internal frictional losses [109]. The rubber aerogel not only showed better acoustic performance than the commercial foam absorber Basmel®, but also its oil absorption capacity was competitive compared to commercial PP and polyurethane (PU) sorbents [109–111].

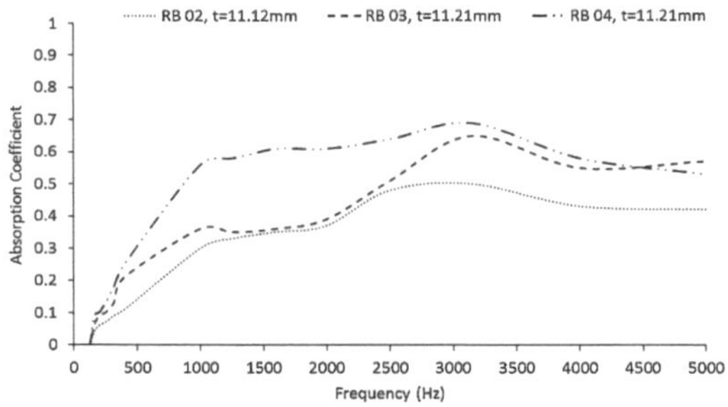

Figure 15. Sound absorption coefficient curves of rubber aerogels at different fiber concentrations (2.0, 3.0, and 4.0 wt.%) [109].

From an environmental point of view, it is also of great importance to develop aerogels for the removal of oil contaminants from oil-polluted waterways to safeguard aquatic and terrestrial lifeforms. Although silica aerogels are widely used to cope with oil spills, organic pollutants and industrial oil wastewater, complex and expensive manufacturing combined with the fragility of these aerogels prevent their widespread application, as high mechanical strength is required for specific applications [112,113]. Recent research on rubber aerogels was devoted to design and produce aerogels via a simple method showing very high oil absorption capacity, about two times more than conventional oil sorbents [114]. Figure 16 shows that water, as an eco-friendly solvent, is used for aerogel fabrication, which is very fast (about 15 h) compared to other sol–gel methods [115], taking more than a week to convert the hydrogels into aerogels [114]. The freeze-drying process was used to fabricate rubber aerogels based on WTTF, PVA and co-crosslinkers (GA) without shrinkage. In addition, methyltrimethoxysilane (MTMS) coating was applied onto the aerogel surface, inducing a very stable super-hydrophobic property (water contact angles of 147.9–153.8°) for over a period of six months, contributing to a high absorption capacity of 25.0 g/g. The super-hydrophobicity and high porosity of rubber aerogel promote its oil absorption by providing extra storage space for oil in the inner pores of the aerogel [114]. The addition of fiber up to 1.0 wt.% led to the lowest density (0.020 g/cm^3) and highest porosity (98.3%) of the aerogels, resulting in a maximum oil absorption capacity (25.0 g/g) similar to cellulose-based aerogels [116], and 2.6 times more important than polyacrylonitrile fiber-reinforced silica aerogels [117]. Furthermore, further increases in fiber content up to 5 wt.% decreased the absorption capacity to only 6.4 g/g due to a more compact network and denser structure, leading to less air pockets and decreasing the porosity and providing less space for oil penetration [114].

In general, the aerogel fabrication process controls the gel formation, while the drying steps determine the final structure and properties of aerogels [115,118]. Contrary to cellulose-based aerogels or silica aerogels that are highly brittle, the addition of scrap tire textile fibers is expected to improve the mechanical properties of rubber aerogels to undergo high deformation, especially for bending and folding without failure of the porous structure [33,105]. The presence of an interconnected network between the tire fibers and PVA enables the rubber aerogels to withstand the applied stresses and prevents overloading of the fibrous structure, resulting in significant geometric deformation under compression. As shown in compressive strain–stress curves (Figure 17a,b), rubber aerogels present three regions of compressive deformation including elastic deformation (strain $\varepsilon < 2\%$), compaction ($2\% < \varepsilon < 60\%$) and densification ($\varepsilon > 60\%$) [105]. The plateau area beyond the yield point is related to the rapidly collapsing porous structure during compressive testing, as the stress slowly increases with increasing compressive strain ($\varepsilon < 60\%$), while the specimens experience a sharp increase in stress due to densification at $\varepsilon > 60\%$ [105,119]. As shown in Figure 17c, deformation of the larger pores ($\varepsilon > 60\%$) and densification of the smaller pores beyond 60% strain resulted in a high level of strain absorption. As a function of WTTF content (1–5 wt.%), incorporation of 5 wt.% rubber fiber led to a maximum Young's modulus of 965.6 kPa, which is much higher than that of Styrofoam (commercial aerogel) and within the range of silica aerogels (0.1–10 MPa) [33,105]. It was observed that increasing the WTTF content increased the compression resistance of aerogels to undergo large deformations without fracture. Figure 17d shows that the introduction of WTTF into the porous structure substantially increased the recovery performance by up to 2.7 times as a result of reduced porosity, thereby decreasing shrinking and preserving the structural integrity of the aerogels [105].

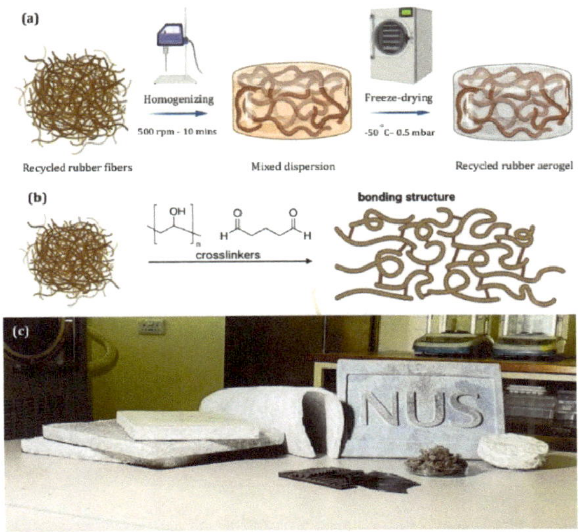

Figure 16. (**a**) Fabrication steps for the recycled rubber aerogels. (**b**) Schematic representation of the bonding structure formation. (**c**) Image presenting typical recycled rubber aerogel samples of different dimensions [114].

Thermal insulation is an important parameter in the design and fabrication of rubber aerogels, reflecting their overall performance by aiming at energy savings and minimizing carbon emissions. The thermal conductivity of the solid and gas phases play dominant roles on the thermal insulation efficiency of aerogels as lightweight structures. The heat transported via convection in the gas phase due to the displacement of molecules under a pressure gradient should be minimized by partially closed cells and pores smaller than the mean free path of gas (<70 nm) [33,120]. According to the literature, the thermal conductivity of rubber aerogel increases with increasing WTTF content because of more filled up volume at high fiber loading, resulting in lower porosities where the total volume of the voids within the aerogel structures is decreased. Therefore, the thermal conductivity of the rubber aerogels increased due to less air being trapped in the voids [33]. For instance, Thai et al. [109] reported increasing thermal conductivity of rubber aerogels based on WTTF and PVA from 0.035 to 0.047 W/m·K with increasing fiber concentration (1–5 wt.%) as a result of lower porosity (97.1% to 88.0%). Increasing the PVA content from 1 to 2 wt.% at fixed WTTF content (2 wt.%) contributed to lower thermal conductivity (from 0.049 to 0.038 W/m·K) because of increased porosity of the rubber aerogels (93.1%

to 95.7%) [109]. The Pham and Phan-Thien [121] model can predict the low thermal conductivity of rubber aerogels indicating that the thermal conductivity of a three-phase composite aerogel (WTTF, PVA, air) is controlled by the volume fraction of each component and the thermal conductivity of air (0.026 W/m·K) [122], WTTF (0.45 W/m·K) [105] and PVA (0.40 W/m·K) [122]. Thus, the low thermal conductivity of rubber aerogels depending on the volume fraction of air can be minimized by varying the tire textile fibers and PVA contents and increasing the overall aerogel porosity [105].

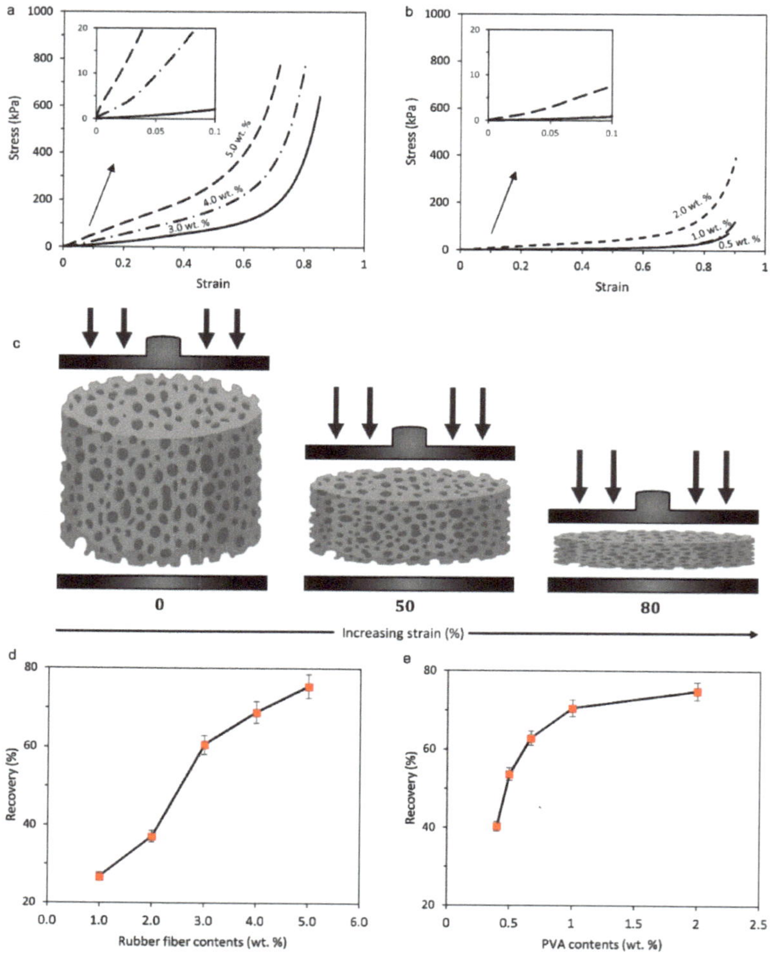

Figure 17. (**a**) Compressive stress–strain curves of rubber aerogels with various contents of WTTF and constant PVA content (1 wt.%). (**b**) Compressive stress–strain curves of rubber aerogels at various PVA contents and constant WTTF content (2 wt.%). (**c**) Schematic description of the changes in the cellular structure with compressive deformation. (**d**,**e**) Recovery performance of rubber aerogels after compression with varying contents of WTTF and PVA [105].

6. Blending with Polymers

Polymer composites comprising polymer matrices and fibers (natural/synthetic) are of high interest to achieve good physical, mechanical, thermal and dynamic properties at affordable cost. Fiber-reinforced polymer composites find a wide range of applications in construction, automotive, electronic, packaging and biomedical sectors [123]. Among the different short fibers available, a wide range of natural fibers are now investigated: hemp [124], flax [125], bamboo [126], jute [127], rice husk [128], coconut husk [129], and pineapple leaf [130]. They are extensively being used as reinforcements in composites, especially as alternatives to petroleum-based materials. Natural fibers have significant advantages over synthetic reinforcements (glass, carbon, aramid, etc.) such as low cost, low density, non-toxicity, renewable and biodegradable, with limited waste disposal problems [123]. However, several challenges are associated with natural fibers such as high moisture content/absorption, poor compatibility with most matrices, low fire resistance, and low impact strength [131]. Several reviews are available on the conventional processing routes and critical issues associated with natural fiber-reinforced composites [132], chemical interaction between natural fibers and matrices [133], and effects of interfacial adhesion on the strength of short fiber-reinforced composites [134]. Although a large body of literature is available on hybrid fiber-reinforced composites, there is a limited number of studies on WTTF as reinforcements into polymer composites [43,44,135].

Marconi et al. [136] investigated the environmental effects of reusing tire textile fibers as a second-life material to reinforce plastic compounds. In terms of climate change and fossil depletion, reusing WTTF in plastics compounding is the best scenario to prevent the emissions of toxic substances through fiber incineration and saving fossil fuels (coal, oil, etc.), as well as decreasing the amount of virgin plastics consumed. However, due to a lack of interest for scrap tire textile fibers contaminated with crumb rubber, they are rarely blended with polymer matrices (thermoplastic, thermoset and rubber) [43,44,135]. Additional cleaning processes (centrifugal separation) are required to remove most of the rubber particles (vulcanized) that may act as defects in polymer composites, degrading the performance and durability of the resulting products, as well as making the processing more difficult (extrusion and injection) for a polymer compound [137]. In recent years, WTTF (about 10% by weight of ELT) has gained more interest to develop sustainable polymer composites with potential applications (geotechnical engineering and construction materials, insulation products, polymer compounds, etc.) and to reduce the effect of waste polymers in landfills and toxic gas emission. For example, Zhang et al. [135] used ELT-derived textile short fibers to produce rubber composites from GTR through mechanical milling. They developed a system of stress-induced mechanochemical devulcanization of waste tire rubber mixed with textile fibers through pulverization at ambient temperature. The introduction of 5 wt.% WTTF through co-milling of fibers mixed with GTR resulted in higher tensile strength (9.5 MPa) than pan milling of the compound (6 MPa) due to improved interfacial adhesion minimizing voids formation, possibility at the ends of the fibers during straining, and hence retarding crack initiation [135]. Strong adhesion between the fibers and devulcanized rubber is responsible for better stress transfer from the matrix to the fibers, leading to increased tensile strength and less probability of specimen failure [44]. Rubber compounds with longitudinal fiber direction showed higher tensile properties than in the transverse direction at low fiber content (5 wt.%) due to higher overall strain resistance and growing crack inhibition, while further increases in WTTF content up to 20 wt.% led to fiber entanglement, preventing unidirectional orientation in the composites [135].

Irrespective of fiber cleaning efforts, tire-derived fibers always contain some impurities and rubber crumbs which are major obstacles to their recycling and are used in polymer composites. Nevertheless, Moghaddamzadeh and Rodrigue [43] reported significant amounts of rubber particles sticking on the fluffy fibers with high fiber entanglement. The average diameter of the fibers was about 20 +/− 2 mm (Figure 18c) with an average length of 2500 +/− 1500 mm (Figure 18a,b), while the rubber particles size distribution was roughly 90 +/− 10 mm (Figure 18d). In addition, energy dispersive spectroscopy (EDS) analyses confirmed the presence of impurities such as metal atoms (Al, Cu and Zn) and other additives (processing/vulcanization package) or polymeric materials used in tire formulations [43]. For example, aluminum silicates are common reinforcing fillers in tire structures, while sulfur and zinc oxides are used for rubber curing [138].

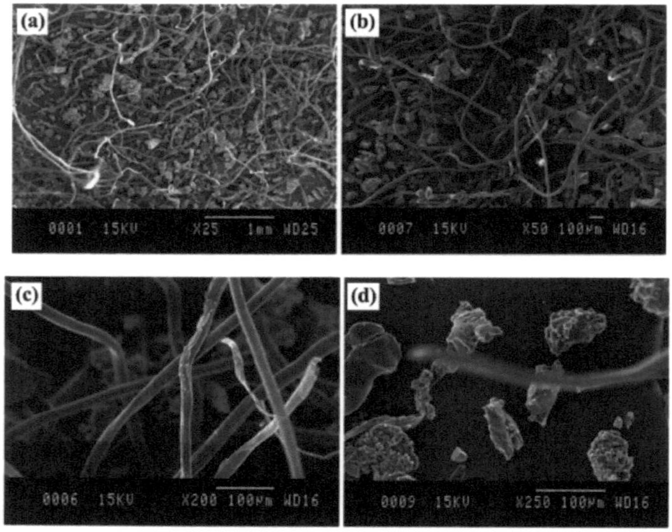

Figure 18. Typical SEM images of WTTF mixed with GTR at different magnifications (**a**–**d**) [43].

Landi et al. [10] used TGA on WTTF (Figure 19A) to compare the curve with Nylon 6.6 (Figure 19B) and confirmed the composition of the textile fibers. Furthermore, Moghaddamzadeh and Rodrigue [43] claimed that polyester was the main component of their WTTF using DTG and DSC, as well as FTIR showing the presence of specific peaks for carbonyl (C=O) (1711 cm^{-1}), hydroxyl (O–H) (3410 cm^{-1}), aromatic ring (1632 and 1409 cm^{-1}), aliphatic C–H stretching (2917 and 2850 cm^{-1}), and C(O)–O stretching (1239 cm^{-1}) of the ester groups in PET (Figure 19C). In a similar work, Onuaguluchi and Banthia used infrared spectra of the as-received WTTF and known polyester fabric to confirm by matching the spectra that both materials were polyester fibers [42]. Thus, the temperature to process WTTF as reinforcement in thermoplastic matrices needs to be lower than the melting point of Nylon 6.6 (259 °C) [46] or PET (253 °C) [43] as the main components of fiber residues. Nevertheless, the thermal degradation of the residual rubber particles on the fiber's surface is around 180–185 °C [139]. Thus, polyolefins (PP, PE, etc.) having good mechanical properties, excellent processability, low cost and high availability are appropriate matrices for melt blending with WTTF [136].

It is expected that melt blending of a polymer matrix with fiber residues contaminated by rubber particles as vulcanized materials with crosslinked network (not melting) would lead to poor mechanical properties due to low compatibility/interactions between the phases [140]. This is why Ferreira et al. [141] investigated the mechanical recycling of tire textile fibers by using a hybrid system based on recycled polyethylene terephthalate (rPET) reinforced by glass fibers and waste polyamide (W-PA) fibers from scrap tires. They observed low tensile properties of the composites attributed to the presence of the remaining rubber particles in W-PA despite the purification performed in the polyamide fiber wastes.

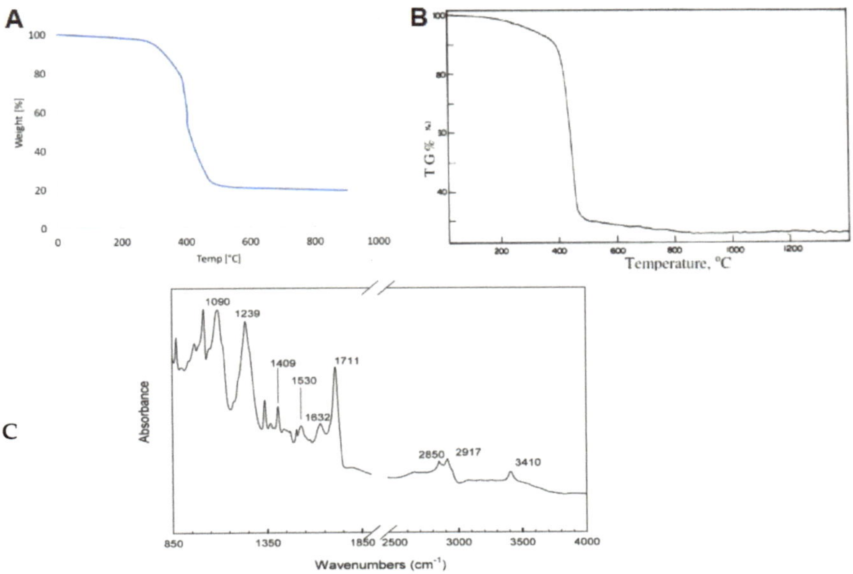

Figure 19. (**A**) TGA of textile materials and (**B**) of Nylon 6.6 [10], as well as (**C**) the FTIR spectrum of WTTF [43].

In agreement with the results of several independent research groups, Fazli and Rodrigue also reported low affinity between thermoplastic polyolefin matrices and GTR particles due to the crosslinked structure of tire rubber (not melting), resulting in poor compatibility and low interfacial adhesion [142,143]. Thus, the tensile strength of recycled high-density polyethylene (rHDPE) decreased from 19.0 to 10.6 MPa with the addition of 20 wt.% GTR, in agreement with similar reports [124,144,145]. Although GTR particles can undergo devulcanization (break-up of the crosslinked network via C-S and/or S-S bonds scission) to promote rubber chain mobility/interaction with the polymer matrix, the rupture of rubber hydrocarbon chains during devulcanization may lead to a drop in molecular weight (MW), thus decreasing the mechanical properties of the resulting composites [2,144]. Therefore, it is of great importance to find a solution for incompatibility issues and performance of tire textile fiber-reinforced composites to compensate for the loss of mechanical properties caused by rubber contamination (stress concentration point). Moghaddamzadeh and Rodrigue used styrene–ethylene–butylene–styrene-grafted maleic anhydride (SEBS-g-MA) as a compatibilizer having a similar structure with polyolefin and fibers to modify the interfacial interactions and mechanical properties of WTTF-reinforced composites [140,146]. Adding 10 wt.% SEBS-g-MA increased the tensile strength of linear low-density polyethylene (LLDPE)/WTTF (75/25) by 14% (10.3 MPa to 11.7 Mpa), which was attributed to improved stress transfer from LLDPE to the reinforcements and good compatibility between styrene (SEBS) and PET (main component of tire textile fibers), ethylene–butylene blocks of the compatibilizer with PE, and maleic anhydride groups with PE, hydroxyl and carboxylic end groups of PET [140]. Figure 20 shows that the introduction of SEBS-g-MA contributed to higher elasticity of the compatibilized samples compared to uncompatibilized ones as a consequence of reduced surface tension between the phases, thus improving interfacial stress transfer from the matrix to fibers. However, the processing conditions also have an important effect on the final properties of multiphase systems. As illustrated in Figure 20b, at constant fiber content (50 wt.%), increasing the extruder screw speed from 110 rpm (L-501) to 180 rpm (L-508) and 250 rpm (L-502) decreased the fiber's aspect ratio due to increased applied shear, leading to higher surface area and elasticity [146]. Figure 20 also reports on the effect of temperature on the final composites as determined via dynamic rheological measurements, which are very sensitive to the composition and structure on the samples.

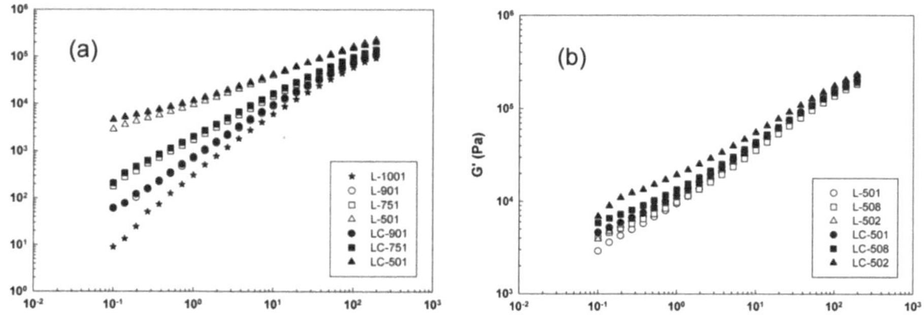

Figure 20. Dynamic elastic modulus as a function of angular frequency to determine the effect of: (**a**) RTF content, and (**b**) extruder screw speed [146].

According to the literature, incorporation of short fibers into plastics results in significant increases in tensile and flexural modulus (high rigidity) at a cost of reduced elasticity and impact strength, resulting in brittle composites [131,147]. Kakroodi et al. [124] reported that adding 30% hemp fibers to maleic anhydride-grafted polyethylene (MAPE) as the matrix increased the tensile modulus by 67% (from 104 to 184 MPa) and tensile strength by 30% (9.1–11.7 MPa) compared to the neat matrix, which did not break under Charpy impact testing (room temperature). However, the presence of hemp fibers (10–60%) significantly decreased the impact strength (from 369.2 to 127.5 J/m).

In recent years, the introduction of recycled tire rubber into fiber-reinforced composites is considered as an effective approach for impact strength modification of composites. However, this improvement strongly depends on the blend composition and mixing conditions. Kakroodi and Rodrigue [148] observed higher impact strength of composites (15% flax fiber) by 38% with the addition of only 15% GTR. However, the tensile strength substantially decreased from 17.5 to 11.6 MPa (34% decrease). Fazli and Rodrigue proposed a new approach for impact modification of polymer composites reinforced with RTF through incorporation of a surface-coated RR with compatibilizer [44]. As illustrated in Figure 21, a masterbatch of reclaimed rubber and MAPE was produced prior to melt blending of RTF and the thermoplastic matrix to ensure good surface coverage of rubber particles with the compatibilizer. Table 5 presents the tension and flexion properties of rHDPE-based composites filled with RR, and a mixture of RR/WTTF. The results suggest that melt blending of recycled tire rubber (35–80%) with thermoplastic (rHDPE) led to lower tensile strength (13–4.7 MPa) compared to the matrix (19 MPa) due to very low compatibility between both phases, resulting in voids around rubber particles (stress concentration points) and promoting interfacial debonding. The addition of RTF (20 wt.%) increased the tensile and flexural moduli of fiber-reinforced rubberized composites up to 246.5 and 405.6 MPa, respectively, imparting higher stiffness and resistance to deformation. The interaction of the maleic anhydride group (MAPE) with the hydroxyl group on the carbon black surface or carboxyl groups of rubber particles contributed to improved stress transfer and increased the tensile strength of the compatibilized composites [44,145]. Successful application of the methodology proposed by Fazli and Rodrigue resulted in a substantial increase in impact strength for composites filled with 20% RTF and 45% or 60% of RR/MAPE (70/30) compound leading to 49% (from 246.5 to 368.2 J/m) and 44% (from 275.6 to 398.7 J/m) improvements. The presence of MAPE surface-coated RR delayed crack growth and propagation due to a uniform filler dispersion in the matrix via thick and soft interphase-surrounded rubber, resulting in more energy dissipation [44].

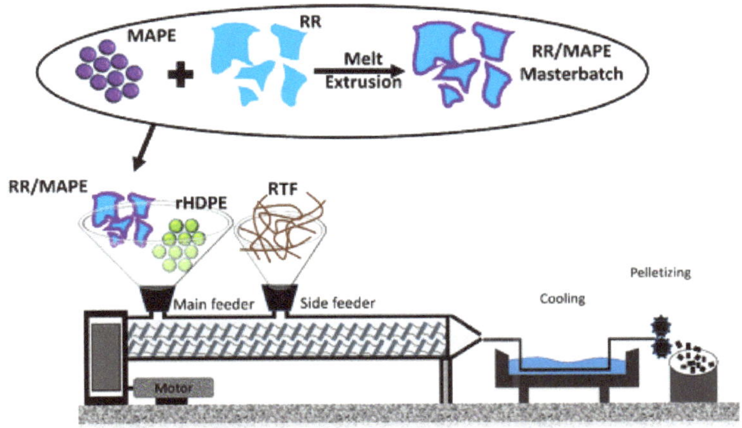

Figure 21. Melt extrusion steps for different rHDPE/(RR/MAPE)/RTF samples [44].

Table 5. Mechanical properties of the samples produced [44].

Sample	Tensile Strength (MPa)	Young's Modulus (MPa)	Tensile Strain at Break (%)	Flexural Modulus (MPa)
rHDPE	19.0 (0.3)	427.1 (14.9)	949.2 (26.4)	594.4 (11.3)
HR35	13.0 (0.3)	191.2 (4.3)	38.1 (4.8)	384.1 (3.5)
HR50	9.2 (0.3)	152.3 (3.2)	44.2 (7.2)	281.8 (5.4)
HR65	7.7 (0.1)	99.3 (4.2)	56.7 (5.3)	189.4 (3.8)
HR80	4.7 (0.4)	32.5 (5.4)	77.9 (8.6)	103.6 (4.7)
HR15F	9.5 (0.1)	246.5 (6.1)	30.2 (6.1)	405.6 (2.1)
HR30F	9.2 (0.3)	170.5 (6.6)	36.4 (4.9)	308.5 (3.8)
HR45F	7.4 (0.2)	109.3 (4.7)	45.3 (6.4)	202.7 (3.5)
HR60F	4.9 (0.1)	45.8 (5.2)	65.2 (5.7)	134.7 (2.9)
HR15F *	13.2 (0.2)	277.3 (4.9)	64.5 (8.2)	437.9 (3.4)
HR30F *	12.1 (0.2)	212.2 (5.3)	87.6 (7.9)	384.6 (4.5)
HR45F *	9.8 (0.1)	126.5 (3.6)	138.2 (7.6)	262.5 (4.2)
HR60F *	8.8 (0.4)	80.9 (4.5)	172.3 (8.3)	182.7 (5.1)

R*x*: RR wt.% (*x*) blended with rHDPE (wt.%). R*x*F: RR wt.% (*x*) blended with 20 wt.% of FF (F) and rHDPE (wt.%). R*y*F *: RR/MPAE wt.% (*y*) blended with 20 wt.% of FF (F) and rHDPE (wt.%).

7. Conclusions

This review presented a comprehensive overview on the recent progress in waste management of WTTF produced as a waste in ELT treatment. The complexity of the recovery and recycling of WTTF is due to fiber contamination with rubber particles, metal wires and impurities that must be separated during the recycling process to improve the performance of the recycled waste for subsequent reuse as a secondary raw material in future applications. The main conclusions of this review can be classified as:

1. Introduction of WTTF as soil reinforcement contributes to increase the load-bearing capacity of expansive soils while decreasing their swelling. The presence of WTTF in expansive subgrade results in the reduction of permanent deformation under repeated loadings and lower volumetric shrinkage strain.
2. Inclusion of WTTF in concrete composites can minimize plastic shrinkage cracking due to the fibers bridging effect across the cracks, preventing their growth. The fiber bridging effect of WTTF can be enhanced by its hybridization with steel fibers to increase the compressive strength of concrete composites.

3. WTTF have high potential to increase the toughness and fracture resistance of HMA, as well as serving as stabilizers to hinder drain down of the asphalt binder (porous asphalt and SMA mixtures).
4. Tire-derived textile fibers are potential candidates to produce ultralight and highly porous aerogels with strong mechanical performance (to be bent and folded without failure) through simple, cost-effective and time-saving fabrication methods showing great acoustic performance (sound insulation) and oil absorption capacity competitive with commercial sorbents.
5. A fiber cleaning step is highly recommended for melt blending of WTTF with polymers to avoid difficult processing (increasing viscosity) and poor mechanical properties (low compatibility). Recent progress in processing WTTF-reinforced polymer composites reported a substantial increase in impact strength by the introduction of surface-coated fillers to postpone failure due to the presence of a thick and soft interphase surrounding the rubber particles and fibers, resulting in more energy being dissipated before complete failure.

For future developments on WTTF recycling, the next steps should focused on better fiber cleaning efficiency (above 65% fiber purity) using simple processes (mechanical), develop new applications (floor mats, dampers, containers (recycling bins), automotive sector parts, wheels, gaskets, sports equipment, etc.) and find new markets/innovative products made from 100% recycled materials to reduce our environmental impact even more and find the best performance/cost ratio for value-added products.

Author Contributions: Conceptualization, A.F. and D.R.; methodology, A.F.; resources, D.R.; writing—original draft preparation, A.F.; writing—review and editing, D.R.; supervision, D.R.; funding acquisition, D.R. All authors have read and agreed to the published version of the manuscript.

Funding: This research was funded by the RECYC-QUÉBEC (Quebec Society for Recovery and Recycling).

Institutional Review Board Statement: Not applicable.

Data Availability Statement: Not applicable.

Acknowledgments: The authors gratefully acknowledge the technical and material support of Royal Mat Inc.

Conflicts of Interest: The authors declare no conflict of interest.

References

1. Sahajwalla, V.; Zaharia, M.; Rahman, M.; Khanna, R.; Saha-Chaudhury, N.; O'Kane, P.; Dicker, J.; Skidmore, C.; Knights, D. Recycling rubber tyres and waste plastics in EAF steelmaking. *Steel Res. Int.* **2011**, *82*, 566–572. [CrossRef]
2. Ramarad, S.; Khalid, M.; Ratnam, C.; Chuah, A.L.; Rashmi, W. Waste tire rubber in polymer blends: A review on the evolution, properties and future. *Prog. Mater. Sci.* **2015**, *72*, 100–140. [CrossRef]
3. Archibong, F.N.; Sanusi, O.M.; Médéric, P.; Hocine, N.A. An overview on the recycling of waste ground tyre rubbers in thermoplastic matrices: Effect of added fillers. *Resour. Conserv. Recycl.* **2021**, *175*, 105894–105906. [CrossRef]
4. Zhang, F.; Zhao, Y.; Wang, D.; Yan, M.; Zhang, J.; Zhang, P.; Ding, T.; Chen, L.; Chen, C. Current technologies for plastic waste treatment: A review. *J. Clean. Prod.* **2021**, *282*, 124523. [CrossRef]
5. Tang, Z.; Li, W.; Tam, V.W.; Xue, C. Advanced progress in recycling municipal and construction solid wastes for manufacturing sustainable construction materials. *Resour. Conserv. Recycl. X* **2020**, *6*, 100036. [CrossRef]
6. United States Tire Manufacturers Association. *US Scrap Tire Management Summary*; U. T. M. Association: Washington, DC, USA, 2018.
7. Canadian Association of Tire Recycling Agencies. *CATRA Annual Report*; Canadian Association of Tire Recycling Agencies: Toronto, ON, Canada, 2017.
8. European Tyre and Rubber Manufacturers Association. *End-of-life Tyre Report Europe*; European Tyre and Rubber Manufacturers Association: Brussels, Belgium, 2015.
9. Mountjoy, E.; Hasthanayake, D.; Freeman, T. *Stocks & Fate of End of Life Tyres-2013–2014 Study*; Hyder Consulting Pty Ltd.: Melbourne, Australia, 2015.
10. Landi, D.; Gigli, S.; Germani, M.; Marconi, M. Investigating the feasibility of a reuse scenario for textile fibres recovered from end-of-life tyres. *Waste Manag.* **2018**, *75*, 187–204. [CrossRef]

11. Fazli, A.; Rodrigue, D. Waste Rubber Recycling: A Review on the Evolution and Properties of Thermoplastic Elastomers. *Materials* **2020**, *13*, 782. [CrossRef] [PubMed]
12. Chen, J.; Chen, K.; Tong, L. On the pyrolysis kinetics of scrap automotive tires. *J. Hazard. Mater.* **2001**, *84*, 43–55. [CrossRef]
13. Fazli, A.; Rodrigue, D. Recycling waste tires into ground tire rubber (GTR)/rubber compounds: A review. *J. Compos. Sci.* **2020**, *4*, 103. [CrossRef]
14. Forrest, M.J. *Recycling and Re-Use of Waste Rubber*, 2nd ed.; De Gruyter: Berlin, Germany, 2019.
15. Yerezhep, D.; Tychengulova, A.; Sokolov, D.; Aldiyarov, A. A multifaceted approach for cryogenic waste tire recycling. *Polymers* **2021**, *13*, 2494. [CrossRef]
16. Awolusi, T.F.; Oke, O.L.; Atoyebi, O.D.; Akinkurolere, O.O.; Sojobi, A.O. Waste tires steel fiber in concrete: A review. *Innov. Infrastruct. Solut.* **2021**, *6*, 1–12. [CrossRef]
17. Ferdous, W.; Manalo, A.; Siddique, R.; Mendis, P.; Zhuge, Y.; Wong, H.S.; Lokuge, W.; Aravinthan, T.; Schubel, P. Recycling of landfill wastes (tyres, plastics and glass) in construction–A review on global waste generation, performance, application and future opportunities. *Resour. Conserv. Recycl.* **2021**, *173*, 105745. [CrossRef]
18. Karger-Kocsis, J.; Mészáros, L.; Bárány, T. Ground tyre rubber (GTR) in thermoplastics, thermosets, and rubbers. *J. Mater. Sci.* **2013**, *48*, 1–38. [CrossRef]
19. Levendis, Y.A.; Atal, A.; Carlson, J.; Dunayevskiy, Y.; Vouros, P. Comparative study on the combustion and emissions of waste tire crumb and pulverized coal. *Environ. Sci. Technol.* **1996**, *30*, 2742–2754. [CrossRef]
20. Williams, P.T. Pyrolysis of waste tyres: A review. *Waste Manag.* **2013**, *33*, 1714–1728. [CrossRef]
21. Sathiskumar, C.; Karthikeyan, S. Recycling of waste tires and its energy storage application of by-products–a review. *Sustain. Mater. Technol.* **2019**, *22*, 00125–00130. [CrossRef]
22. Tang, L.; Huang, H. An investigation of sulfur distribution during thermal plasma pyrolysis of used tires. *J. Anal. Appl. Pyrolysis* **2004**, *72*, 35–40. [CrossRef]
23. Medina, N.F.; Garcia, R.; Hajirasouliha, I.; Pilakoutas, K.; Guadagnini, M.; Raffoul, S. Composites with recycled rubber aggregates: Properties and opportunities in construction. *Constr. Build. Mater.* **2018**, *188*, 884–897. [CrossRef]
24. Zanchet, A.; Carli, L.; Giovanela, M.; Brandalise, R.; Crespo, J. Use of styrene butadiene rubber industrial waste devulcanized by microwave in rubber composites for automotive application. *Mater. Des.* **2012**, *39*, 437–443. [CrossRef]
25. Siddika, A.; al Mamun, M.A.; Alyousef, R.; Amran, Y.M.; Aslani, F.; Alabduljabbar, H. Properties and utilizations of waste tire rubber in concrete: A review. *Constr. Build. Mater.* **2019**, *224*, 711–731. [CrossRef]
26. Bilema, M.; Aman, M.; Hassan, N.; Haloul, M.; Modibbo, S. Influence of crumb rubber size particles on moisture damage and strength of the hot mix asphalt. *Mater. Today Proc.* **2021**, *42*, 2387–2391. [CrossRef]
27. Milad, A.; Ahmeda, A.G.; Taib, A.M.; Rahmad, S.; Solla, M.; Yusoff, N.I.M. A review of the feasibility of using crumb rubber derived from end-of-life tire as asphalt binder modifier. *J. Rubber Res.* **2020**, *23*, 203–216. [CrossRef]
28. Alfayez, S.A.; Suleiman, A.R.; Nehdi, M.L. Recycling tire rubber in asphalt pavements: State of the art. *Sustainability* **2020**, *12*, 9076. [CrossRef]
29. Moreno-Navarro, F.; Sol-Sánchez, M.; Rubio-Gámez, M. Reuse of deconstructed tires as anti-reflective cracking mat systems in asphalt pavements. *Constr. Build. Mater.* **2014**, *53*, 182–189. [CrossRef]
30. Naseem, A.; Mumtaz, W.; de Backer, H. Stabilization of expansive soil using tire rubber powder and cement kiln dust. *Soil. Mech. Found. Eng.* **2019**, *56*, 54–58. [CrossRef]
31. Fernández, A.; Barriocanal, C.; Alvarez, R. Pyrolysis of a waste from the grinding of scrap tyres. *J. Hazard. Mater.* **2012**, *203*, 236–243. [CrossRef] [PubMed]
32. Adhikari, J.; Das, A.; Sinha, T.; Saha, P.; Kim, J.K. Grinding of Waste Rubber. In *Grinding of Waste Rubber: Royal Society of Chemistry*; Kim, K.J., Saha, P., Thomas, S., Haponiuk, J.T., Aswathi, M.K., Eds.; Royal Society of Chemistry (RSC): London, UK, 2018; pp. 1–23. [CrossRef]
33. Thai, Q.B.; Le-Cao, K.; Nguyen, P.T.; Le, P.K.; Phan-Thien, N.; Duong, H.M. Fabrication and optimization of multifunctional nanoporous aerogels using recycled textile fibers from car tire wastes for oil-spill cleaning, heat-insulating and sound absorbing applications. *Colloids Surf. Physicochem. Eng. Asp.* **2021**, *628*, 127363–127378. [CrossRef]
34. Maderuelo-Sanz, R.; Nadal-Gisbert, A.V.; Crespo-Amorós, J.E.; Parres-García, F. A novel sound absorber with recycled fibers coming from end of life tires (ELTs). *Appl. Acoust.* **2012**, *73*, 402–408. [CrossRef]
35. Medina, N.F.; Medina, D.F.; Hernández-Olivares, F.; Navacerrada, M. Mechanical and thermal properties of concrete incorporating rubber and fibres from tyre recycling. *Constr. Build. Mater.* **2017**, *144*, 563–573. [CrossRef]
36. Onuaguluchi, O.; Banthia, N. Value-added reuse of scrap tire polymeric fibers in cement-based structural applications. *J. Clean. Prod.* **2019**, *231*, 543–555. [CrossRef]
37. Abbaspour, M.; Aflaki, E.; Nejad, F.M. Reuse of waste tire textile fibers as soil reinforcement. *J. Clean. Prod.* **2019**, *207*, 1059–1071. [CrossRef]
38. Chen, M.; Zhong, H.; Zhang, M. Flexural fatigue behaviour of recycled tyre polymer fibre reinforced concrete. *Cem. Concr. Compos.* **2020**, *105*, 103441–103483. [CrossRef]
39. Baričević, A.; Rukavina, M.J.; Pezer, M.; Štirmer, N. Influence of recycled tire polymer fibers on concrete properties. *Cem. Concr. Compos.* **2018**, *91*, 29–41. [CrossRef]

40. Narani, S.; Abbaspour, M.; Hosseini, S.M.M.; Aflaki, E.; Nejad, F.M. Sustainable reuse of Waste Tire Textile Fibers (WTTFs) as reinforcement materials for expansive soils: With a special focus on landfill liners/covers. *J. Clean. Prod.* **2020**, *247*, 119151. [CrossRef]
41. Bekhiti, M.; Trouzine, H.; Rabehi, M. Influence of waste tire rubber fibers on swelling behavior, unconfined compressive strength and ductility of cement stabilized bentonite clay soil. *Constr. Build. Mater.* **2019**, *208*, 304–313. [CrossRef]
42. Onuaguluchi, O.; Banthia, N. Durability performance of polymeric scrap tire fibers and its reinforced cement mortar. *Mater. Struct.* **2017**, *50*, 1–10. [CrossRef]
43. Moghaddamzadeh, S.; Rodrigue, D. The effect of polyester recycled tire fibers mixed with ground tire rubber on polyethylene composites. Part I: Morphological analysis. *Prog. Rubber Plast. Recycl.* **2018**, *34*, 200–220. [CrossRef]
44. Fazli, A.; Rodrigue, D. Phase morphology, mechanical, and thermal properties of fiber-reinforced thermoplastic elastomer: Effects of blend composition and compatibilization. *J. Reinf. Plast. Compos.* **2022**, *41*, 267–283. [CrossRef]
45. Acevedo, B.; Fernández, A.; Barriocanal, C. Identification of polymers in waste tyre reinforcing fibre by thermal analysis and pyrolysis. *J. Anal. Appl. Pyrolysis* **2015**, *111*, 224–232. [CrossRef]
46. Wong, S.; Sui, G.; Yue, C.; Mai, Y.-W. Characterization of microstructures and toughening behavior of fiber-containing toughened nylon 6, 6. *J. Mater. Sci.* **2002**, *37*, 2659–2667. [CrossRef]
47. Abbaspour, M.; Narani, S.; Aflaki, E.; Nejad, F.M.; Hosseini, S.M.M. Strength and swelling properties of a waste tire textile fiber-reinforced expansive soil. *Geosynth. Int.* **2020**, *27*, 476–489. [CrossRef]
48. Putman, B.J.; Amirkhanian, S.N. Utilization of waste fibers in stone matrix asphalt mixtures. *Resour. Conserv. Recycl.* **2004**, *42*, 265–274. [CrossRef]
49. Bockstal, L.; Berchem, T.; Schmetz, Q.; Richel, A. Devulcanisation and reclaiming of tires and rubber by physical and chemical processes: A review. *J. Clean. Prod.* **2019**, *236*, 117574. [CrossRef]
50. Firoozi, A.A.; Olgun, C.G.; Firoozi, A.A.; Baghini, M.S. Fundamentals of soil stabilization. *Int. J. Geo-Eng.* **2017**, *8*, 1–16. [CrossRef]
51. Kianimehr, M.; Shourijeh, P.T.; Binesh, S.M.; Mohammadinia, A.; Arulrajah, A. Utilization of recycled concrete aggregates for light-stabilization of clay soils. *Constr. Build. Mater.* **2019**, *227*, 116792. [CrossRef]
52. Phanikumar, B.; Shankar, M.U. Studies on hydraulic conductivity of fly ash-stabilised expansive clay liners. *Geotech. Geol. Eng.* **2016**, *34*, 449–462. [CrossRef]
53. Cabalar, A.F.; Karabash, Z.; Mustafa, W.S. Stabilising a clay using tyre buffings and lime. *Road Mater. Pavement Des.* **2014**, *15*, 872–891. [CrossRef]
54. Zhang, M.; Guo, H.; El-Korchi, T.; Zhang, G.; Tao, M. Experimental feasibility study of geopolymer as the next-generation soil stabilizer. *Constr. Build. Mater.* **2013**, *47*, 1468–1478. [CrossRef]
55. Khemissa, M.; Mahamedi, A. Cement and lime mixture stabilization of an expansive overconsolidated clay. *Appl. Clay Sci.* **2014**, *95*, 104–110. [CrossRef]
56. Tyagi, A.; Soni, D. Effects of Granulated Ground Blast Furnace Slag and Fly Ash on Stabilization of Soil. Agnihotri, A., Reddy, K., Bansal, A., Eds.; Springer: Singapore, 2019; pp. 79–90. [CrossRef]
57. Thyagaraj, T.; Soujanya, D. Polypropylene fiber reinforced bentonite for waste containment barriers. *Appl. Clay Sci.* **2017**, *142*, 153–162. [CrossRef]
58. Chaduvula, U.; Viswanadham, B.; Kodikara, J. A study on desiccation cracking behavior of polyester fiber-reinforced expansive clay. *Appl. Clay Sci.* **2017**, *142*, 163–172. [CrossRef]
59. Yao, X.; Huang, G.; Wang, M.; Dong, X. Mechanical properties and microstructure of PVA fiber reinforced cemented soil. *KSCE J. Civ. Eng.* **2021**, *25*, 482–491. [CrossRef]
60. Mirzababaei, M.; Mohamed, M.; Arulrajah, A.; Horpibulsuk, S.; Anggraini, V. Practical approach to predict the shear strength of fibre-reinforced clay. *Geosynth. Int.* **2018**, *25*, 50–66. [CrossRef]
61. Rabab'ah, S.; al Hattamleh, O.; Aldeeky, H.; Alfoul, B.A. Effect of glass fiber on the properties of expansive soil and its utilization as subgrade reinforcement in pavement applications. *Case Stud. Constr. Mater.* **2021**, *14*, 00485–00496. [CrossRef]
62. Bao, X.; Huang, Y.; Jin, Z.; Xiao, X.; Tang, W.; Cui, H.; Chen, X. Experimental investigation on mechanical properties of clay soil reinforced with carbon fiber. *Constr. Build. Mater.* **2021**, *280*, 122517. [CrossRef]
63. Ghadakpour, M.; Choobbasti, A.J.; Kutanaei, S.S. Investigation of the Kenaf fiber hybrid length on the properties of the cement-treated sandy soil. *Transp. Geotech.* **2020**, *22*, 100301. [CrossRef]
64. Karatai, T.R.; Kaluli, J.W.; Kabubo, C.; Thiong'o, G. Soil stabilization using rice husk ash and natural lime as an alternative to cutting and filling in road construction. *J. Constr. Eng. Manag.* **2017**, *143*, 04016127. [CrossRef]
65. Syed, M.; GuhaRay, A. Effect of natural fiber reinforcement on strength response of alkali activated binder treated expansive soil: Experimental investigation and reliability analysis. *Constr. Build. Mater.* **2021**, *273*, 121743. [CrossRef]
66. Dang, L.C.; Fatahi, B.; Khabbaz, H. Behaviour of expansive soils stabilized with hydrated lime and bagasse fibres. *Procedia Eng.* **2016**, *143*, 658–665. [CrossRef]
67. Kafodya, I.; Okonta, F. Effects of natural fiber inclusions and pre-compression on the strength properties of lime-fly ash stabilised soil. *Constr. Build. Mater.* **2018**, *170*, 737–746. [CrossRef]
68. Yixian, W.; Panpan, G.; Shengbiao, S.; Haiping, Y.; Binxiang, Y. Study on strength influence mechanism of fiber-reinforced expansive soil using jute. *Geotech. Geol. Eng.* **2016**, *34*, 1079–1088. [CrossRef]

69. Yoon, Y.W.; Heo, S.B.; Kim, K.S. Geotechnical performance of waste tires for soil reinforcement from chamber tests. *Geotext. Geomembr.* **2008**, *26*, 100–107. [CrossRef]
70. Valipour, M.; Shourijeh, P.T.; Mohammadinia, A. Application of recycled tire polymer fibers and glass fibers for clay reinforcement. *Transp. Geotech.* **2021**, *27*, 100474. [CrossRef]
71. Olgun, M. Effects of polypropylene fiber inclusion on the strength and volume change characteristics of cement-fly ash stabilized clay soil. *Geosynth. Int.* **2013**, *20*, 263–275. [CrossRef]
72. Wang, Y.-X.; Guo, P.-P.; Ren, W.-X.; Yuan, B.-X.; Yuan, H.-P.; Zhao, Y.-L.; Shan, S.-B.; Cao, P. Laboratory investigation on strength characteristics of expansive soil treated with jute fiber reinforcement. *Int. J. Geomech.* **2017**, *17*, 04017101. [CrossRef]
73. Tabakouei, A.R.; Narani, S.; Abbaspour, M.; Aflaki, E.; Siddiqua, S. Coupled specimen and fiber dimensions influence measurement on the properties of fiber-reinforced soil. *Measurement* **2022**, *188*, 110556. [CrossRef]
74. Narani, S.; Abbaspour, M.; Hosseini, S.M.M.; Nejad, F.M. Long-term dynamic behavior of a sandy subgrade reinforced by Waste Tire Textile Fibers (WTTFs). *Transp. Geotech.* **2020**, *24*, 100375. [CrossRef]
75. Abbaspour, M.; Narani, S.S.; Aflaki, E.; Nejad, F.M. Behavior of a subgrade soil reinforced by waste tire textile fibers under static and cyclic loading. *J. Mater. Civ. Eng.* **2020**, *32*, 04020208. [CrossRef]
76. Abbaspour, M.; Narani, S.S.; Aflaki, E.; Nejad, F.M. Dynamic characteristics of a sandy subgrade reinforced by waste tire textile fibres. *Int. J. Pavement Eng.* **2020**, *23*, 2293–2308. [CrossRef]
77. Mohammadifar, L.; Miraki, H.; Rahmani, A.; Jahandari, S.; Mehdizadeh, B.; Rasekh, H.; Samadi, P.; Samali, B. Properties of Lime-Cement Concrete Containing Various Amounts of Waste Tire Powder under Different Ground Moisture Conditions. *Polymers* **2022**, *14*, 482. [CrossRef]
78. Angelin, A.F.; da Silva, F.M.; Barbosa, L.A.; Lintz, R.C.; de Carvalho, M.A.; Franco, R.A.S. Voids identification in rubberized mortar digital images using K-Means and Watershed algorithms. *J. Clean. Prod.* **2017**, *164*, 455–464. [CrossRef]
79. Onuaguluchi, O. Effects of surface pre-coating and silica fume on crumb rubber-cement matrix interface and cement mortar properties. *J. Clean. Prod.* **2015**, *104*, 339–345. [CrossRef]
80. Mucsi, G.; Szenczi, Á.; Nagy, S. Fiber reinforced geopolymer from synergetic utilization of fly ash and waste tire. *J. Clean. Prod.* **2018**, *178*, 429–440. [CrossRef]
81. Song, W.; Yin, J. Hybrid effect evaluation of steel fiber and carbon fiber on the performance of the fiber reinforced concrete. *Materials* **2016**, *9*, 704. [CrossRef]
82. Zhao, Q.; Yu, J.; Geng, G.; Jiang, J.; Liu, X. Effect of fiber types on creep behavior of concrete. *Constr. Build. Mater.* **2016**, *105*, 416–422. [CrossRef]
83. Wang, J.; Dai, Q.; Si, R.; Guo, S. Mechanical durability, and microstructural properties of macro synthetic polypropylene (PP) fiber-reinforced rubber concrete. *J. Clean. Prod.* **2019**, *234*, 1351–1364. [CrossRef]
84. Yuan, T.-F.; Lee, J.-Y.; Min, K.-H.; Yoon, Y.-S. Experimental Investigation on Mechanical Properties of Hybrid Steel and Polyethylene Fiber-Reinforced No-Slump High-Strength Concrete. *Int. J. Polym. Sci.* **2019**, *2019*, 4737384. [CrossRef]
85. Banthia, N.; Onuaguluchi, O. Recycling Scrap Tire-Derived Fibers in Concrete. *Trans. Indian Natl. Acad. Eng.* **2022**, *7*, 207–217. [CrossRef]
86. Yu, Z.; Wang, Y.; Li, J. Performance Investigation and Cost–Benefit Analysis of Recycled Tire Polymer Fiber-Reinforced Cemented Paste Backfill. *Polymers* **2022**, *14*, 708. [CrossRef]
87. Banthia, N.; Gupta, R. Influence of polypropylene fiber geometry on plastic shrinkage cracking in concrete. *Cem. Concr. Res.* **2006**, *36*, 1263–1267. [CrossRef]
88. Pacheco-Torres, R.; Cerro-Prada, E.; Escolano, F.; Varela, F. Fatigue performance of waste rubber concrete for rigid road pavements. *Constr. Build. Mater.* **2018**, *176*, 539–548. [CrossRef]
89. Serdar, M.; Baričević, A.; Rukavina, M.J.; Pezer, M.; Bjegović, D.; Štirmer, N. Shrinkage behaviour of fibre reinforced concrete with recycled tyre polymer fibres. *Int. J. Polym. Sci.* **2015**, *215*, 145918. [CrossRef]
90. Chen, M.; Zhong, H.; Chen, L.; Zhang, Y.; Zhang, M. Engineering properties and sustainability assessment of recycled fibre reinforced rubberised cementitious composite. *J. Clean. Prod.* **2021**, *278*, 123996–124010. [CrossRef]
91. Ranjbar, N.; Zhang, M. Fiber-reinforced geopolymer composites: A review. *Cem. Concr. Compos.* **2020**, *107*, 103498. [CrossRef]
92. Walraven, J.C. High performance fiber reinforced concrete: Progress in knowledge and design codes. *Mater. Struct.* **2009**, *42*, 1247–1260. [CrossRef]
93. Ríos, J.D.; Cifuentes, H.; Yu, R.C.; Ruiz, G. Probabilistic flexural fatigue in plain and fiber-reinforced concrete. *Materials* **2017**, *10*, 767. [CrossRef]
94. Pais, J.C.; Santos, C.R.; Presti, D.L. Application of textile fibres from tire recycling in asphalt mixtures. *Road Mater. Pavement Des.* **2021**, *23*, 1–22. [CrossRef]
95. Jin, D.; Ge, D.; Zhou, X.; You, Z. Asphalt Mixture with Scrap Tire Rubber and Nylon Fiber from Waste Tires: Laboratory Performance and Preliminary ME Design Analysis. *Buildings* **2022**, *12*, 160. [CrossRef]
96. Ling, S.; Yu, F.; Sun, D.; Sun, G.; Xu, L. A comprehensive review of tire-pavement noise: Generation mechanism, measurement methods, and quiet asphalt pavement. *J. Clean. Prod.* **2021**, *287*, 125056. [CrossRef]
97. Wong, C.C.; Wong, W.-G. Effect of crumb rubber modifiers on high temperature susceptibility of wearing course mixtures. *Constr. Build. Mater.* **2007**, *21*, 1741–1745. [CrossRef]

98. Valdés-Vidal, G.; Calabi-Floody, A.; Duarte-Nass, C.; Mignolet, C.; Díaz, C. Development of a New Additive Based on Textile Fibers of End-of-Life Tires (ELT) for Sustainable Asphalt Mixtures with Improved Mechanical Properties. *Polymers* **2022**, *14*, 3250. [CrossRef] [PubMed]
99. Landi, D.; Marconi, M.; Bocci, E.; Germani, M. Comparative life cycle assessment of standard, cellulose-reinforced and end of life tires fiber-reinforced hot mix asphalt mixtures. *J. Clean. Prod.* **2020**, *248*, 119295. [CrossRef]
100. Bocci, E.; Prosperi, E. Recycling of reclaimed fibers from end-of-life tires in hot mix asphalt. *J. Traffic Transp. Eng.* **2020**, *7*, 678–687. [CrossRef]
101. Moreno, F.; Rubio, M.; Martinez-Echevarria, M. Analysis of digestion time and the crumb rubber percentage in dry-process crumb rubber modified hot bituminous mixes. *Constr. Build. Mater.* **2011**, *25*, 2323–2334. [CrossRef]
102. Chen, H.; Xu, Q. Experimental study of fibers in stabilizing and reinforcing asphalt binder. *Fuel* **2010**, *89*, 1616–1622. [CrossRef]
103. Rubio, M.C.; Martínez, G.; Baena, L.; Moreno, F. Warm mix asphalt: An overview. *J. Clean. Prod.* **2012**, *24*, 76–84. [CrossRef]
104. Schiavoni, S.; Bianchi, F.; Asdrubali, F. Insulation materials for the building sector: A review and comparative analysis. *Renew. Sust. Energ. Rev.* **2016**, *62*, 988–1011. [CrossRef]
105. Thai, Q.B.; Chong, R.O.; Nguyen, P.T.; Le, D.K.; Le, P.K.; Phan-Thien, N.; Duong, H.M. Recycling of waste tire fibers into advanced aerogels for thermal insulation and sound absorption applications. *J. Environ. Chem. Eng.* **2020**, *8*, 104279. [CrossRef]
106. Cao, L.; Fu, Q.; Si, Y.; Ding, B.; Yu, J. Porous materials for sound absorption. *Compos. Commun.* **2018**, *10*, 25–35. [CrossRef]
107. Baetens, R.; Jelle, B.P.; Gustavsen, A. Aerogel insulation for building applications: A state-of-the-art review. *Energy Build.* **2011**, *43*, 761–769. [CrossRef]
108. Gesser, H.; Goswami, P. Aerogels and related porous materials. *Chem. Rev.* **1989**, *89*, 765–788. [CrossRef]
109. Thai, Q.B.; Siang, T.E.; Le, D.K.; Shah, W.A.; Phan-Thien, N.; Duong, H.M. Advanced fabrication and multi-properties of rubber aerogels from car tire waste. *Colloids Surf. Physicochem. Eng. Asp.* **2019**, *577*, 702–708. [CrossRef]
110. Nguyen, S.T.; Feng, J.; Ng, S.K.; Wong, J.P.; Tan, V.B.; Duong, H.M. Advanced thermal insulation and absorption properties of recycled cellulose aerogels. *Colloids Surf. Physicochem. Eng. Asp.* **2014**, *445*, 128–134. [CrossRef]
111. Cao, L.; Si, Y.; Wu, Y.; Wang, X.; Yu, J.; Ding, B. Ultralight, superelastic and bendable lashing-structured nanofibrous aerogels for effective sound absorption. *Nanoscale* **2019**, *11*, 2289–2298. [CrossRef] [PubMed]
112. Sorour, M.H.; Hani, H.A.; Al-Bazedi, G.A.; El-Rafei, A. Hydrophobic silica aerogels for oil spills clean-up, synthesis, characterization and preliminary performance evaluation. *J. Porous Mater.* **2016**, *23*, 1401–1409. [CrossRef]
113. Wang, D.; McLaughlin, E.; Pfeffer, R.; Lin, Y. Adsorption of oils from pure liquid and oil–water emulsion on hydrophobic silica aerogels. *Sep. Purif. Technol.* **2012**, *99*, 28–35. [CrossRef]
114. Thai, Q.B.; Le, D.K.; Do, N.H.; Le, P.K.; Phan-Thien, N.; Wee, C.Y.; Duong, H.M. Advanced aerogels from waste tire fibers for oil spill-cleaning applications. *J. Environ. Chem. Eng.* **2020**, *8*, 104016. [CrossRef]
115. Dorcheh, A.S.; Abbasi, M. Silica aerogel, synthesis, properties and characterization. *J. Mater. Process. Technol.* **2008**, *199*, 10–26. [CrossRef]
116. Thai, Q.B.; Nguyen, S.T.; Ho, D.K.; Tran, T.D.; Huynh, D.M.; Do, N.H.; Luu, T.P.; Le, P.K.; Le, D.K.; Phan-Thien, N. Cellulose-based aerogels from sugarcane bagasse for oil spill-cleaning and heat insulation applications. *Carbohydr. Polym.* **2020**, *228*, 115365–115371. [CrossRef]
117. Shi, M.; Tang, C.; Yang, X.; Zhou, J.; Jia, F.; Han, Y.; Li, Z. Superhydrophobic silica aerogels reinforced with polyacrylonitrile fibers for adsorbing oil from water and oil mixtures. *RSC Adv.* **2017**, *7*, 4039–4045. [CrossRef]
118. Chang, C.; Zhang, L. Cellulose-based hydrogels: Present status and application prospects. *Carbohydr. Polym.* **2011**, *84*, 40–53. [CrossRef]
119. Shi, J.; Jiang, K.; Ren, D.; Zou, H.; Wang, Y.; Lv, X.; Zhang, L. Structure and performance of reclaimed rubber obtained by different methods. *J. Appl. Polym. Sci.* **2013**, *129*, 999–1007. [CrossRef]
120. Sakai, K.; Kobayashi, Y.; Saito, T.; Isogai, A. Partitioned airs at microscale and nanoscale: Thermal diffusivity in ultrahigh porosity solids of nanocellulose. *Sci. Rep.* **2016**, *6*, 1–7. [CrossRef]
121. Pham, D.; Phan-Thien, N. On the optimal bounds for the effective conductivity of isotropic quasi-symmetric multiphase media. *J. Appl. Math. Phys.* **1997**, *48*, 744–759. [CrossRef]
122. Xie, X.; Li, D.; Tsai, T.-H.; Liu, J.; Braun, P.V.; Cahill, D.G. Thermal conductivity, heat capacity, and elastic constants of water-soluble polymers and polymer blends. *Macromolecules* **2016**, *49*, 972–978. [CrossRef]
123. Faruk, O.; Bledzki, A.K.; Fink, H.P.; Sain, M. Progress report on natural fiber reinforced composites. *Macromol. Mater. Eng.* **2014**, *299*, 9–26. [CrossRef]
124. Kakroodi, A.R.; Kazemi, Y.; Rodrigue, D. Mechanical, rheological, morphological and water absorption properties of maleated polyethylene/hemp composites: Effect of ground tire rubber addition. *Compos. B. Eng* **2013**, *51*, 337–344. [CrossRef]
125. Ghasemzadeh-Barvarz, M.; Duchesne, C.; Rodrigue, D. Mechanical, water absorption, and aging properties of polypropylene/flax/glass fiber hybrid composites. *J. Compos. Mater.* **2015**, *49*, 3781–3798. [CrossRef]
126. Lokesh, P.; Kumari, T.S.; Gopi, R.; Loganathan, G.B. A study on mechanical properties of bamboo fiber reinforced polymer composite. *Mater. Today Proc.* **2020**, *22*, 897–903. [CrossRef]
127. Sever, K.; Sarikanat, M.; Seki, Y.; Erkan, G.; Erdoğan, Ü.H.; Erden, S. Surface treatments of jute fabric: The influence of surface characteristics on jute fabrics and mechanical properties of jute/polyester composites. *Ind. Crop. Prod.* **2012**, *35*, 22–30. [CrossRef]

128. Fávaro, S.L.; Lopes, M.S.; Neto, A.G.V.d.; de Santana, R.R.; Radovanovic, E. Chemical, morphological, and mechanical analysis of rice husk/post-consumer polyethylene composites. *Compos. Part A Appl. Sci.* **2010**, *41*, 154–160. [CrossRef]
129. Fernandes, E.M.; Correlo, V.M.; Mano, J.F.; Reis, R.L. Novel cork–polymer composites reinforced with short natural coconut fibres: Effect of fibre loading and coupling agent addition. *Compos. Sci. Technol.* **2013**, *78*, 56–62. [CrossRef]
130. Saha, A.; Kumar, S.; Kumar, A. Influence of pineapple leaf particulate on mechanical, thermal and biodegradation characteristics of pineapple leaf fiber reinforced polymer composite. *J. Polym. Res.* **2021**, *28*, 1–23. [CrossRef]
131. Chimeni, D.Y.; Toupe, J.L.; Dubois, C.; Rodrigue, D. Effect of hemp surface modification on the morphological and tensile properties of linear medium density polyethylene (LMDPE) composites. *Compos. Interfaces* **2016**, *23*, 405–421. [CrossRef]
132. Balla, V.K.; Kate, K.H.; Satyavolu, J.; Singh, P.; Tadimeti, J.G.D. Additive manufacturing of natural fiber reinforced polymer composites: Processing and prospects. *Compos. Part B Eng.* **2019**, *174*, 106956. [CrossRef]
133. Mahmud, S.; Hasan, K.; Jahid, M.; Mohiuddin, K.; Zhang, R.; Zhu, J. Comprehensive review on plant fiber-reinforced polymeric biocomposites. *J. Mater. Sci.* **2021**, *56*, 7231–7264. [CrossRef]
134. DiBenedetto, A. Tailoring of interfaces in glass fiber reinforced polymer composites: A review. *Mater. Sci. Eng. A* **2001**, *302*, 74–82. [CrossRef]
135. Zhang, X.X.; Lu, C.H.; Liang, M. Preparation of rubber composites from ground tire rubber reinforced with waste-tire fiber through mechanical milling. *J. Appl. Polym. Sci.* **2007**, *103*, 4087–4094. [CrossRef]
136. Marconi, M.; Landi, D.; Meo, I.; Germani, M. Reuse of tires textile fibers in plastic compounds: Is this scenario environmentally sustainable? *Procedia Cirp* **2018**, *69*, 944–949. [CrossRef]
137. Anthony, W.S. Separation of Crumb and Fiber in Tire Recycling Operations. In *ASAE Annual Meeting*; American Society of Agricultural and Biological Engineers: St. Joseph, MI, USA, 2005.
138. Fazli, A.; Rodrigue, D. Effect of ground tire rubber (GTR) particle size and content on the morphological and mechanical properties of recycled high-density polyethylene (rHDPE)/GTR blends. *Recycling* **2021**, *6*, 44. [CrossRef]
139. Fazli, A.; Rodrigue, D. Thermoplastic elastomers based on recycled high-density polyethylene/ground tire rubber/ethylene vinyl acetate: Effect of ground tire rubber regeneration on morphological and mechanical properties. *J. Thermoplast. Compos. Mater.* **2022**, *35*, 08927057221095388. [CrossRef]
140. Moghaddamzadeh, S.; Rodrigue, D. The effect of polyester recycled tire fibers mixed with ground tire rubber on polyethylene composites. Part II: Physico-mechanical analysis. *Prog. Rubber Plast. Recycl.* **2018**, *34*, 128–142. [CrossRef]
141. Ferreira, C.T.; Perez, C.A.; Hirayama, D.; Saron, C. Recycling of polyamide (PA) from scrap tires as composites and blends. *J. Environ. Chem. Eng.* **2013**, *1*, 762–767. [CrossRef]
142. Fazli, A.; Rodrigue, D. Thermoplastic Elastomer based on Recycled HDPE/Ground Tire Rubber Interfacially Modified with an Elastomer: Effect of Mixing Sequence and Elastomer Type/Content. *Polym. Plast. Technol. Eng.* **2022**, *61*, 1021–1038. [CrossRef]
143. Fazli, A.; Stevanovic, T.; Rodrigue, D. Recycled HDPE/Natural Fiber Composites Modified with Waste Tire Rubber: A Comparison between Injection and Compression Molding. *Polymers* **2022**, *14*, 3197. [CrossRef]
144. Fazli, A.; Rodrigue, D. Morphological and Mechanical Properties of Thermoplastic Elastomers Based on Recycled High Density Polyethylene and Recycled Natural Rubber. *Int. Polym. Proc.* **2021**, *36*, 156–164. [CrossRef]
145. Simon-Stőger, L.; Varga, C. PE-contaminated industrial waste ground tire rubber: How to transform a handicapped resource to a valuable one. *Waste Manag.* **2021**, *119*, 111–121. [CrossRef]
146. Moghaddamzadeh, S.; Rodrigue, D. Rheological characterization of polyethylene/polyester recycled tire fibers/ground tire rubber composites. *J. Appl. Polym. Sci.* **2018**, *135*, 46563–46571. [CrossRef]
147. Verma, D.; Goh, K.L.; Vimal, V. Interfacial Studies of Natural Fiber-Reinforced Particulate Thermoplastic Composites and Their Mechanical Properties. *J. Nat. Fibers* **2020**, *19*, 2299–2326. [CrossRef]
148. Kakroodi, A.R.; Bainier, J.; Rodrigue, D. Mechanical and morphological properties of flax fiber reinforced high density polyethylene/recycled rubber composites. *Int. Polym. Proc.* **2012**, *27*, 196–204. [CrossRef]

polymers

MDPI

Article

The Preservative Action of Protein Hydrolysates from Legume Seed Waste on Fresh Meat Steak at 4 °C: Limiting Unwanted Microbial and Chemical Fluctuations

Eman T. Abou Sayed-Ahmed [1], Karima Bel Hadj Salah [2,3], Rasha M. El-Mekkawy [4], Nourhan A. Rabie [1], Mada F. Ashkan [2], Soha A. Alamoudi [2], Mohammed H. Alruhaili [5], Soad K. Al Jaouni [6], Mohammed S. Almuhayawi [5], Samy Selim [7,*], Ahmed M. Saad [8,*] and Mohammad Namir [1]

1 Department of Food Science, Faculty of Agriculture, Zagazig University, Zagazig 44511, Egypt
2 Biological Sciences Department, College of Science & Arts, King Abdulaziz University, Rabigh 21911, Saudi Arabia
3 Laboratory of Transmissible Diseases and Biologically Active Substances, Faculty of Pharmacy, University of Monastir, Monastir 5089, Tunisia
4 Department of Botany and Microbiology, Faculty of Science, Zagazig University, Zagazig 44511, Egypt
5 Medical Microbiology and Parasitology Department, Faculty of Medicine, King AbdulAziz University, Jeddah 21589, Saudi Arabia
6 Department of Hematology/Oncology, Yousef Abdulatif Jameel Scientific Chair of Prophetic Medicine Application, Faculty of Medicine, King Abdulaziz University, Jeddah 21589, Saudi Arabia
7 Department of Clinical Laboratory Sciences, College of Applied Medical Sciences, Jouf University, Sakaka 72388, Saudi Arabia
8 Biochemistry Department, Faculty of Agriculture, Zagazig University, Zagazig 44511, Egypt
* Correspondence: sabdulsalam@ju.edu.sa (S.S.); amsaadm@zu.edu.eg (A.M.S.)

Citation: Sayed-Ahmed, E.T.A.; Salah, K.B.H.; El-Mekkawy, R.M.; Rabie, N.A.; Ashkan, M.F.; Alamoudi, S.A.; Alruhaili, M.H.; Jaouni, S.K.A.; Almuhayawi, M.S.; Selim, S.; et al. The Preservative Action of Protein Hydrolysates from Legume Seed Waste on Fresh Meat Steak at 4 °C: Limiting Unwanted Microbial and Chemical Fluctuations. *Polymers* **2022**, *14*, 3188. https://doi.org/10.3390/polym14153188

Academic Editors: Cristina Cazan and Mihaela Cosnita

Received: 16 June 2022
Accepted: 28 July 2022
Published: 4 August 2022

Abstract: Valorizing agricultural wastes to preserve food or to produce functional food is a general trend regarding the global food shortage. Therefore, natural preservatives were developed from the seed waste of the cluster bean and the common bean to extend the shelf life of fresh buffalo meat steak and boost its quality via immersion in high-solubility peptides, cluster bean protein hydrolysate (CBH), and kidney bean protein hydrolysate (RCH). The CBH and the RCH were successfully obtained after 60 min of pepsin hydrolysis with a hydrolysis degree of 27–30%. The SDS-PAGE electropherogram showed that at 60 min of pepsin hydrolysis, the CBH bands disappeared, and RCH (11–48 kD bands) nearly disappeared, assuring the high solubility of the obtained hydrolysates. The CBH and the RCH have considerable antioxidant activity compared to ascorbic acid, antimicrobial activity against tested microorganisms compared to antibiotics, and significant functional properties. The CBH and the RCH (500 μg/mL) successfully scavenged 93 or 89% of DPPH radicals. During the 30-day cold storage (4 °C), the quality of treated and untreated fresh meat steaks was monitored. Protein hydrolysates (500 g/g) inhibited lipid oxidation by 130–153% compared to the control and nisin and eliminated 31–55% of the bacterial load. The CBH and the RCH (500 μg/g) significantly enhanced meat redness (a^* values). The protein maintained 80–90% of the steak's flavor and color ($p < 0.05$). In addition, it increased the juiciness of the steak. CBH and RCH are ways to valorize wastes that can be safely incorporated into novel foods.

Keywords: legume; wastes; enzymatic hydrolysis; protein; antioxidant; antimicrobial; buffalo meat steak; cold storage

1. Introduction

A short life distinguishes fresh products as they are an excellent medium for microbial growth and prone to lipid oxidation. The increase in microbial load and oxidation rate in food causes considerable economic losses and ethical problems worldwide. The deterioration in color and sensory traits of fresh meat is an obvious consequence of these problems [1]. Cutting fresh meat in steak form can effectively and efficiently ease the incorporation of additives that can enhance storage stability.

The Food and Agriculture Organization (FAO) revealed that food waste and food loss occur when "the quantity or quality of food decreases along the food supply chain" [2].

Food-spoiling microorganisms are the leading cause of food waste and loss. Approximately 20% of global meat production is lost due to microbial contamination. Controlling the population of microorganisms in animal products is one of the most important ways to reduce food waste [3]. In theory, increasing the levels of preservatives in food could be a solution to this problem. However, consumers dislike chemical preservatives that extend food shelf life [4]. As a result, there is a demand for natural alternatives to chemical preservatives.

Antimicrobial peptides (AMPs) are compounds that represent the first line of defense in plants, animals, and microbes against pathogenic microorganisms [5].

Biologically active peptides derived from vegetable wastes, especially legumes by enzymatic hydrolysis, are promising alternatives to chemical preservatives as they can extend the shelf life of food by inhibiting microbial contamination and food components' oxidation. They are also biocompatible and biodegradable in vivo [6].

Enzymatic hydrolysis increased the human diet's long-term bioavailability of amino acids, besides enhancing their absorbance through the small intestine [7]. Additionally, the protein hydrolysates have various activities. Recently, chickpea protein hydrolysate by chymotrypsin was isolated as an antimicrobial peptide against various foodborne pathogens, and it can be used as a food preservative [8]. A 2 h Flavourzyme-gram bean hydrolysate is characterized by a distinct structure and significant antioxidant activity [9]. Therefore, the application of these enzymatic hydrolysates has a significant impact on food formulation. Many trials are proceeding to incorporate various protein hydrolysates as potential natural preservatives into meat products and juices [10–12].

The antimicrobial mechanism of AMPs is briefed on reacting with certain parts of the bacterial membrane, such as anionic phospholipids and lipopolysaccharides, which break down the membrane and kill the bacteria [13]. The reaction may depend on peptides' hydrophobicity by binding the hydrophobic groups in the membrane. Additionally, in ionic/electrostatic interactions, the peptides are deposited on the bilayer surface, causing cellular membrane flux and disintegration [14,15]. Furthermore, amphipathicity, when peptides contain hydrophobic and hydrophilic residues, both previous mechanisms may be functioning. In addition, peptide length plays a critical role; short peptide has a good amphipathic structure with powerful antimicrobial activity [16,17].

Legume wastes, i.e., broken seeds, which consumers do not prefer, are precious sources of nutrients, especially protein [18]. However, many of these biomaterials do not get used and wind up in municipal landfills, causing major environmental problems and negative economic impacts. Therefore, managing massive amounts of various degradable materials is challenging [19].

Several studies highlighted the extraction of protein from food wastes (FW). They concerned the extraction of protein from FW to strengthen the concept of recycling and to utilize valuable extracted protein from FW as an equally valuable recycled ingredient and product to induce sustainability. So, the extraction processes often target protein yield, and they must be environmentally greener. Enzymatic hydrolysis in our study achieved this equation by excluding harsh chemicals. The extraction steps of protein from food waste are of prime importance to maximize protein yield and quality, where polysaccharide removal may affect protein solubility. Solubility is a marker of protein extractability. Tabal et al. [20] provided a new protein hydrolysate from pigeon pea milling waste (26% protein) for use in food formulation. In addition, Tassoni et al. [21] highlighted the use of pea, bean, and chickpea agro-industrial wastes in preparing protein for the formulation of feed, food, cosmetic, and packaging products.

However, there are no studies on valorizing the unwanted cluster bean (52% protein) and common red bean (25% protein) seeds to produce novel, eco-friendly protein hydrolysates and incorporate them in food formulation.

Cluster bean or guar (*Cyamopsis tetragonoloba*) is a valuable legume. Its seeds contain 52.6% protein and high mineral and vitamin content, mainly Fe and Vitamin C [22]. Green or dried seeds have numerous medicinal and industrial applications for humans and animals [23]. Additionally, red kidney beans (*Phaseolus vulgaris* L.) are an excellent source of protein (20–40%), consumed as an inexpensive protein source in many developing countries [24].

In this study, we explored novel pepsin protein hydrolysates from cluster bean and common red bean wastes (CBH and RCH) which have not been used in previous studies—that were chemically characterized by SDS-PAGE. Functional properties, e.g., the hydrolysate's solubility, water-holding capacity, and oil holding capacity were evaluated and associated with their activities (antioxidant and antimicrobial). Buffalo meat steak was covered with CBH and RCH hydrolysates and stored for 30 days under cold conditions, while continuously monitoring the chemical and microbial changes.

2. Materials and Methods

2.1. Protein Hydrolysates Isolation

The seed wastes of the cluster (*Cyamopsis tetragonoloba*) and kidney beans (*Phaseolus vulgaris* L.) were finely ground by a Moulinex blender (FP823125, France). The powder was homogenized in hexane (1:3, w:v). The protein was isolated from defatted powder as per Millan-Linares et al., [25]

The cluster bean and the common bean protein isolates were blended with pepsin (0.5%), homogenized in acidic phosphate buffer pH 2, and put in a heat bath (37 °C) for intervals of 0, 30, 45, and 60 min. The enzyme was inhibited at 90 °C for 15 min. The solution was centrifuged at (14,000× *g*, 5 min) to obtain CBH and RCH, lyophilized, and kept for further analysis [11,26].

2.2. The Degree of Hydrolysis (DH)

The %DH of cluster bean and common bean protein isolates after 0, 30, 45, and 60 min were determined by Holye and Merrltt [27]. A total of 100 μL of Trichloroacetic acid (10%) was added to 100 μL of protein isolates, then centrifuged under cooling at 12,298× *g* for 10 min. The total nitrogen in the supernatant protein and TCA was measured by the Kjeldahl method [28].

2.3. Characterization of Cluster Bean and Common Bean Hydrolysates

2.3.1. SDS-PAGE

After pepsin hydrolysis for 30 and 60 min, the protein hydrolysates were separated by discontinuous SDS-PAGE (18%) (Arabian Group for Integrated Technologies "Agitech", New Cairo, Egypt). The buffer system was (Tris HCl, pH 6.8 for staking gel, and Tris HCl, pH 8.8 for resolving gel) following Laemmli [29]. A total of 5 μL of protein in loading sample buffer was loaded in each well. A 5 to 245 kDa Tris-Glycine marker was embedded to configure the detected bands in electropherogram, which was stained by Coomassie brilliant blue.

2.3.2. Physicochemical Analysis of Hydrolysates

The suspensions of protein hydrolysates and seed residues were served for the following analysis. Moisture content was determined using AOAC method 925.10 [28]. The protein content was evaluated by elemental microanalysis as % nitrogen content × 6.25, using a NB9830 full automatic kjeldahl protein analyzer by AOAC method 920.87 [28]. The ash content was evaluated using the direct ignition method (550 °C for 25 h), AOAC method 923.03 [28]. The fat content was determined using AOAC method 945.16 [28]. Carbohydrate content was determined by difference. Carbohydrate was calculated by subtracting the sum percentage of moisture, protein, fat, ash, crude and dietary fiber. The protein hydrolysates used in the following analysis were obtained after 60 min of pepsin hydrolysis.

2.3.3. Functional Properties of Protein Hydrolysates

Solubility

The solubility of CBH and RCH (60 min of pepsin hydrolysis) was estimated at different pH (2–10) according to the method described in Saad et al. [12], with some modification. The CBH and the RCH (0.1 mL) were suspended in 25 mL of distilled water, stirred for 45 min at 45 °C, while adjusting pH, and then centrifuged under cooling at (5000× *g*, 10 min) to estimate the total nitrogen in the protein supernatant using the Kjeldahl method [28], which was then applied in Equation (2).

$$Solubility\ (\%) = \frac{Amount\ of\ protein\ in\ the\ supernantent}{Amount\ of\ protein\ in\ sample} \times 100 \quad (1)$$

Water Absorption Capacity

In weighted test tubes, 100 mg of CBH and RCH were stirred with 10 mL of sterilized distilled water for 30 min. The tubes were centrifuged at 6000× *g* for 30 min. The supernatant was suspended, then the tubes remained tilted at 45° for 30 min until the surface water was broken, then reweighted [30].

$$water\ absorbing\ capacity = \frac{absorbed\ water\ (g)}{sample\ weight\ (g)} \quad (2)$$

Oil Absorption Capacity

In weighted test tubes, 500 mg of CBH and RCH were homogenized in 10 mL of oil for 30 min. The tubes were centrifuged for 30 min at 6000× *g*, the supernatant was discarded, and the tubes were left upside down for 30 min to remove surface oil before being reweighted [30].

$$oil\ absorbing\ capacity = \frac{absorbed\ oil\ (g)}{sample\ weight\ (g)} \quad (3)$$

2.4. Total Phenolic Compounds (TPC) in the Hydrolysates

Total phenolic content was determined by the Folin–Ciocalteu reagent according to the method of Müller et al. [31]. In brief, in a 96-well microplate, 20 microliters of CBH and RCH obtained after 60 min pepsin hydrolysis were added to 100 microliters of Folin–Ciocalteu reagent and 75 microliters of Na_2CO_3 solution (7.5%), then incubated for 60 min in the dark. The absorbance was read at 765 nm by a microplate reader. The total phenolic content was expressed as g gallic acid equivalent/mL of protein by applying the following equation: y = 0.004x + 0.1257.

2.5. Biological Activity of Pepsin Protein Hydrolysates

2.5.1. Antioxidant Activity

The scavenging potential of the DPPH free radical was determined as per Gali and Bedjou [32]. In brief, 160 µL of DPPH solution was added to 40 µL of CBH and RCH levels (50, 100, 200, 300, and 500 µg/mL), and ascorbic acid (500 µg/mL) was used as a reference. After 30 min of incubation in a dark place at 25 °C, the absorbance was read at 517 nm. The ability of protein hydrolysates to scavenge the DPPH radical was applied in this Equation (5):

$$Radical\ scavenging\ activity\ (\%) = \frac{Abs.\ control - Abs.\ sample}{Abs.control} \times 100 \quad (4)$$

2.5.2. Antimicrobial Activity

The microbial strains, *Bacillus cereus* (BC), *Listeria monocytogenes* (LM), *Staphylococcus aureus* (SA), *Escherichia coli* (EC), *Pseudomonas aeruginosa* (PA), *Salmonella typhi* (ST), *Aspergillus niger* (AN), *Aspergillus flavus* (AF), *Candida gelbeta* (CG), *Candida tropicalis* (CT), and *Candida albicans* (CA) were taken from the microbial culture collection (MIRCN) in the Faculty of Agriculture, Ain Shams University, Egypt. The bacterial inoculum was prepared at 1×10^8 CFU/mL, and the fungal inoculum was prepared at 1×10^5 CFU/mL.

Antibacterial

The disc diffusion method was used to assess antibacterial activity. Discs (6 mm) were saturated with CBH and RCH at concentrations of (25, 50, 100, 200, 400, and 500 μg/mL) and placed on the surface of bacterial strains-inoculated Muller Hinton agar (MHA) plates and incubated at 37 °C for a day. The positive control was penicillin (500 μg/mL), and water was used as negative control. The diameters of the inhibition zones (mm) were calculated [12,33].

The minimum inhibitory concentration (MIC) and minimum bactericidal concentration (MBC) were estimated [12,33]. For MIC, 50 μL of CBH or RCH concentrations were added to Muller Hinton broth (MHB) tubes inoculated with bacterial stains. The tubes were incubated at 37 °C for 1 day before the turbidity was measured at 600 nm. For MBC, 100 μL of MIC tubes were spread on new MHA plates that were incubated at 37 °C, and bacterial population was observed after 24 h.

Antifungal

The disc diffusion technique was used to assess the antifungal activity of CBH and RCH [34,35]. The *Candida* and fungal inoculum were prepared in Sabouraud dextrose (SD) broth at a concentration of 10^5 CFU/mL. The prepared inoculum was spread over Sabouraud dextrose Agar (SDA) plates. CBH and RCH saturated 6 mm discs with different concentrations were placed on SDA plates, while clotrimazole (500 μg/mL) saturated discs were used as a positive control. The SDA plates were incubated for 2 days at 37 °C and 5 days at 30 °C for *Candida* and fungi, respectively. The distances of the inhibition zones (mm) were estimated. The MIC and MFC were determined with microdilution broth and MFC with spreading plates as mentioned in antibacterial activity, considering the incubation conditions of *Candida* and fungi.

2.6. Preservation of Fresh Meat Steak

Two factors were studied in this experiment: 1. Concentrations of CBH and RCH (0, 100, 250, and 500 μg/g meat); and 2. Storage time: 0, 10, 20, and 30 days.

Meat steaks were immersed in 100 mL of CBH and RCH suspension concentrations (0, 100, 250, and 500 μg/mL) for 24 h at 4 °C before being packed in polyethylene bags and stored in refrigeration at 4 °C for analysis.

Meat steaks were divided into 8 equal proportions and mixed with different concentrations of CBH, RCH, and nisin according to the following formulations: Using the immersion method, CBH 0, 100, 250, and 500 μg/g were T1-T4; RCH 0, 100, 250, and 500 μg/g were T5-T7, and nisin (500 μg/g) was T8, respectively. The samples were packaged and stored in a refrigerator at 4 °C. A total of 8 random samples were taken for analysis during the storage period (0, 10, 20, and 30 days).

2.6.1. Physicochemical Analysis of Meat Sample

pH and Glycogen Content Estimation

The pH of steak samples was measured by Ibrahim et al. [36]. Steak samples (10 g) were homogenized in 100 mL of distilled water for 1 min. The pH was then measured by a pH meter. Glycogen content in meat samples was measured according to Dreiling et al. [37].

Metmyoglobin (MetMb) Analysis

The MetMb content in meat samples was estimated by Krzywicki [38]. The treated and untreated steak samples (1 g) were blended for 10 s in a magnetic stirrer with 10 mL of ice-cold 0.04 M phosphate buffer at pH 6.8 before being centrifuged at 3000× *g* for 30 min at 4 °C. The supernatant was further clarified by filtration through Whatman No. 1 filter paper. The absorbance of the filtrate was measured at 525, 572, and 700 nm using a UV-VIS spectrophotometer (Shimadzu, Nakagyo-ku, Kyoto, Japan). The %Met-Mb concentration was estimated by Equation (6) [38].

$$MetMb\ (\%) = \left[1.395 - \frac{A572 - A700}{A525 - A700}\right] \times 100 \tag{5}$$

The Percentage of Inhibition in Lipid Peroxidation (LPI)

The Witte et al. [39] method was used to estimate the percentage of LPI. The meat steak samples were suspended in cold 50 mM phosphate buffer (pH 7) and centrifuged at high speed (14,000× *g*, 30 min, 4 °C). The obtained supernatant (100 µL) was mixed with barbituric acid (2 mL) and boiled for 30 min before cooling. A spectrophotometer (Shimadzu, Nakagyo-ku, Kyoto, Japan) was used to measure the sample absorbance at 530 nm. The percentage of LPI was calculated in Equation (6):

$$Lipid\ oxidation\ inhibition\ (\%) = \left[1 - \frac{Sample\ absorbance}{control\ absorbance}\right] \times 100 \tag{6}$$

2.6.2. Sensorial Properties and Color Measurement

The steak samples were evaluated for sensory characteristics, including color, flavor and aroma, tenderness, juiciness, and overall acceptability. Steak samples (after a 30-day storage period) were cooked in an oven at 176 °C for 8.5 min until the internal temperature reached 70 °C, then served warm at 60 °C to eight trained panelists [40]. Steak samples from different treatments were randomized and evaluated within the session. Water was mounted after each sample assessment. Panelists rated each sample attribute using a 9-point hedonic scale. The higher score values indicate a greater preference.

Concerning the meat color, the Hunter color analyzer (Hunter Lab color Flex EZ, USA) was used to measure the color parameters (L^*, a^*, and b^*) of meat steak samples [41].

2.6.3. Microbiological Analysis

The microbial load of buffalo meat steak was calculated as per Saad et al. [42]; 10 g of meat steak samples were homogenized with 90 mL sterilized buffer peptone water for 10 min to prepare a 10^{-1} concentration; decimal dilutions were prepared up to 10^{-6}; in one-use Petri-dishes, 1 mL of each dilution was added, followed by the appropriate media; on plate count agar (PCA), the total bacterial count (TBC) was determined after a 24-h incubation period at 30 °C. Additionally, after a 10-day incubation period at 7 °C, psychrophilic bacteria counts (PBC) were counted at PCA [43]. The microbial counts were transformed to logarithms (CFU/g).

2.7. Statistics

All experiments were done three times. The average of the replicated data was analyzed by One Way ANOVA at a probability level of 5%, followed by an LSD test to define the significant differences between means using Microsoft Excel (v. 2019).

3. Results and Discussion

3.1. The Approximate Composition of Protein Isolates and Hydrolysates

Table 1 presents the proximate analysis of the seed wastes and their protein hydrolysates of cluster and kidney beans. Cluster bean seed wastes and their protein hydrolysate contain high protein, i.e., 57.2 and 93.2%, respectively, compared to 23.35 and

88.9% in red kidney beans. Furthermore, red common bean seeds have the highest carbohydrate content, recording 64.1%. Ash content was higher in cluster bean hydrolysates, reaching 7.3%. In a previous study, the approximate composition of black beans was carbohydrates (71.4%), protein (23.1%), ash (4.3%), and fat (1.2%). However, the protein content in its protein isolates increased by 253% compared to black bean seed, while carbohydrates decreased by 11.4% [44].

Table 1. Approximate analysis of seed wastes and protein hydrolysate of cluster and red kidney beans.

Proximate Composition (%)	Beans Seed Wastes			
	Cluster		Red Kidney	
Material Status	**Seed**	**Hydrolysate**	**Seed**	**Hydrolysate**
Protein	57.2 c	93.2 a	23.35 d	88.9 b
Carbohydrates	30.8 b	0.13 d	64.1 a	4.6 c
Fat	2.25 a	-	2.3 a	-
Ash	4.3 c	7.3 a	4.19 c	6.5 b
Moisture	6.1 a	-	6.3 a	-

The lowercase letters next to values means in the same row indicate significant differences at $p \leq 0.05$. Protein hydrolysate is obtained after 60 min of hydrolysis with pepsin at 37 °C.

Generally, ash content increased with pepsin hydrolysis in CBH rather than RCH. The results indicated that CBH protein content was significantly increased by 63%. However, the increase was 282% in RCH because of the increase of 70% in ash content in CBH. A prior study found a comparable decrease in protein level in Alcalase-black kidney bean protein hydrolysate, probably because of the increase in ash content, which may be due to the addition of NaOH to maintain the pH during hydrolysis [44]. In Alcalase-hydrolyzed chickpea protein, a comparable reduction in protein content was observed, along with an increase in ash content [45].

3.2. *Physiochemical Characterization of the Hydrolysates*

3.2.1. SDS-PAGE Electropherogram

Figure 1 shows an SDS-PAGE electropherogram of the obtained CBH and RCH after pepsin hydrolysis for 30 min and 60 min. In total, 10 bands in the range of 17–100 kD in RCH were detected in lane 1. However, 5 bands of 17–48 kD in CBH were detected in lane 3. These 15 bands correspond to storage proteins, where 47–75 kD refers to vicilin (7S), and 40–48 kD refers to phaseolin (8S), following Los et al. [46], the 8S and the 7S bands in RCH after 30 min hydrolysis are more intense than those in CBH. After 60 min of pepsin hydrolysis, 7S bands (50–63 kD) still existed in lane 2 (RCH), but no bands were detected in lane 4, indicating the complete hydrolysis of CBH, which agreed with Saad et al. [11], who found the total disappearance of white kidney bean protein bands after 6 h of pepsin hydrolysis.

3.2.2. The Degree of Hydrolysis (DH)

Figure 2 depicts the DH (%) of cluster and red common bean protein isolates after 60 min of pepsin hydrolysis at 37 °C. In terms of hydrolysis time dependence, the DH was significantly increased. After 60 min of pepsin hydrolysis, the maximum DH of CBH and RCH was recorded, reaching 30 and 27%, respectively. The CBH had a high DH by a relative rise of nearly 11% over the RCH. Saad et al. [11] found the degree of hydrolysis was 33.3% for white kidney bean protein after hydrolysis by pepsin 1% for 6 h. Additionally, pepsin black bean protein was hydrolyzed by 27% with 2 h of pepsin hydrolysis [47]. In addition, when Bumrungsart and Duangmal [9] used Flavourzyme® (6%) for 6 h to hydrolyze black gram bean protein isolate, they obtained a high DH (75%).

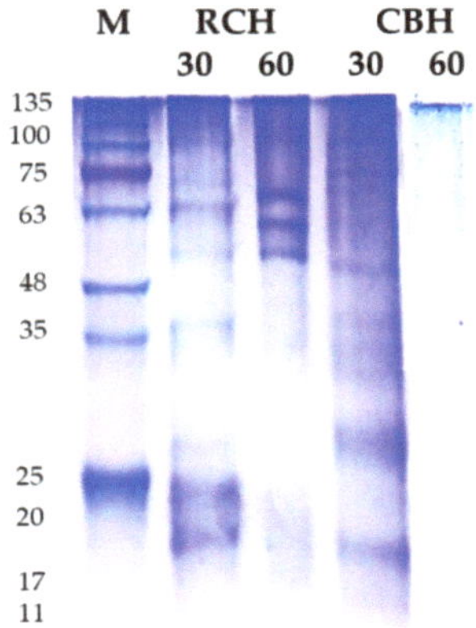

Figure 1. SDS electropherogram (18%) of RCH and CBH isolated from cluster and red kidney bean seeds wastes. Lane 1, M, molecular marker (Tris-Glycine SDS-PAGE, 4–20%). Lanes 2 and 4 represented protein bands of protein isolates after 30 min pepsin hydrolysis; lanes 3 and 5, represented 60-min pepsin hydrolysis at 37 °C of protein isolates. Buffer system, discontinuous SDS-PAGE buffer system (Tris HCl, pH 6.8 for staking gel, and Tris HCl, pH 8.8 for resolving gel). 5 μL of protein in loading sample buffer was loaded in each well.

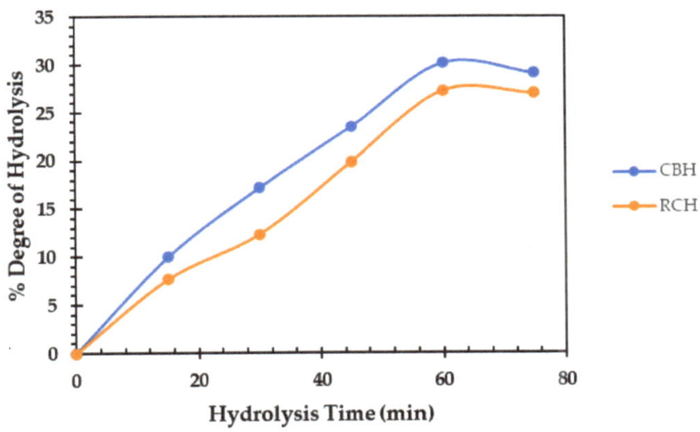

Figure 2. Degree of hydrolysis (DH) of cluster bean hydrolysate (CBH) and red bean protein hydrolysates (RCH) (60 min pepsin hydrolysis at 37 °C).

3.2.3. The pH-Protein Solubility

Incorporating functional protein into food formulation is depends on its solubility, influencing protein foaming and emulsifying properties [48]. The CBH and the RCH had isoelectric pH (lowest solubility) of 4–6 similar to the intact mother protein (Table 2). Protein solubility increased significantly ($p < 0.05$) when the pH shifted away from the isoelectric point. After 60 min of pepsin hydrolyzed, the solubility of CBH and RCH were 80 and 75% at pH 3, which increased to 100 and 90% at pH 11, respectively. The order of solubility level is CBH > RCH, which is related to the degree of hydrolysis and means that the solubility is improved. The fact that the basic side has a higher solubility than the acidic side is consistent with Los et al. [46], who found the solubility of carioca bean and soybean protein hydrolysate at pH 3.0 was 35.13%; but, when pH was raised to 10, the hydrolysates dissolved. Furthermore, the solubility of papain-kidney bean protein hydrolysate at pH 10 was 78% [49].

Table 2. Functional properties (water-absorbing capacity, oil-absorbing capacity, and solubility) of CBH and RCH were obtained after 60 min of pepsin hydrolysis at 37 °C.

Protein Hydrolysate	HT (min)	Functional Properties						
		WAC (g/g)	OAC (g/g)	Solubility (%)				
				pH 3	pH 5	pH 7	pH 9	pH 11
CBH	0	6.3 e	6.9 e	20 b	5 d	11 c	22 b	35 a
	30	8.7 c	10.22 c	50 ab	12 d	24 c	41 b	64 a
	60	11.6 a	13.66 a	80 ab	20 d	55 c	72 b	100 a
RCH	0	6.1 e	6.5 e	14 bc	5 d	9 c	19 b	29 a
	30	7.6 d	8.3 d	35 b	10 d	22 c	38 b	55 a
	60	9.4 b	11.96 b	75 ab	14 d	41 c	60 b	90 a

Means with different lowercase letters in the same raw indicate significant differences between hydrolysates solubility at $p \leq 0.05$. WAC, water-absorbing capacity; HT, hydrolysis time; OAC, oil-absorbing capacity. Means with different lowercase letters in the same column indicate significant differences between WAC and OAC values.

3.2.4. Functional Properties

Table 2 shows the water-holding capacity (WHC) and oil-absorbing capacity (OAC) of CBH and RCH after 0-, 30-, and 60-min pepsin-hydrolysis. The CBH has the highest WHC, with a solubility improvement of 22% over the RCH. After 60 min of pepsin hydrolysis, the OAC of CBH was increased by 11% compared to RCH. In addition, Eckert et al. [50] observed an increase from 6.12 to 8.21% in OAC of faba bean protein after pepsin hydrolysis. Enzymatic hydrolysis is commonly used to improve the functional properties of proteins. Incorporating plant proteins instead of animal proteins in food formulation is a new trend in the food industry [51,52]. Plant proteins' nutritional value and functional properties are critical [53]. High-solubility proteins improve the technical qualities of fortified foods and are required in many food applications [54].

3.2.5. Total Phenolic Content

The total phenolic content of different protein hydrolysates is presented in Figure 3A. The content of polyphenols grew in a concentration-dependent manner. The CBH (500 µg/mL) had higher phenolic compound values with 75.4 mg GAE/g, which increased 10% above RCH. Dark (44.3 mg gallic acid/g) and red (38.89 mg GAE/g) bean protein hydrolysates have comparable phenolic contents [55]. Protein–phenolic interactions may affect protein physicochemical properties; peptide activity may be increased by hindering specific residues of amino acids, thereby increasing the polyphenol absorption and activity [56]. Enzymatic hydrolysis increased the total polyphenols by breaking down the protein–polyphenol complexes and releasing some polyphenols entrapped in the peptide fragments [55,57].

3.2.6. Biological Activity

Antioxidant Activity

The DPPH-scavenging ability of protein hydrolysates is presented in Figure 3B. Most of the DPPH radical (93%) was scavenged by CBH (500 µg/mL). The antioxidant activity of protein hydrolysates depends on phenolic compounds in a concentration-dependent manner. The high polyphenol content in CBH and RCH is accountable for the higher scavenging activity of these hydrolysates. The scavenging power of pepsin-kidney bean hydrolysate was 85%, and for papain-kidney bean hydrolysate, it was 89% [11,49]. The scavenging power of CBH was stronger than intact protein or ascorbic acid (500 µg/mL), which is used commercially in the food industry [55]. The mode of action of antioxidant hydrolysates depends on making free radicals stable by donating electrons or transferring protons from aromatic amino acids in hydrolysates. Furthermore, the acidic amino acids may stabilize free radicals by sharing a proton with the NH_2 and COOH residues [58].

Antioxidant peptides are critical in the food industry because they prevent the oxidation of protein, lipid, and nucleic acid, ensuring product quality [59].

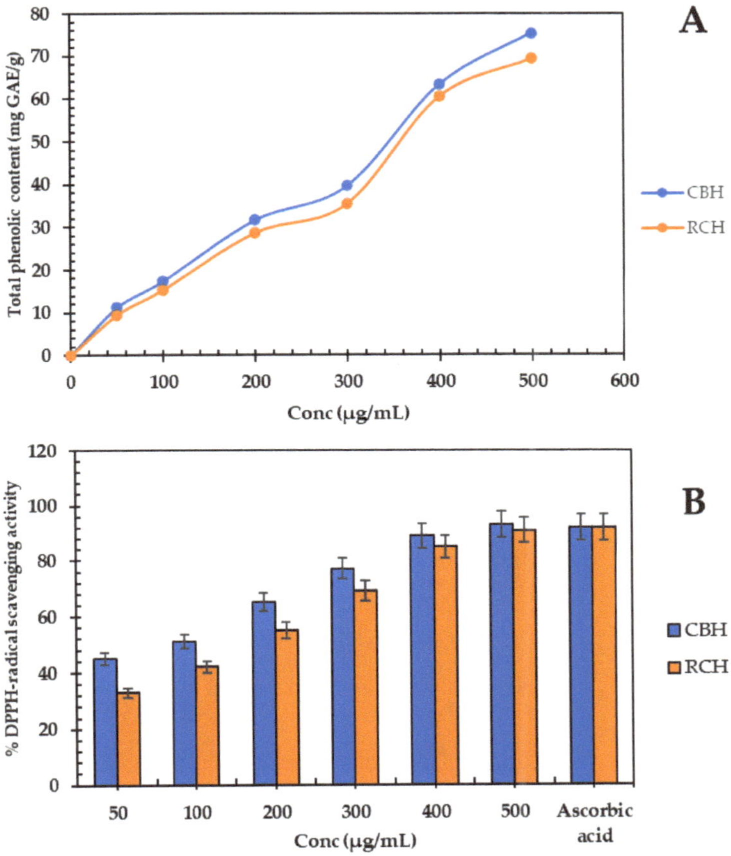

Figure 3. (**A**), Total phenolic content (mg gallic acid equivalent/g) of CBH and RCH concentrations (50–500 μg/mL). (**B**), DPPH˙ scavenging activity of CBH and RCH concentrations compared to ascorbic acid (500 μg/mL).

Antimicrobial Activity

Table 3 illustrates the inhibition zone (DIZ, mm) distances of the mentioned meat-borne bacteria and fungi when exposed to different concentrations of CBH and RCH (25, 50, 100, 200, 400, and 500 μg/mL). CBH (25–500 μg/mL) caused the biggest DIZs of 13–35 mm, followed by RCH (12–32 mm). The bacteria most vulnerable to the protein hydrolysates were *S. aureus* and *E. coli*, demonstrating 26 and 35 mm DIZ, respectively. However, *P. aeruginosa* and *L. monocytogenes* were the most resistant isolates to the hydrolysates, with an estimated DIZ of 28 and 23 mm, respectively.

The resistant Gram-negative bacteria have a lower DIZ than the Gram-positive. The unique membrane structure of Gram-negative bacteria is attributable to their resistance to antibacterial drugs through their lipopolysaccharide layer and certain enzymes in the periplasmic area [60,61].

Table 3. The diameters of inhibition zones (mm) of CBH and CBH at different concentrations of 25–500 μg/mL against Gram-positive, Gram-negative bacteria and fungi.

Bacteria	CBH						RCH					
	25	50	100	200	300	400	25	50	100	200	300	400
G+												
B. cereus (BC)	-	14 ab	20 b	22 b	25 b	27 b	-	13 ab	18 b	21 b	23 b	26 b
L. monocytogenes (LM)	-	13 b	18 c	20 c	24 c	26 c	-	12 b	17 c	20 c	22 c	25 c
S. aureus (SA)	-	15 a	22 a	25 a	28 a	32 a	-	14 a	21 a	23 a	26 a	30 a
G-												
E. coli (EC)	-	12 a	17 a	19 a	22 a	24 a	-	-	15 a	18 a	20 a	23 a
P. aeruginosa (PA)	-	9 c	13 c	15 b	19 c	21 c	-	-	11 c	14 c	17 c	19 c
S. typhi (ST)	-	10 b	15 b	18 ab	20 b	22 b	-	-	13 b	16 b	18 b	20 b
Fungi												
A. niger (AN)	-	8 b	12 d	13 d	17 d	22 d	-	-	9 c	11 b	15 d	20 d
A. flavus (AF)	-	9 ab	13 c	14 c	19 c	23 cd	-	-	11 ab	13 ab	17 bc	21 c
C. gelbeta (CG)	-	8 b	11 bc	14 c	17 b	24 c	-	-	9 c	11 b	15 d	21 c
C. tropicalis (CT)	-	10 a	14 a	18 a	21 a	29 a	-	-	12 a	14 a	19 a	26 a
C. albicans (CA)	-	9 ab	12 b	16 b	20 ab	27 b	-	-	10 b	13 ab	18 b	24 b

The lowercase letters next to values means indicate significant differences $p < 0.05$.

The lowest MIC values (20–45 μg/mL) were recorded in CBH-treated MHA plates, with a relative decrease of 40% compared to RCH (Table 4). The lowest CBH concentration that could kill the tested bacteria was 45–80 μg/mL, compared to 65–145 μg/mL for RCH at higher doses. There are similar values of DIZ in red kidney bean hydrolysate; 19.23 mm and 20.26 mm against *P. aeruginosa* and *E. coli*, respectively. Additionally, the MIC of papain-kidney bean protein hydrolysate was 70–90 μg/mL [49].

Table 4. The lowest concentration (μg/mL) of CBH, RCH, and antibiotic, inhibiting microbial strains, (MIC), and the lowest concentration (μg/mL) killing bacterial strains, (MBC), and fungal strains, (MFC).

Bacterial Strain	CBH (μg/mL)		RCH (μg/mL)		Antibiotic * (μg/mL)	
	MIC	MBC	MIC	MBC	MIC	MBC
BC	30 d	60 e	35 e	70 e	25 d	50 d
LM	35 c	65 d	40 d	75 d	30 c	60 c
SA	30 d	50 f	35 e	65 f	20 e	45 e
EC	40 b	70 c	65 c	125 c	35 b	70 b
PA	45 a	90 a	80 a	155 a	40 a	80 a
ST	40 b	75 b	70 b	130 b	35 b	70 b
Fungal strain	**MIC**	**MFC**	**MIC**	**MFC**	**MIC**	**MFC**
AN	40 b	80 b	80 b	165 a	40 a	85 a
AF	30 c	75 c	70 d	140 c	35 b	70 b
CG	45 a	85 a	85 a	150 b	40 a	85 a
CT	30 c	70 d	65 e	125 e	30 c	65 c
CA	30 c	75 c	75 c	135 d	35 b	70 b

Different lowercase letters next to means indicate significant differences at probability level 5%. * Penicillin with Bacteria, clotrimazole with fungi. Minimum inhibitory concentration (MIC), minimum bactericidal concentration (MBC), minimum fungicidal concentration (MFC).

The type of bacteria determines the antibacterial effect of peptides. The peptides may bind electrostatically to the bacterial membranes, and the membrane rigidity and cell components could be impaired [62].

Table 3 shows the antifungal activity of CBH and RCH. The DIZs of tested hydrolysates (25, 50, 100, 200, 300, 400, and 500 μg/mL) were in the range of (8–35 mm) against tested

Candida and fungi. The results showed no significant differences between the antifungal activities of hydrolysates on fungi or *Candida*. The fungi vulnerable to CBH (500 µg/mL) were Candida tropicalis *(CT)* and Candida albicans *(CA)*, with DIZs of 32–35 mm. However, the resistant fungi, *C. gleberta*, and *A. niger* recorded DIZs of approximately 24 and 27 mm, respectively.

Table 4 shows that the MIC and MFC of CBH are lower than those of RCH. The CBH suppressed fungal growth at MICs of 35–40 mm and MFCs of 75–90 mm; MIC and MFC against fungi and MIC and MBC against bacteria; the fungi had higher values (Table 4), indicating that hydrolysates have more powerful antibacterial than antifungal activity. The proposed mechanisms suggest that hydrolysates are more suited for bacteria than fungi, which have a more complex structure than bacteria. Heymich et al. [63] stated that chickpea protein hydrolysate exhibited potent antibacterial activity against *Escherichia coli* and *Bacillus subtilis* with minimum inhibitory concentrations of 31.3–62.5 µM. Additionally, it displayed antifungal activity with minimum inhibitory concentrations of 125–500 µM against *Saccharomyces cerevisiae* and *Zygosaccharomyces bailii*. In addition, the purified peptide from the seeds of the cowpea plant displayed higher MIC values against fungal spoilers, with MICs of 50 µg/mL against *F. culmorum* and > 500 µg/mL against *Penicillium expansum* [64].

3.3. Fresh Meat Steak Preservation by Cluster and Common Bean Protein Hydrolysates

3.3.1. Physiochemical Alternation during Cold Storage

The physiochemical alternations of treated and untreated fresh meat steaks during storage of 0, 10, 20, and 30 days at 4 °C are presented in Table 5. The studied parameters included tracking the changes in pH, metmyoglobin (%), and lipid oxidation inhibition (%) in response to CBH and RCH addition (0, 100, 250, and 500 µg/g) with a preference for CBH. All parameters' values considerably increased in storage period dependence, but lipid oxidation inhibition (%) significantly decreased in concentration dependence. The pH values of the control fresh steak increased by 52% from the start of storage until day 30 of cold storage.

The pH in supplemented meat samples with RCH and CBH (500 µg/g) significantly decreased by 30–33% compared to the control at the end of the storage period. The gradual increase in the pH value with storage time is probably a result of food spoilage microorganisms that can hydrolyze the proteins into NH_3 [65]. The addition of CBH and RCH (500 µg/g) to the stored meat significantly reduced this value in a concentration-dependent manner. This action appears to be due to the previously demonstrated antimicrobial activity of the protein hydrolysates. Similar results were observed by Saad et al. [11] and Saad et al. [12] on using common bean protein hydrolysates in preserving minced beef or chicken meat.

The estimated glycogen content in meat samples is 1.1 µmol/g in the control sample, 0.8 µmol/g in RCH-supplemented meat, 0.5 µmol/g in CBH-supplemented meat, and 0.6 µmol/g in nisin-supplemented meat. The low glycogen content didn't affect the pH value of meat and that agrees with England et al. [66].

The Met-myoglobin levels in the untreated samples dramatically increased by 416% at the end of storage, exceeding the acceptable level in meat (40%). The treated meat samples with CBH and RCH (500 µg/g) were characterized by low metmyoglobin levels of 21–26%, which decreased by 96–142% compared to the untreated samples. The considerable decrease (142%) was achieved by CBH (500 µg/g). Increasing metmyoglobin levels in stored meat significantly affect the meat color because of the oxidation of myoglobin and the auto-oxidation of protein in the meat [67].

Table 5. Physicochemical fluctuation in raw buffalo steak supplemented with cluster bean and red common bean protein hydrolysate (CBH and RCH) compared to nisin during 0–30 days cold storage at 4 °C.

Sample	Conc (μg/g)	Cold storage Time (days)											
		pH				Metmyoglobin (%)				Lipid Oxidation Inhibition (%)			
		0	10	20	30	0	10	20	30	0	10	20	30
Control	0.0	5.8 a	6.33 a	7.50 a	8.80 a	10.00 a	27.00 a	43.00 a	51.60 a	38 d	30 d	22 d	13 g
CBH	100	5.7 b	5.89 c	6.25 c	7.00 c	9.1 ab	19.00 c	21.00 b	30.45 d	37 ab	33 c	30 c	20 e
	250	5.65 b	5.74 c	6.20 c	6.55 d	8.2 bc	14.50 cd	16.00 c	26.00 e	37 ab	35 b	33 ab	28 c
	500	5.49 c	5.67 c	5.88 d	6.14 e	7.67 c	11.99 d	15.00 c	21.65 f	38 a	37 a	34 a	33 a
RCH	100	5.89 a	6.00 b	6.54 b	7.35 b	9.8 a	21.30 b	25.90 b	36.00 b	36 b	31 d	28 d	17 f
	250	5.72 b	5.9 c	6.32 c	6.87 cd	8.7 b	16.55 c	23.10 b	32.50 c	36 b	34 c	31 bc	24 d
	500	5.56 c	5.79 c	6.25 c	6.44 d	7.9 c	14.33 cd	16.80 c	26.35 e	37 ab	36 ab	32 b	30 b
Nisin	500	5.51 c	5.70 c	6.12 c	6.20 e	7.7 c	12.54 d	15.56 c	24.66 ef	37 ab	36 ab	33 ab	32 ab

Means with different lowercase letters in the same column indicate significant differences.

The antioxidant activity of added hydrolysates (Figure 3) is probably responsible for the met-myoglobin reduction. A decrease in met-myoglobin was observed when treating beef burgers with cinnamon, rosemary, and thyme extracts [68]. Lipid oxidation is one of the main factors that affect meat quality by producing hydroperoxides and aldehydes [1]. Therefore, adding RCH and CBH (500 μg/g) to meat steak dramatically reduces lipid oxidation by approximately 130–153% compared to the control.

The antioxidant activity of CBH and RCH is probably responsible for scavenging the radicals produced by lipid oxidation [1]. Moreover, Pineul et al. [69] found inhibition of lipid oxidation by 82% in zebrafish meat treated with red bean hydrolysate. Additionally, Aslam et al. [70] discovered that storing supplemented chicken breast with fish protein hydrolysate delayed lipid oxidation and other undesirable changes.

3.3.2. The Fluctuation in Color Parameters and Sensory Properties of Stored Buffalo Steak for 30 Days at 4 °C

Table 6 and Figure 4A present color parameter fluctuations in fresh meat steak treated or untreated with CBH and RCH during a storage period of 30 days. The lightness values (L^*) of meat steak decreased with CBH and RCH (500 μg/g) in concentration dependence by 6–10% compared to the control. The dark color of hydrolysates may cause this decrease. However, a^* value increased by 4–6% compared to the control. Additionally, blueness values increased from 3.08 to −2.50 in CBH-treated samples and from 3.08 to −0.5 in RCH-treated steak. In general, color components deteriorated at the end of storage, but protein hydrolysate inclusion significantly preserved approximately 80–90% of color attributes based on the protein hydrolysates compared to nisin and the control.

Table 6. The alternation in color parameters in buffalo steak supplemented with CBH and RCH at graded concentrations (0, 100, 250, and 500μg/g) during 0–30 days cold storage.

Sample	Conc	Storage (Day)						
	(μg/g)	0	30	0	30	0	30	ΔE
		L^*		a^*		b^*		
Control	0	49.55 a	48.66 a	14.00 c	13.70 c	4.00 a	3.80 a	1.34 a
CBH	100	48.70 b	47.92 b	15.10 a	14.80 b	1.90 c	1.10 c	1.29 b
	250	46.54 c	46.00 c	14.90 b	14.30 bc	−0.90 e	−1.80 ef	1.30 ab
	500	44.65 d	44.22 d	15.20 a	15.00 a	−1.50 f	−2.50 f	1.22 c
RCH	100	49.00 ab	48.12 ab	14.20 c	13.98 c	2.30 b	2.00 b	1.34 a
	250	48.22 bc	47.57 b	14.40 c	14.00 bc	0.50 d	0.20 d	1.32 ab
	500	46.33 c	45.88 cd	14.80 b	14.20 bc	−0.50 e	−1.10 e	1.27 b

Means with different lowercase letters in the same column indicate significant differences at $p \leq 0.05$. Cont.: control. The Lightness (L^*) [(0–100) lightness to darkness], redness (a^*) [(− to +) redness to greenness], and b^* value reflected (+) yellowness to (−) blueness.

No studies shed light on the alternations in color components and sensory characteristics of fresh meat steak enriched with CBH. However, the obtained results were in agreement with Arshad et al. [71], who studied the beneficial effects of bioactive peptides on the oxidative stability and functional properties of beef nuggets supplemented with milk casein protein hydrolysates at levels of 0, 2, 4, 6, and 8% during cold storage of 15 days.

The Hunter color test also revealed a significant difference between groups, where lightness values of beef nuggets were decreased in 8% supplemented samples and during the storage period. In addition, yellowness decreased, but the redness of beef nuggets increased. On the other hand, the addition of CBH and RCH considerably kept the meat quality parameters at high levels (Figure 4B–F). The highest values of tenderness and juiciness, 8.6 and 8.8, were observed in CBH-supplemented steak because of the highest scores in water-holding capacity in CBH (Table 2).

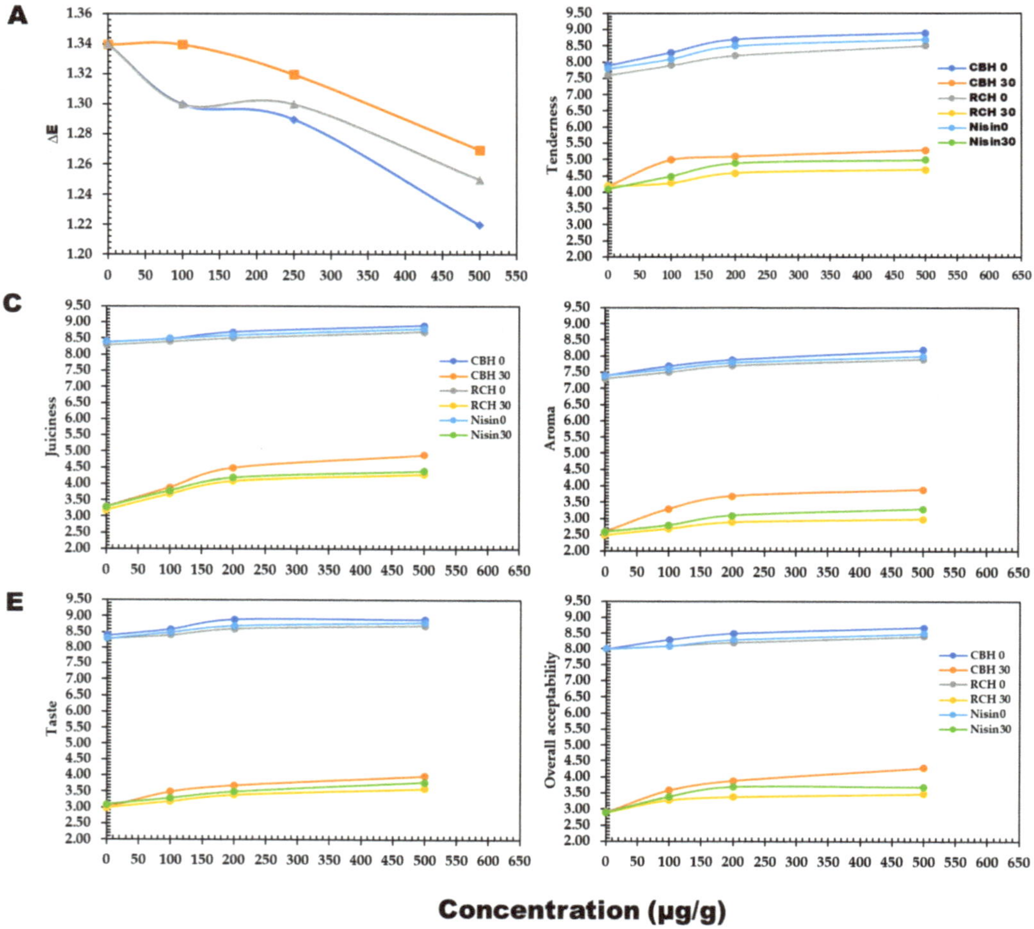

Figure 4. The changes in sensorial quality of buffalo meat steak supplemented with CBH and RCH (100, 250, and 500 μg/g) and nisin 500 μg/g persevered at 4 °C during 0–30 days storage. The sensorial parameters; Color (**A**), Tenderness (**B**), Juiciness (**C**), Aroma (**D**), Taste (**E**), and Overall Acceptability (**F**).

CBH enhanced cooked steak flavor and taste, followed by RCH supplementation. All quality parameters decreased at the end of cold storage, but the precious roles of CBH and RCH significantly reduced the unwanted changes by 55–60% (Figure 4B–F) compared to nisin and the control.

The above-mentioned sensorial scores affected the meat's overall acceptability scores, where CBH-supplemented fresh steak was safe for approximately 28–29 days of cold storage, while the RCH-supplemented steak (500 μg/g) was safe for 20 days.

Similar results have been noticed in the sensorial traits of raw buffalo meat supplemented with pea, and red kidney bean hydrolysates were maintained and were highly acceptable compared to the control concerning shelf-life [72].

3.3.3. Microbial Alternation in Fresh Meat Steak during 30 Days Storing at 4 °C

Figure 5 shows a considerable ($p < 0.05$) increase in bacterial load during the storage period at refrigeration. At the end of the storage period, the total bacterial count declined in supplemented fresh buffalo steak with RCH, CBH, and nisin at a high concentration (500 µg/g). As a result, the total bacterial count in supplemented samples was reduced by 31–55% compared to controls.

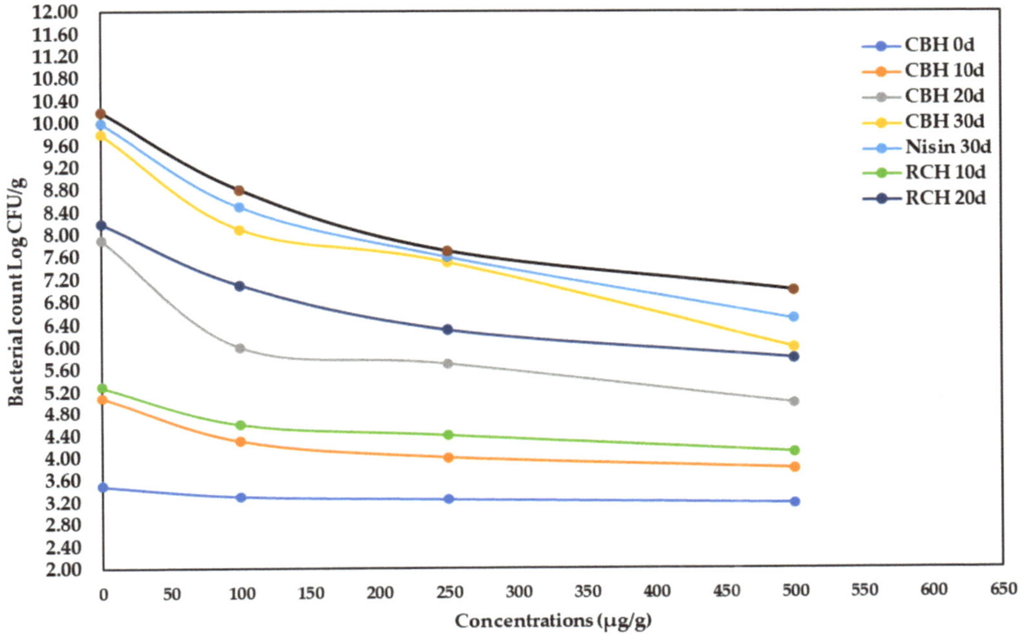

Figure 5. The alteration of total bacterial count Log CFU/g of stored buffalo steaks at 4 °C during 0- and 30-days storage period, as supplemented with CBH, RCH (100, 250, and 500 µg/g), and Nisin at 500 µg/g.

The antibacterial activity of CBH could extend the secure cold storage of fresh meat for approximately 20–30 days, when kept refrigerated. Saad et al. [11] found that found that adding kidney bean protein hydrolysate to minced beef lowered the microbial load by 22%. Additionally, Sharma et al. [73] discovered that adding natural preservatives to chicken sausages, such as a herbal blend, considerably reduced the microbial population. The allowable bacterial count in fresh buffalo meat must be less than 1×10^6 CFU/g, according to Egyptian Standards No. 4334 [74] and the International Commission on Microbiological Specification (ICMS, 1982).

4. Conclusions

Food waste harms the environment, but legumes are rich in nutrients, so maximizing these wastes is critical. Enzymatic hydrolysis and microbial fermentation can quickly and precisely produce bioactive peptides. CBH and RCH can be exploited as possible antioxidants and efficient antibacterial in food systems. They can extend the cold lifetime of preserved fresh steak to 28–29 days while sustaining acceptable sensory qualities.

Author Contributions: Conceptualization, E.T.A.S.-A., M.F.A., S.A.A. and A.M.S.; methodology, E.T.A.S.-A., R.M.E.-M.; A.M.S., N.A.R. and M.N.; formal analysis, A.M.S. and M.N.; investigation, E.T.A.S.-A., K.B.H.S., M.F.A., S.A.A. and A.M.S.; data curation, K.B.H.S., M.F.A. and S.A.A.; writing—original draft preparation, A.M.S. and M.N.; software, M.H.A. and S.K.A.J.; resources, M.S.A. and S.S.; investigation and formal analysis, M.H.A., S.K.A.J., M.S.A. and S.S.; writing—review and editing, A.M.S.; visualization, A.M.S. and M.N. All authors have read and agreed to the published version of the manuscript.

Funding: This work received no funding.

Institutional Review Board Statement: Not applicable.

Informed Consent Statement: Not applicable.

Data Availability Statement: Not applicable.

Conflicts of Interest: The authors declare no conflict of interest.

Sample Availability: Samples of the compounds are not available from the authors.

References

1. Bing, S.; Zang, Y.; Li, Y.; Zhang, B.; Mo, Q.; Zhao, X.; Yang, C. A combined approach using slightly acidic electrolyzed water and tea polyphenols to inhibit lipid oxidation and ensure microbiological safety during beef preservation. *Meat Sci.* **2022**, *183*, 108643. [CrossRef] [PubMed]
2. FAO. Moving Forward on Food Loss and Wast Reduction Food and Agriculture. 2019. Available online: http://www.fao.org/publications (accessed on 1 October 2019).
3. Saucier, L. Microbial spoilage, quality and safety within the context of meat sustainability. *Meat Sci.* **2016**, *120*, 78–84. [CrossRef] [PubMed]
4. Bedale, W.; Sindelar, J.J.; Milkowski, A.L. Dietary nitrate and nitrite: Benefits, risks, and evolving perceptions. *Meat Sci.* **2016**, *120*, 85–92. [CrossRef] [PubMed]
5. Corrêa, J.A.F.; Evangelista, A.G.; de Melo Nazareth, T.; Luciano, F.B. Fundamentals on the molecular mechanism of action of antimicrobial peptides. *Mater* **2019**, *8*, 100494.
6. Deng, M.; Lin, Y.; Dong, L.; Jia, X.; Shen, Y.; Liu, L.; Chi, J.; Huang, F.; Zhang, M.; Zhang, R. Physicochemical and functional properties of dietary fiber from pummelo (*Citrus grandis* L. Osbeck) and grapefruit (*Citrus paradisi* Mcfad) cultivars. *Food Biosci.* **2021**, *40*, 100890. [CrossRef]
7. Noman, A.; Xu, Y.; AL-Bukhaiti, W.Q.; Abed, S.M.; Ali, A.H.; Ramadhan, A.H.; Xia, W. Influence of enzymatic hydrolysis conditions on the degree of hydrolysis and functional properties of protein hydrolysate obtained from Chinese sturgeon (*Acipenser sinensis*) by using papain enzyme. *Process Biochem.* **2018**, *67*, 19–28. [CrossRef]
8. Heymich, M.-L.; Friedlein, U.; Trollmann, M.; Schwaiger, K.; Böckmann, R.A.; Pischetsrieder, M. Generation of antimicrobial peptides Leg1 and Leg2 from chickpea storage protein, active against food spoilage bacteria and foodborne pathogens. *Food Chem.* **2021**, *347*, 128917. [CrossRef]
9. Bumrungsart, N.; Duangmal, K. Optimization of enzymatic hydrolysis condition for producing black gram bean (*Vigna mungo*) hydrolysate with high antioxidant activity. *Food Appl. Biosci. J.* **2019**, *7*, 105–117.
10. El-Saadony, M.T.; Saad, A.M.; Elakkad, H.A.; El-Tahan, A.M.; Alshahrani, O.A.; Alshilawi, M.S.; El-Sayed, H.; Amin, S.A.; Ahmed, A.I. Flavoring and extending the shelf life of cucumber juice with aroma compounds-rich herbal extracts at 4 °C through controlling chemical and microbial fluctuations. *Saudi J. Biol. Sci.* **2022**, *29*, 346–354. [CrossRef]
11. Saad, A.M.; Osman, A.O.M.; Mohamed, A.S.; Ramadan, M.F. Enzymatic hydrolysis of *Phaseolus vulgaris* protein isolate: Characterization of hydrolysates and effect on the quality of minced beef during cold storage. *Int. J. Pept. Res. Ther.* **2020**, *26*, 567–577. [CrossRef]
12. Saad, A.M.; Sitohy, M.Z.; Ahmed, A.I.; Rabie, N.A.; Amin, S.A.; Aboelenin, S.M.; Soliman, M.M.; El-Saadony, M.T. Biochemical and functional characterization of kidney bean protein alcalase-hydrolysates and their preservative action on stored chicken meat. *Molecules* **2021**, *26*, 4690. [CrossRef] [PubMed]
13. Seyfi, R.; Kahaki, F.A.; Ebrahimi, T.; Montazersaheb, S.; Eyvazi, S.; Babaeipour, V.; Tarhriz, V. International Journal of Peptide Research and Therapeutics. Antimicrobial peptides (AMPs): Roles, functions and mechanism of action. *Int. J. Pept. Res. Ther.* **2020**, *26*, 1451–1463. [CrossRef]
14. Lei, J.; Sun, L.; Huang, S.; Zhu, C.; Li, P.; He, J.; Mackey, V.; Coy, D.H.; He, Q. The antimicrobial peptides and their potential clinical applications. *Am. J. Transl. Res.* **2019**, *11*, 3919. [PubMed]
15. Trabulo, S.; Cardoso, A.L.; Mano, M.; De Lima, M.C.P. Cell-penetrating peptides—mechanisms of cellular uptake and generation of delivery systems. *Pharmacetucals* **2010**, *3*, 961–993. [CrossRef]
16. Hollmann, A.; Martínez, M.; Noguera, M.E.; Augusto, M.T.; Disalvo, A.; Santos, N.C.; Semorile, L.; Maffía, S.B. Role of amphipathicity and hydrophobicity in the balance between hemolysis and peptide–membrane interactions of three related antimicrobial peptides. *Colloids Surf. B Biointerfaces.* **2016**, *141*, 528–536. [CrossRef]
17. Lyu, Y.; Yang, Y.; Lyu, X.; Dong, N.; Shan, A. Antimicrobial activity, improved cell selectivity and mode of action of short PMAP-36-derived peptides against bacteria and *Candida*. *Sci. Rep.* **2016**, *6*, 27258. [CrossRef]
18. Lau, K.Q.; Sabran, M.R.; Shafie, S.R. Utilization of Vegetable and Fruit By-products as Functional Ingredient and Food. *Front. Nutr.* **2021**, *8*, 261. [CrossRef]

19. Torres-León, C.; Ramírez-Guzman, N.; Londoño-Hernandez, L.; Martinez-Medina, G.A.; Díaz-Herrera, R.; Navarro-Macias, V.; Aguilar, C.N. Food waste and byproducts: An opportunity to minimize malnutrition and hunger in developing countries. *Front. Sustain. Food Syst.* **2018**, *52*, 1–17. [CrossRef]
20. Tapal, A.; Vegarud, G.E.; Sreedhara, A.; Tiku, P.K. Nutraceutical protein isolate from pigeon pea (*Cajanus cajan*) milling waste by-product: Functional aspects and digestibility. *Food Func.* **2019**, *10*, 2710–2719. [CrossRef]
21. Tassoni, A.; Tedeschi, T.; Zurlini, C.; Cigognini, I.M.; Petrusan, J.-I.; Rodríguez, Ó.; Neri, S.; Celli, A.; Sisti, L.; Cinelli, P. State-of-the-art production chains for peas, beans and chickpeas—Valorization of agro-industrial residues and applications of derived extracts. *Molecules* **2020**, *25*, 1383. [CrossRef]
22. Ahmed, M.; Hamed, R.; Ali, M.; Hassan, A.; Babiker, E. Proximate composition, antinutritional factors and protein fractions of guar gum seeds as influenced by processing treatments. *Pak. J. Nutr.* **2006**, *5*, 340–345.
23. Arora, R.; Pahuja, S. Mutagenesis in guar (*Cyamopsis tetragonoloba* (L.) Taub.). *Plant Mutat. Rep.* **2008**, *2*, 7–9.
24. Ahmed, J.; Al-Ruwaih, N.; Mulla, M.; Rahman, M.H. Effect of high pressure treatment on functional, rheological and structural properties of kidney bean protein isolate. *LWT Food Sci. Technol.* **2018**, *91*, 191–197. [CrossRef]
25. Millan-Linares, M.C.; Lemus-Conejo, A.; Yust, M.M.; Pedroche, J.; Carrillo-Vico, A.; Millan, F.; Montserrat-de la Paz, S. Gpetaflr, a novel bioactive peptide from *Lupinus angustifolius* L. protein hydrolysate, reduces osteoclastogenesis. *J. Func. Food.* **2018**, *47*, 299–303. [CrossRef]
26. El-Saadony, M.T.; Khalil, O.S.; Osman, A.; Alshilawi, M.S.; Taha, A.E.; Aboelenin, S.M.; Shukry, M.; Saad, A.M. Bioactive peptides supplemented raw buffalo milk: Biological activity, shelf life and quality properties during cold preservation. *Saudi J. Biol. Sci.* **2021**, *28*, 4581–4591. [CrossRef]
27. Holye, N.T.; Merrltt, J.H. Quality of fish protein hydrolysates from herring (*Clupea harengus*). *J. Food Sci.* **1994**, *59*, 76–79. [CrossRef]
28. AOAC. *Official Methods of Analysis*, 18th ed.; AOAC International: Gaithersburg, MD, USA, 2012.
29. Laemmli, U.K. Cleavage of structural proteins during the assembly of the head of bacteriophage T4. *Nature* **1970**, *227*, 680–685. [CrossRef]
30. Wani, I.A.; Sogi, D.S.; Gill, B.S. Physicochemical and functional properties of flours from three Black gram (*Phaseolus mungo* L.) cultivars. *Inter. J. Food Sci. Technol.* **2013**, *48*, 771–777. [CrossRef]
31. Müller, L.; Gnoyke, S.; Popken, A.M.; Böhm, V. Antioxidant capacity and related parameters of different fruit formulations. *LWT-Food. Sci.Technol.* **2010**, *43*, 992–999. [CrossRef]
32. Gali, L.; Bedjou, F. Antioxidant and anticholinesterase effects of the ethanol extract, ethanol extract fractions and total alkaloids from the cultivated *Ruta chalepensis*. *S. Afr. J. Bot.* **2019**, *120*, 163–169. [CrossRef]
33. Abdel-Moneim, A.M.E.; El-Saadony, M.T.; Shehata, A.M.; Saad, A.M.; Aldhumri, S.A.; Ouda, S.M.; Mesalam, N.M. Antioxidant and antimicrobial activities of Spirulina platensis extracts and biogenic selenium nanoparticles against selected pathogenic bacteria and fungi. *Saudi J. Biol. Sci.* **2022**, *29*, 1197–1209. [CrossRef] [PubMed]
34. El-Saadony, M.T.; Elsadek, M.F.; Mohamed, A.S.; Taha, A.E.; Ahmed, B.M.; Saad, A.M. Effects of chemical and natural additives on cucumber juice's quality, shelf life, and safety. *Foods* **2020**, *9*, 639. [CrossRef] [PubMed]
35. Elgorban, A.; Aref, S.; Seham, S.; Elhindi, K.; Bahkali, A.; Sayed, S.; Manal, M. Extracellular synthesis of silver nanoparticles using *Aspergillus versicolor* and evaluation of their activity on plant pathogenic fungi. *Mycosphere* **2016**, *7*, 844–852. [CrossRef]
36. Ibrahim, H.M.; Abou-Arab, A.A.; Salem, F.M.A. Effects on lamb patties quality. *J. Food Technol.* **2010**, *8*, 134–142.
37. Dreiling, C.; Brown, D.; Casale, L.; Kelly, L. Muscle glycogen: Comparison of iodine binding and enzyme digestion assays and application to meat samples. *Meat Sci.* **1987**, *20*, 167–177. [CrossRef]
38. Krzywicki, K. The determination of haem pigments in meat. *Meat Sci.* **1982**, *7*, 29–36. [CrossRef]
39. Witte, V.C.; Krause, G.F.; Bailey, M.E. A new extraction method for determining 2-thiobarbituric acid values of pork and beef during storage. *J. Food Sci.* **1970**, *35*, 582–585. [CrossRef]
40. Murphy, M.A.; Zerby, H.N. Prerigor infusion of lamb with sodium chloride, phosphate, and dextrose solutions to improve tenderness. *Meat Sci.* **2004**, *66*, 343–349. [CrossRef]
41. Hunter, R. Scales for the measurements of color difference. In *The Measurement of Appearance*; John Willey & Sons: New York, NY, USA, 1975; pp. 133–140.
42. Saad, A.M.; El-Saadony, M.T.; Mohamed, A.S.; Ahmed, A.I.; Sitohy, M.Z. Impact of cucumber pomace fortification on the nutritional, sensorial and technological quality of soft wheat flour-based noodles. *Int. J. Food Sci. Technol.* **2021**, *56*, 3255–3268. [CrossRef]
43. Milton, D.L.; Johnson, T.W.; Johnson, K.; Murphy, B.; Carter, H.; Hurt, R.T.; Mundi, M.S.; Epp, L.; Spurlock, A.Y.; Hussey, J. Accepted safe food-handling procedures minimizes microbial contamination of home-prepared blenderized tube-feeding. *Nutr. Clin. Pr.* **2020**, *35*, 479–486. [CrossRef]
44. Evangelho, J.A.d.; Berrios, J.d.J.; Pinto, V.Z.; Antunes, M.D.; Vanier, N.L.; Zavareze, E.d.R. Antioxidant activity of black bean (*Phaseolus vulgaris* L.) protein hydrolysates. *Food Sci. Technol.* **2016**, *36*, 23–27. [CrossRef]
45. del Mar Yust, M.; Pedroche, J.; del Carmen Millán-Linares, M.; Alcaide-Hidalgo, J.M.; Millán, F. Improvement of functional properties of chickpea proteins by hydrolysis with immobilised Alcalase. *Food Chem.* **2010**, *122*, 1212–1217. [CrossRef]
46. Los, F.G.B.; Demiate, I.M.; Dornelles, R.C.P.; Lamsal, B. Enzymatic hydrolysis of Carioca bean (*Phaseolus vulgaris* L.) protein as an alternative to commercially rejected grains. *LWT-Food Sci. Technol.* **2020**, *125*, 109191. [CrossRef]

47. do Evangelho, J.A.; Vanier, N.L.; Pinto, V.Z.; De Berrios, J.J.; Dias, A.R.G.; da Rosa Zavareze, E. Black bean (*Phaseolus vulgaris* L.) protein hydrolysates: Physicochemical and functional properties. *Food Chem.* **2017**, *214*, 460–467. [CrossRef] [PubMed]
48. Moure, A.; Sineiro, J.; Domínguez, H.; Parajó, J.C. Functionality of oilseed protein products: A review. *Food Res. Int.* **2006**, *39*, 945–963. [CrossRef]
49. Wahdan, K.; Saad, A. Antibacterial and Antioxidant Activities of an Enzymatic Hydrolysate Kidney Bean (*Phaseolus vulgaris* L.) Protein Isolates. *J. Agric. Chem. Biotechnol.* **2018**, *9*, 85–89. [CrossRef]
50. Eckert, E.; Han, J.; Swallow, K.; Tian, Z.; Jarpa-Parra, M.; Chen, L. Effects of enzymatic hydrolysis and ultrafiltration on physicochemical and functional properties of faba bean protein. *Cereal Chem.* **2019**, *96*, 725–741. [CrossRef]
51. Chandra-Hioe, M.V.; Wong, C.H.; Arcot, J. The potential use of fermented chickpea and faba bean flour as food ingredients. *Plant Foods Hum. Nutr.* **2016**, *71*, 90–95. [CrossRef]
52. Multari, S.; Stewart, D.; Russell, W.R. Potential of fava bean as future protein supply to partially replace meat intake in the human diet. *Compr. Rev. Food Sci. Food Saf.* **2015**, *14*, 511–522. [CrossRef]
53. Aydemir, L.Y.; Yemenicioğlu, A. Potential of Turkish Kabuli type chickpea and green and red lentil cultivars as source of soy and animal origin functional protein alternatives. *LWT-Food Sci. Technol.* **2013**, *50*, 686–694. [CrossRef]
54. Faustino, M.; Veiga, M.; Sousa, P.; Costa, E.M.; Silva, S.; Pintado, M. Agro-food byproducts as a new source of natural food additives. *Molecules* **2019**, *24*, 1056. [CrossRef]
55. Sarker, A.; Chakraborty, S.; Roy, M. Dark red kidney bean (*Phaseolus vulgaris* L.) protein hydrolysates inhibit the growth of oxidizing substances in plain yogurt. *J. Agric. Food Res.* **2020**, *2*, 100062. [CrossRef]
56. Jakobek, L. Interactions of polyphenols with carbohydrates, lipids and proteins. *Food Chem.* **2015**, *175*, 556–567. [CrossRef] [PubMed]
57. Dias, D.R.; Abreu, C.M.P.d.; Silvestre, M.P.C.; Schwan, R.F. In vitro protein digestibility of enzymatically pre-treated bean (*Phaseolus vulgaris* L.) flour using commercial protease and *Bacillus* sp. protease. *Food Sci. Technol.* **2010**, *30*, 94–99. [CrossRef]
58. Esfandi, R.; Walters, M.E.; Tsopmo, A. Antioxidant properties and potential mechanisms of hydrolyzed proteins and peptides from cereals. *Heliyon* **2019**, *5*, e01538. [CrossRef]
59. Nwachukwu, I.D.; Aluko, R.E. Structural and functional properties of food protein-derived antioxidant peptides. *J. Food Biochem.* **2019**, *43*, e12761. [CrossRef]
60. Breijyeh, Z.; Jubeh, B.; Karaman, R. Resistance of Gram-negative bacteria to current antibacterial agents and approaches to resolve it. *Molecules* **2020**, *25*, 1340. [CrossRef]
61. El-Saadony, M.T.; Zabermawi, N.M.; Zabermawi, N.M.; Burollus, M.A.; Shafi, M.E.; Alagawany, M.; Abd El-Hack, M.E. Nutritional aspects and health benefits of bioactive plant compounds against infectious diseases: A review. *Food Rev. Int.* **2021**, 1–23. [CrossRef]
62. Benkerroum, N. Antimicrobial peptides generated from milk proteins: A survey and prospects for application in the food industry. A review. *Int. J. Dairy Technol.* **2010**, *63*, 320–338. [CrossRef]
63. Heymich, M.-L.; Nißl, L.; Hahn, D.; Noll, M.; Pischetsrieder, M. Antioxidative, antifungal and additive activity of the antimicrobial peptides Leg1 and Leg2 from chickpea. *Foods* **2021**, *10*, 585. [CrossRef]
64. Schmidt, M.; Arendt, E.K.; Thery, T.L. Isolation and characterisation of the antifungal activity of the cowpea defensin Cp-thionin II. *Food Microbiol.* **2019**, *82*, 504–514. [CrossRef] [PubMed]
65. Karabagias, I.; Badeka, A.; Kontominas, M. Shelf life extension of lamb meat using thyme or oregano essential oils and modified atmosphere packaging. *Meat Sci.* **2011**, *88*, 109–116. [CrossRef]
66. England, E.M.; Matarneh, S.K.; Oliver, E.M.; Apaoblaza, A.; Scheffler, T.L.; Shi, H.; Gerrard, D.E. Excess glycogen does not resolve high ultimate pH of oxidative muscle. *Meat Sci.* **2016**, *114*, 95–102. [CrossRef] [PubMed]
67. Chaijan, M. Lipid and myoglobin oxidations in muscle foods. *Songklanakarin J. Sci. Technol.* **2008**, *30*, 47–53.
68. Hashemi Gahruie, H.; Hosseini, S.M.H.; Taghavifard, M.H.; Eskandari, M.H.; Golmakani, M.-T.; Shad, E. Lipid oxidation, color changes, and microbiological quality of frozen beef burgers incorporated with Shirazi thyme, cinnamon, and rosemary extracts. *J. Food Qual.* **2017**, *2017*, 6350156. [CrossRef]
69. Piñuel, M.L.; Vilcacundo, E.; Boeri, P.A.; Barrio, D.A.; Morales, D.; Pinto, A.; Moran, R.; Samaniego, I.; Carrillo, W. Extraction of protein concentrate from red bean (*Phaseolus vulgaris* L.): Antioxidant activity and inhibition of lipid peroxidation. *J. Appl. Pharm. Sci.* **2019**, *9*, 1–14.
70. Aslam, S.; Shukat, R.; Khan, M.I.; Shahid, M. Effect of dietary supplementation of bioactive peptides on antioxidant potential of broiler breast meat and physicochemical characteristics of nuggets. *Food Sci. Anim. Resour.* **2020**, *40*, 55–73. [CrossRef]
71. Arshad, M.S.; Hina, G.; Anjum, F.M.; Suleria, H.A.R. Effect of milk-derived bioactive peptides on the lipid stability and functional properties of beef nuggets. *Sci. Rep.* **2022**, *12*, 1242. [CrossRef]
72. El-Saadony, M.T.; Abd El-Hack, M.E.; Swelum, A.A.; Al-Sultan, S.I.; El-Ghareeb, W.R.; Hussein, E.O.; Ba-Awadh, H.A.; Akl, B.A.; Nader, M.M. Enhancing quality and safety of raw buffalo meat using the bioactive peptides of pea and red kidney bean under refrigeration conditions. *Ital. J. Anim. Sci.* **2021**, *20*, 762–776. [CrossRef]
73. Sharma, H.; Mendiratta, S.; Agarwal, R.K.; Gurunathan, K. Bio-preservative effect of blends of essential oils: Natural anti-oxidant and anti-microbial agents for the shelf life enhancement of emulsion based chicken sausages. *J. Food Sci. Technol.* **2020**, *57*, 3040–3050. [CrossRef]
74. *4334/2004*; Fresh Meat. Egyptian Organization for Standardization and Quality Control: Cairo, Egypt, 2004.

polymers

Article

The Dependence of the Properties of Recycled PET Electrospun Mats on the Origin of the Material Used for Their Fabrication

Ewa Kijeńska-Gawrońska [1,2,*], Katarzyna Wiercińska [2] and Monika Bil [1]

1 Centre for Advanced Materials and Technologies CEZAMAT, Warsaw University of Technology, Poleczki 19, 02-822 Warsaw, Poland; monika.bil@pw.edu.pl
2 Faculty of Materials Science and Engineering, Warsaw University of Technology, Woloska 141, 02-507 Warsaw, Poland; katarzynka.wiercinska@gmail.com
* Correspondence: ewa.kijenska@pw.edu.pl

Abstract: Plastic materials are one of the significant components of construction materials omnipresent in all areas of the industry and everyday life. One of these plastics is polyethylene terephthalate (PET). Due to its processing properties, with a simultaneous low production cost, PET has been used in many industrial applications, including the production of various types of bottles. Moreover, the high consumption of PET bottles causes the accumulation of large amounts of their waste and necessitates finding an effective way to recycle them. Electrospinning is a well-known non-complicated method for the fabrication of nonwovens from polymers and composites, which can be utilized in many fields due to their outstanding properties. In addition, it might be a promising technique for the recycling of plastic materials. Therefore, in this study, the electrospinning approach for the recycling of two types of PET bottle wastes—bottles made of virgin PET and bottles made of recycled PET (PET bottles) has been utilized, and a comparison of the properties of the obtained materials have been performed. The fibers with diameters of 1.62 ± 0.22, 1.64 ± 0.18, and 1.89 ± 0.19 have been produced from solutions made of virgin PET granulate, PET bottles, and PET bottles made of recycled bottles, respectively. Obtained fibers underwent morphological observation using a scanning electron microscope. Physico-chemical properties using FTIR, gel chromatography, and differential scanning calorimetry have been evaluated, and mechanical properties of obtained mats have been investigated. Cytotoxicity tests using the L929 mouse fibroblast cell line revealed no cytotoxicity for all tested materials.

Keywords: polyethylene terephthalate (PET); recycling; electrospinning; fibers

check for updates

Citation: Kijeńska-Gawrońska, E.; Wiercińska, K.; Bil, M. The Dependence of the Properties of Recycled PET Electrospun Mats on the Origin of the Material Used for Their Fabrication. *Polymers* **2022**, *14*, 2881. https://doi.org/10.3390/polym14142881

Academic Editors: Cristina Cazan and Mihaela Cosnita

Received: 23 June 2022
Accepted: 14 July 2022
Published: 16 July 2022

1. Introduction

Polymer materials, sometimes called the *material of the millennium* [1], are lightweight, pliable, durable materials, easily convertible to many various forms with a low cost of production. Therefore, a broad spectrum of applications was found in industry and everyday life. Their invention and progress in processing have revolutionized the modern world and pushed the limits of human capabilities, as wood, ceramics, glass, copper, bronze, iron or steel did in earlier stages of our civilization [2]. Modern science has not yet found substitutes that can replace plastics in terms of their multi-functionality, features, and properties. However, the main drawback of their utilization is the accumulation of plastic waste in the environment and, of course, the human factor in its improper storage, processing, and littering. The production of the polymers worldwide reached 359 MTs in 2018, with an estimated 18% involvement of polyethylene terephthalate (PET). Only a small amount of its waste, mainly bottles, has been reported to be recycled [3]. Therefore, many actions have been undertaken to reduce its waste and develop alternative recycling methods for PET waste reduction.

PET material used for bottle production is a semi-crystalline thermoplastic with good mechanical properties and is highly resistant to environmental conditions and many chemical reagents. Moreover, it can be recycled multiple times, and recently plastic bottles can be recycled into other bottles, sleeping bags, t-shirts, clothes insulation, etc. [4]. Nonwovens obtained by processing PET waste could also be of great potential as a substrate for various kinds of fibers applied for water filtration and treatment and air filtration. Furthermore, it may also find application in hygienic membranes utilized in the cosmetic industry and beautician services or fabrics for respiratory and medical protection. Their porous structure and high tensile strength allow it for many applications, and its main advantage is low cost and easy access to the material (wastes).

Electrospinning is a versatile, scalable (from laboratory to industry) method for fiber fabrications having diameters ranging from nanometers to sub microns, with controlled morphology, orientation and structure features [5]. It is frequently used for the production of nonwoven for biomedical application as tissue engineering scaffolds [6], wound dressings [7], multisource catalysis [8], filtration [9], water treatment [10] or bio-active agent delivery carrier [11]. On the other hand, virgin PET granulate has been utilized to produce wound dressings [12] and blood vessel grafts [13]. Attempts have also been made to utilize recycled PET bottles for the production of filtering membranes for air and water filtration [14,15]. The mentioned studies have drawn attention to the importance of using post-consumer PET bottles to create new materials, which seems to be the right course of action from the circular economy perspective. These researches focus mainly on the utilization of bottles made of virgin PET (which might be correlated with the inexistence of bottles made out of 100% recycled PET by the time that the studies have been conducted), and comparison to pristine material has not been performed. Moreover, most of the studies regarding the production of fibrous materials from recycled PET use trifluoroacetic acid (TFA) or a mixture of dichloromethane with trifluoroacetic acid for electrospinning of the fibers [16]. When using the PET alone, the choice of TFA does not influence material at a high level. However, if the designed material aims to be a composite with metal oxides, the use of TFA is inadvisable. Therefore, in the presented studies, 1,1,1,3,3,3-hexafluoro-2-propanol (HFP) has been employed to prepare the solution for electrospinning. Though its toxicity is known, no remaining HFP stays in resultant fibers, and its industrial application is possible since it is non-flammable. This property is usually crucial for the utilization of solvents in the industry. It is also utilized to prepare a composite solution containing metallic nanoparticles [17].

This study aimed to use the waste PET bottles made of virgin PET granulate and PET bottles made out of recycled bottles to fabricate electrospun nonwovens and to compare their properties with each other, along with a comparison to electrospun mats produced from PET granulate. In this way, the dependence of mats' properties on the origin of the materials used for their production could be evaluated, and the question of possible recycling of already recycled bottles could be answered. Moreover, by characterizing the main properties of PET mats, we were able to investigate whether the obtained materials might be promising candidates for application for hygienic membranes, air and water filters or other products that stay in contact with human tissues, such as skin, without causing cytotoxic reactions.

To the best of the authors' knowledge, this is the first time that bottles made entirely of recycled PET bottles have been utilized for fibers fabrication, and detailed studies on the comparison of properties of PET fibers made of PET of different origins have been performed.

2. Materials and Methods

2.1. Materials

Polyethylene terephthalate in the form of granules for water bottle manufacturing (LighterTM C93- Equipolymers) was purchased from RESINEX Poland Sp. z o.o. (Warszawa, Poland). Recycled bottles utilized in this study were made from virgin PET (granulate) and entirely (in 100%) from recycled PET. Both types were purchased from the same local supplier, and before use, they contained mineral water. In all experiments, granules of PET were used in the received form, while bottles were firstly cleaned with detergent and deionized water, dried at 37 °C, and cut into square flakes of about 5 × 5 mm. Additionally, 1,1,1,3,3,3-hexafluoro-2-propanol (HFP) utilized as a solvent for electrospinning was obtained from abcr GmbH (Karlsruhe, Germany). HPLC-grade chloroform and absolute ethanol were purchased from Chempur (Piekary Slaskie, Poland). Polystyrene standards utilized for molecular weight measurements and phosphate-buffered saline (PBS) were purchased from Sigma-Aldrich (St. Louis, MO, USA). L929 murine fibroblast cells were obtained from American Type Culture Collection (ATCC) (Manassas, VA, USA). RPMI 1640 medium, fetal bovine serum (FBS), penicillin-streptomycin solution (PS), and trypsin-EDTA were bought from Thermo Fisher Scientific (Waltham, MA, USA). CellTiter 96 Aqueous One solution was purchased from Promega (Madison, WI, USA). Sodium chloride 0.9% solutions were obtained from Baxter (Deerfield, IL, USA).

2.2. *Characterization of Pristine Unprocessed Materials*

2.2.1. Gel Permeation Chromatography (GPC)

Gel permeation chromatograph (Agilent Technologies, Singapore) using a refractive index detector has been utilized to determine the average molecular weights (Mw, Mn) of the PET granulate and PET flakes derived from both types of bottles. First, 4 mg of PET granules were added to 100 μL HFP, heated to 50 °C, and mixed for 30 min to enable dissolution. Then, the solution was mixed with 1.9 mL HPLC-grade chloroform and mixed for another 24 h. In the case of PET bottles, flakes were added to 100 μL HFP and mixed for 30 min, followed by the addition of 1900 μL of HPLC-graded chloroform. Each solution was mixed to obtain the concentration of 2 mg/mL (*w*/*v*). The final step of preparation was filtration through a 0.22 μm porous filter. During testing, 100 μL aliquots of each type of PET solution were inoculated into chloroform and further separated on two linear coupled SEC columns (PLgel 5 mm MIXED-C, Agilent Technologies, Cheadle, UK, 300 × 7.5 mm) at 35 °C and a flow rate of about 0.7 mL/min. Calibration of the system was performed using three polystyrene standards with known molecular weights (Mp ranging from 500 to 1,800,000 g/mol).

2.2.2. Differential Scanning Calorimetry (DSC)

The crystallinity of the pristine materials before processing was evaluated using a differential scanning calorimeter Q2000 (TA Instruments, New Castle, DE, USA) under a nitrogen flow of 50 mL/min. Specimens with the weight of 6–8 mg were heated in aluminum pans from 30 to 300 °C with a rate of 10 °C/min, then cooled back to 30 °C and heated back to 300 °C. Collected data from the second heating run were analyzed using software provided by the equipment manufacturer (TA Universal Analysis).

2.2.3. Surface Hydrophilicity

The surface hydrophilicity of three types of unprocessed materials was assessed through water contact angle (WCA) measurements, and WCA was measured at RT by a sessile drop method using the OCA 20a goniometer (DataPhysics, Filderstadt, Germany). The water droplet size was equal to 1 μL with a dosing rate of 1 μL/s. Images of the droplet shape in contact with material were taken using a CCD camera running in real-time and saved for further analysis.

2.3. *Optimization of PET Mats Electrospinning*

To evaluate the optimal concentration of virgin PET granulate and flakes made out of two types of bottles to obtain uniform fibers, the optimization of the paramaters of the electrospinning process with the usage of different concentrations of the solutions was performed. The summarized list of the utilized solution and parameters has been presented in Table 1. For the preparation of the virgin PET granulate solution first pellets were mixed with 1,1,1,3,3,-hexafluoro-2-propanol at concentrations of 10, 15, and 20% (*w*/*v*) and stirred for 30 min at 50 °C. Furthermore, the solution was stirred for another 24 h at room temperature. In the case of both types of bottle flakes, the desired amount of flakes has been added to HFP to obtain concentrations of 10, 15, and 20% and stirred at RT for 24 h without previous heating. Electrospinning of all types of PET was conducted under 12 kV, with a 1.0 mL/h flow rate using a 27 G flattened needle to a collector distance of 13 cm. A steel plate covered with aluminum foil has been used for collecting the fibers during all the experiments. All processes were carried out under environmental conditions with 23–33% humidity and a temperature of 22–24 °C.

Table 1. Parameters for three types of PET optimization.

	PET Granulate			PET Bottles			Bottles from Recycled PET		
Concentration [% (*w*/*v*)]	10	15	20	10	15	20	10	15	20
Voltage [kV]	12	12	12	12	12	12	12	12	12
Flow rate [mL/h]	1	1	1	1	1	1	1	1	1
Needle diameter [G]	27	27	27	27	27	27	27	27	27
Needle to collector distance [cm]	13	13	13	13	13	13	13	13	13

After electrospinning, all samples were evaluated under the scanning electron microscope (SEM, PhenomX, Eindhoven, The Netherlands) after sputter coating with a 14 nm layer of gold (sputter coater Leica EM SCD 500, Leica Mikrosysteme GmbH, Wien, Austria), at an accelerating voltage of 10 kV. Analysis of images allows for the selection of optimal concentration for the PET mats preparation for further studies. For the samples revealing uniform morphology, fibers' diameters were examined from the obtained SEM images using ImageJ software (National Institute of Health, Bethesda, MD, USA). The average diameter was measured based on the calculation of 100 randomly selected fibers, and the distribution histograms were prepared.

2.4. *Characterization of Fabricated Mats*

2.4.1. Gel Permeation Chromatography (GPC)

Gel permeation chromatography was performed as described for pristine materials and 4 mg of each type of fabricated mats, namely gPET, PET, and rPET, were used.

2.4.2. Differential Scanning Calorimetry (DSC)

The crystallinity of the PET mats, similarly to granules and flakes, was evaluated during annealing using a differential scanning calorimeter.

2.4.3. Fourier Transform Infrared (FTIR) Spectroscopy

Fourier transform infrared spectrophotometer (Thermo Fisher Scientific model Nicolet 6700) was utilized to collect infrared spectra of fabricated electrospun gPET, PET, and rPET mats. Measurements were carried out using the attenuated total reflectance (ATR) mode, and each sample was scanned 64 times at a resolution of 4 cm^{-1} over the wavenumber range of 4000–400 cm^{-1}.

2.4.4. Surface Hydrophilicity

The surface hydrophilicity of three types of fabricated mats was measured following the procedure used for the granulate and bottles' flakes.

2.4.5. Tensile Testing

The tensile properties of the electrospun PET mats were evaluated using a tensile testing machine Instron 5943 (Instron, Norwood, MA, USA). The samples for mechanical testing were cut to 20 mm in length and 4 mm in width and attached to the hydraulic clamps. The tests were performed at a 5 mm/min crosshead speed at RT and ambient humidity.

2.5. Cytotoxicity Evaluation

The cytotoxicity of the gPET, PET, and rPET mats was evaluated according to ISO 10993-5 standard [18]. First, L929 mouse fibroblasts were cultured in RPMI 1640 supplemented with heat FBS and 1% PS at 37 °C in a 5% CO_2 incubator. Fibroblasts were cultivated for 48–72 h for the attainment of sufficient confluence of the cells. After their detachment using 0.05% trypsin-EDTA, cells were seeded in 96-well plates at a concentration of 10^4 in 100 μL of complete media. At the same time, gPET, PET, and rPET mats were exposed to UV for 30 min on each side and rinsed three times with RPMI 1640 with 1% PS to sterilize the samples. Extracts were derived by soaking the mats separately in a complete medium of RPMI1640, 10% FBS, and 1% PS. Referencing ISO standard guidelines, the mass of specimens within the extraction media was altered to 100 mg/mL (ISO 10993-12 [19]). Furthermore, specimens were incubated for 24 h in an incubator. Obtained 24 h extracts were utilized in dilutions of 1, 2.5, 5, and 10x. After aspiration of cell culture media, 100 μL of each extract or its dissolution was added to cells, and fibroblasts were further incubated for another 24 h. Thereafter, media in each well was replaced with 100 μL of 20% MTS solution in RPMI1640 and incubated for 120 min at 37 °C. Then, absorbance of the aliquots was measured at 490 nm using a microplate reader (Spark, Tecan Austria GmbH, Grodig, Austria). The cytotoxicity results are presented as the percentage of the viability cultured in control condition of complete medium.

2.6. Statistical Analysis

Performed experiments were carried out at least in triplicates, and data are expressed as mean ± standard deviation (SD). Post-hoc one-way ANOVA with a Tukey-Kramer pair-wise comparison have been employed for statistical analysis. A value of $p \leq 0.05$ is considered statistically significant, and additional significance is indicated with ** $p < 0.01$ and *** $p < 0.001$. All statistical analyses were calculated using GraphPad Prism version 9.2.0 for Mac OS X (GraphPad Software, La Jolla, CA, USA).

3. Results and Discussion

3.1. Properties of Pristine Materials

To make a characterization of all three types of polyethylene terephthalate, namely granulate of virgin PET, PET bottles, and PET bottles made of recycled PET, gel permeation chromatography, differential scanning calorimetry, and contact angle measurements were performed. Obtained data allowed for molecular weight and crystallinity evaluation, and the results are summarized in Table 2. It can be seen that no significant differences have been observed in the case of molecular weight of all three forms of PET. Moreover, both types of bottles exhibited higher crystallinity than granulate virgin PET. This can be attributed to the mechanism of PET bottle manufacturing. The process of bottle production takes two stages: (i) Melting of the granulate at around 280 °C and processing them into preforms, and (ii) and heating the preforms to about 110 °C and blowing them into bottles. This "stretch blow-molding process" causes the partial crystallization of PET, which after that improves its durability, thermal and barrier anti-carbon dioxide, and oxygen properties [20].

Hydrophilicity measurements revealed the hydrophobic nature of the surface of all three types of unprocessed materials.

3.2. Optimization of Electrospinning

Optimal concentrations for the solutions of virgin PET granulate and both types of PET bottles' flakes were chosen on the basis of the fiber morphology. Figure 1 presents the morphology of the fibers obtained in the optimization process and the distribution of fiber diameters for 20% solutions. As it can be observed, fibers electrospun from 10% (w/v) solutions, regardless of the utilized type of PET, consisted of large beads connected with fragile fibers. Increasing the solution concentration to 15% (w/v) resulted in more uniform fibers, with spindle-like beads formed within their structure. Furthermore, an increase to 20% (w/v) allowed for the attainment of the homogenous structure of non-beaded fibers. In this way, the optimal parameters have been optimized to produce fibrous mats of uniform fibers, and mats made of 20% solutions were chosen to produce the fibers for further studies. The diameters of the fibers made of virgin PET granulate (gPET), PET bottle flakes (PET), and flakes from bottles made of recycled PET (rPET) were 1.62 ± 0.22 µm, 1.64 ± 0.18 µm, and 1.89 ± 0.19 µm, respectively.

Table 2. Properties of unprocessed virgin PET granulate, PET bottles, and PET bottles made of recycled PET.

Properties	PET Granulate	PET Bottles	Bottles from Recycled PET
M_w [g/mol]	61,800	56,600	58,200
M_n [g/mol]	16,300	17,400	13,400
PDI	3.784	3.250	4.091
Crystallinity [%]	12	22	21.6
T_g [°C]	79.74	82.88	82.23
T_m [°C]	246.91	247.10	246.68
Water contact angle [°]	69 ± 7	81 ± 3	79 ± 7

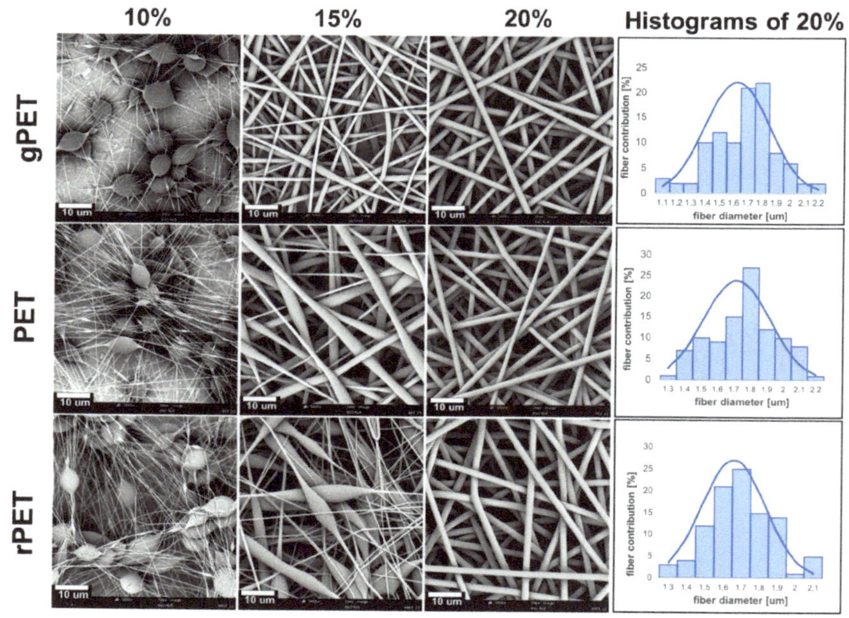

Figure 1. Morphology of obtained fibers during the optimization of electrospinning of PET granulate (gPET), PET bottle flakes (PET), and flakes from bottles made of recycled PET (rPET); histograms obtained for the samples made of 20% PET solutions.

Moreover, the performed Shapiro-Wilk test revealed that obtained distribution of diameters could be described as normal. Based on the properties of normal distribution, it can be concluded that 95% of all diameters in tested mats lie within two standard deviations away from the average. Obtained fibers had larger diameters than fibers obtained by Zander et al. from HFP solutions of PET bottles ranging from 105.5 ± 49 nm to 1039.5 ± 326 nm. However, they used the addition of TBAC salt to decrease bead size and fiber diameters.

3.3. Physico-Chemical and Mechanical Properties of Electrospun Fibrous Mats

To investigate the influence of the electrospinning process on the molecular weight of the poly (ethylene terephthalate), gel permeation chromatography was performed for mats obtained from all types of PET, and molecular weight values have been listed in Table 3. Results obtained for gPET show that the electrospinning process caused a 27% decrease in the M_w, a slight increase of M_n, and a decrease in PDI, compared to PET granulate, which suggests that more uniform macromolecules were observed in the polymer matrix. This might be attributed to random chain scission of PET macromolecules followed by the dissolution of the part of the degraded material having the shortest chain lengths [21]. The DSC results confirm this thesis as the gPET mat is amorphous, which indicates that easier crystallizable short molecules were removed. As stated in the Materials and Methods section, the preparation of PET granulate solution required heating of the granulate in HFP at 50 °C for 30 min during mixing. Moreover, unprocessed PET granulate exhibited only 12% of crystalline phase, which is considerably less compared to the other two types of utilized PET, which showed around 22% crystallinity, and it has been reported that more amorphic materials tend to degrade faster than their more crystalline counterparts due to their increased susceptibility to solvents [22]. Unprocessed PET bottles of both types exhibited similar molecular weight as 56,600 and 58,200 g/mol for PET bottles and bottles made of recycled PET, respectively, and it remained almost unchanged after electrospinning. Therefore, it might be concluded that the fabrication of the mats did not entail any significant change in the polymers' macromolecular structure.

Table 3. Properties of gPET, PET, and rPET mats.

Properties	gPET	PET	rPET
M_w [g/mol]	49,000	57,300	57,900
M_n [g/mol]	17,900	21,000	13,100
PDI	2.747	2.718	4.415
Crystallinity [%]	-	22.2	22.4
T_g [°C]	77.17	80.36	79.03
T_c [°C]	-	195.6	196.18
T_m [°C]	-	247.62	246.7
ΔH_m [J/g]	-	31.1	31.39
Water contact angle [°]	138 ± 6	131 ± 3	132 ± 7

Based on DSC curves obtained for all the tested materials, characteristic temperatures, such as glass transition (T_g), crystallization temperature (T_c) or melting temperature (T_m) with melting enthalpy (ΔHm), were determined. Employing melting enthalpy, the crystallinity of the materials was calculated. In Figure 2A, a comparison of DSC curves obtained for different PET forms has been presented, and Table 3 sums the values of characteristic parameters obtained from these curves. The gPET mat possessed a glass transition temperature of 77.2 °C, which is comparable to other samples and unprocessed materials, as well. There are no peaks from melting of the crystalline phase and crystallization, thus the conclusion is that the mat from gPET was amorphous in structure within the range of temperatures tested. This could have two reasons: As the solvent evaporated rapidly in the electrospinning process performed at the room temperature, macromolecules did not have enough time and energy to arrange appropriately to form a crystalline structure, and as the PET granulate is a pure polymer, there were no nucleating agents, that could start the crystallization process [22]. As can be seen on the graph in Figure 2A, a comparison of PET and rPET revealed that there were no noticeable changes in the shape of DSC curves for these materials. Characteristic temperatures and enthalpy confirm that both mats produced from PET bottles possess almost the same glass transition temperature (T_g at about 80 °C), crystallization temperature (T_c at about 195 °C), melting temperature (T_m at about 247 °C), and melting enthalpy, which indicates that their crystallinity is also very similar (22.2% for

PET and 22.4% for rPET). Both values are almost the same as those of unprocessed PET and rPET (Table 1).

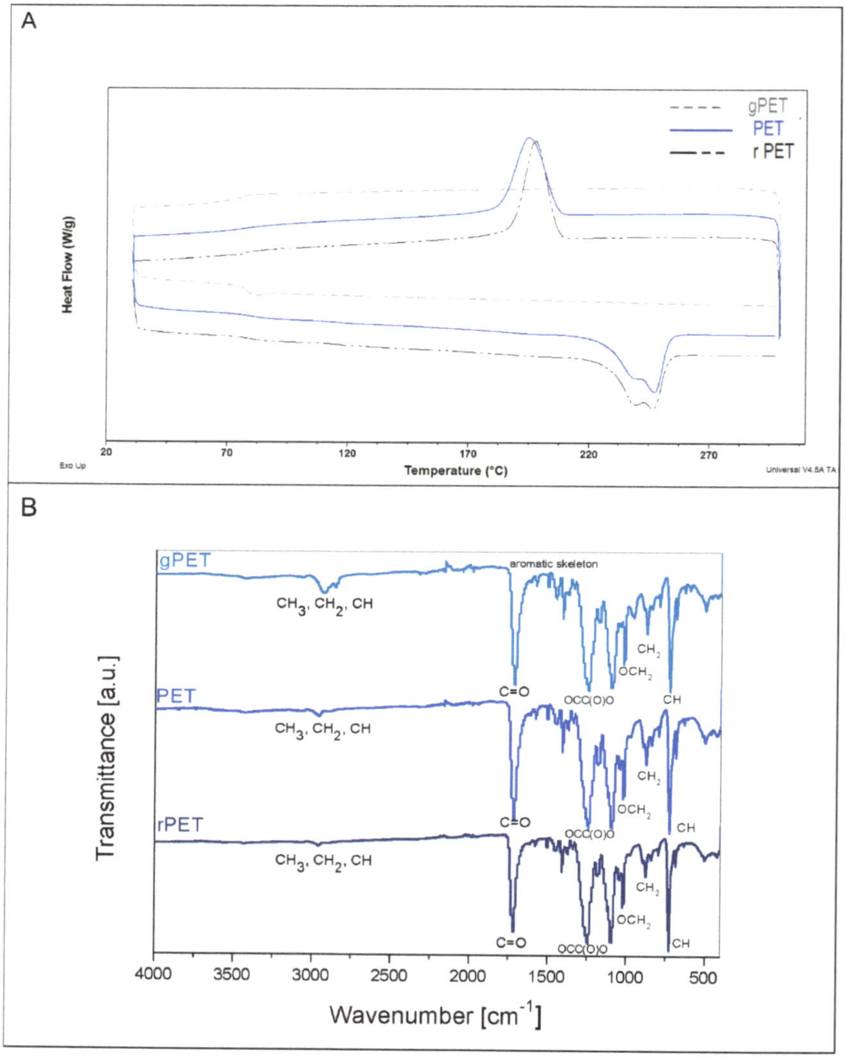

Figure 2. (**A**) Comparison of DSC curves and (**B**) comparison of FTIR spectra for mats made of different types of PET.

FTIR spectra were collected and analyzed to determine if there are any differences between gPET, PET, and rPET chemical structures. Characteristic peaks have been marked, and Figure 2B shows a comparison of spectra obtained for gPET, PET, and rPET mats. Obtained data were compared to the literature finding to identify characteristic bands. Band 724 cm^{-1} was assigned to the interaction of polar ester groups with benzene rings, while 793 cm^{-1} was matched with vibrations of two adjacent aromatic hydrogen atoms in *p*-substituted compounds and aromatic bands. Absorption bands 839, 873, and 1016 cm^{-1} are characteristic of aromatic rings. Other peaks significant for material identification are at 1042 and 1093 cm^{-1} for the methylene group and vibrations of ester C-O bond; 1117 and 1182 cm^{-1} for the terephthalate group. Bands 1340, 1409, and 1452 are assigned to a few different phenomena: Stretching of CO group, deformation of OH group, and bending of the ethylene glycol segment of the molecule. Additionally, 1505 and 1578 cm^{-1} are characteristic bands for vibrations of an aromatic skeleton with stretching of C=C bond, 1714 cm^{-1} is assigned to stretching of C=O bond from carboxylic acid group, and the peak around 2950 cm^{-1} is from stretching of the C-H bond [23]. No significant differences were detected between the gPET, PET, and rPET samples.

To investigate the wettability of electrospun mats, water contact angle measurements have been performed, and the obtained values of WCA are listed in Table 3. Obtained data indicate that gPET, PET, and rPET mats exhibited WCA approximately above 130° (with no significant difference), which indicates their hydrophobic behavior. It is also consistent with the literature, where values of 131 and 133° are reported [24,25].

Tensile tests have been performed to obtain information about the stiffness and the elasticity of the studied mats. Tensile strength [MPa] and strain at break [%] for all electrospun mats were measured and are compared in Figure 3A,B. Mechanical tests revealed how the form of PET influenced tensile strength and strain at the break of PET. The gPET mat exhibited significantly decreased mechanical properties than the remaining materials (tensile strength at ~4,37 MPa and strain at the break at 64%). It can be assigned to its amorphous structure. It can also be noticed that the rPET showed higher values for both studied parameters than the PET. It might result from chain extenders used in the recycling of PET, as they can also cause branching of the polymer, which can increase mechanical properties [26]. Considering the usefulness of the fabricated materials in applications, such as filters or hygienic membranes, their mechanical properties are of great importance. Values of the mechanical strength obtained for mats made of bottles reached a higher level, with a tensile strength of 6.01 $\pm$ 0.43 MPa for PET and 6.67 $\pm$ 1.09 for rPET, than the tensile strength reported for electrospun filters made of PAN or PLLA, possessing tensile strength of 3.8 and 2 MPa, respectively [27,28]. Moreover, electrospun mats made of bottles exhibited very high elongation at break of 235 and 264% for PET and rPET, respectively, which was higher than the elongation of PET electrospun membrane filters obtained by Bonfim et al. [14].

3.4. Cytotoxicity Evaluation

To evaluate the cytotoxicity of produced materials, in vitro studies using L929 mouse fibroblast cell lines have been performed. ISO 10993-5 standard specifies the reduction of cell viability by more than 30% as exhibiting a cytotoxic effect. Therefore, materials demonstrating the viability of the cells cultured with contact with materials' extract of less than 70% of the cell control are considered cytotoxic [17]. No extract from gPET, PET, and rPET mats shows the viability of less than 70% (Figure 4). The lowest viability of 76% has been observed for 1x solute extract of gPET. However, from 2.5x onwards, the viability was around 100%. L929 cells cultured with non-diluted extract of PET and rPET exhibited the viability of 93 and 91%, respectively. As a result, it can be concluded that mats prepared from PET granulate, PET from bottles, and PET from bottles made of recycled PET can be considered non-cytotoxic and might find the application in hygienic products of short intended contact time (i.e., surgical masks [14]).

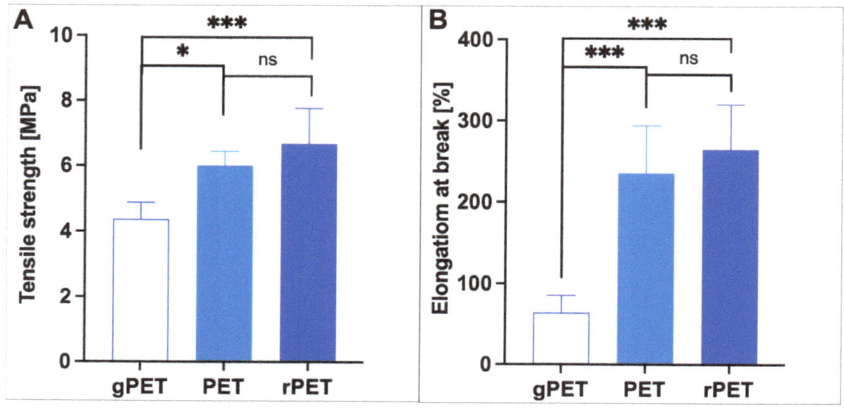

Figure 3. Comparison of (**A**) tensile strength and (**B**) strain at break values for investigated mats. Statistical difference is indicated with * $p \leq 0.05$, *** $p < 0.001$, "ns" indicates no statistical difference.

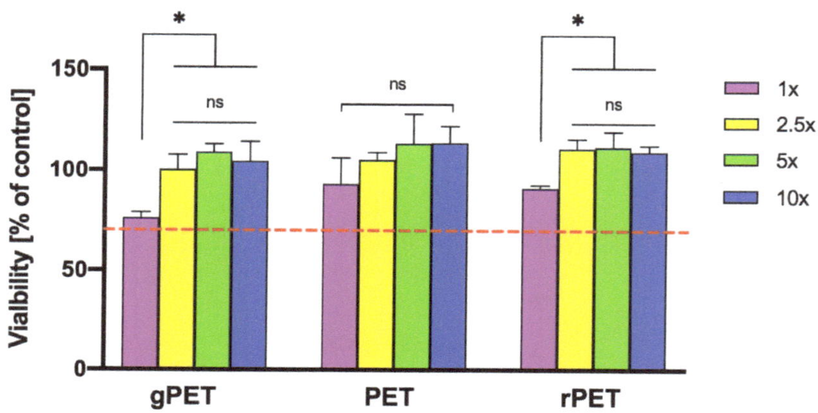

Figure 4. Cytotoxicity (% viability) of L929 fibroblast cell line treated with gPET, PET or rPET mats solute aliquots [1× (undiluted), 2.5, 5, and 10× indicate dilution factor of the solutes]. Red line indicates 70% of the control value. Statistical difference is indicated with * $p \leq 0.05$ and "ns" indicates no statistical difference.

4. Conclusions

In the presented study, electrospun fibers made of PET of different origins have been produced and characterized to investigate the impact of PET substrate on the obtained fibrous mat. Performed optimization of the electrospinning process allowed for the selection of optimal parameters for all three studied types of poly (ethylene terephthalate): PET granulate, PET bottles, and bottles for recycled PET. Physico-chemical characterization and mechanical studies show superior properties of PET and rPET mats compared to mats made of PET granulate. Moreover, cytotoxicity evaluation has shown their non-toxicity, confirming their potential use for materials that can be in contact with human tissue, such as skin with limited contact devices, such as masks.

Future studies on the potential utilization of the fibers made of PET from the bottles should focus on the development of antimicrobial and filtration layered fabrics coupled with particles (i.e., metal particles) or other bio-active agents (i.e., enzymes), which could find application for fabrication of membranes for respiratory protection, air, and water filtration or hygienic materials.

Author Contributions: Conceptualization, E.K.-G.; methodology, E.K.-G.; software, E.K.-G., K.W. and M.B.; validation, E.K.-G. and M.B.; formal analysis, E.K.-G. and K.W.; investigation, E.K.-G. and K.W.; resources, E.K.-G.; data curation, E.K.-G.; writing—original draft preparation, E.K.-G. and K.W.; writing—review and editing, E.K.-G. and M.B.; visualization, E.K.-G., K.W. and M.B.; supervision, E.K.-G.; project administration, E.K.-G.; funding acquisition, E.K.-G. All authors have read and agreed to the published version of the manuscript.

Funding: This work was performed in the frame of the project entitled "The development of antimicrobial and filtration layered fabrics for sanitary and medical protection and fabrication technology based on metal-polymer composites" financially supported by the National Centre for Research and Development (Contract No. POIR.01.01.01-00-1246/20).

Institutional Review Board Statement: Not applicable.

Informed Consent Statement: Not applicable.

Data Availability Statement: The data presented in this study are available on request from the corresponding author.

Conflicts of Interest: The authors declare no conflict of interest.

References

1. Kijeński, J. Tworzywa polimerowe w zrównoważonym rozwoju—od potrzeby użycia do potrzeby zużycia. Cz. I. Nie ma odwrotu od "plastików". *Polimery* **2019**, *64*, 725–738.
2. Kijeński, J. Tworzywa polimerowe w zrównoważonym rozwoju—od potrzeby użycia do potrzeby zużycia. Cz. II. Powrót do monomerów. *Polimery* **2019**, *64*, 740–750.
3. Sadeghi, B.; Marfavi, Y.; AliAkbari, R.; Kowsari, E.; Borbor Ajdari, F.; Ramakrishna, S. Recent Studies on Recycled PET Fibers: Production and Applications: A Review. *Mater. Circ. Econ.* **2021**, *3*, 4. [CrossRef]
4. Sezgin, H.; Yalcin-Enis, I. Turning Plastic Wastes Into Textile Products. In *Handbook of Solid Waste Management: Sustainability through Circular Economy*; Baskar, C., Ramakrishna, S., Baskar, S., Sharma, R., Chinnappan, A., Sehrawat, R., Eds.; Springer: Singapore, 2020; pp. 1–27. [CrossRef]
5. Kijeńska-Gawrońska, E.; Swieszkowski, W. General requirements of electrospun materials for tissue engineering. In *Electrospun Materials for Tissue Engineering and Biomedical Applications: Research, Design and Commercialization*; Elsevier: Amsterdam, The Netherlands, 2017; pp. 3–56. [CrossRef]
6. Dębski, T.; Kijeńska-Gawrońska, E.; Zołocińska, A.; Siennicka, K.; Słysz, A.; Paskal, W.; Włodarski, P.K.; Święszkowski; Pojda, Z.W. Bioactive Nanofiber-Based Conduits in a Peripheral Nerve Gap Management—An Animal Model Study. *Int. J. Mol. Sci.* **2021**, *22*, 5588. [CrossRef]
7. Altan, E.; Karacelebi, Y.; Saatcioglu, E.; Ulag, S.; Sahin, A.; Aksu, B.; Croitoru, A.-M.; Codrea, C.I.; Ficai, D.; Gunduz, O.; et al. Fabrication of Electrospun Juglans regia (Juglone) Loaded Poly(lactic acid) Scaffolds as a Potential Wound Dressing Material. *Polymers* **2022**, *14*, 1971. [CrossRef] [PubMed]
8. Orudzhev, F.; Ramazanov, S.; Sobola, D.; Kaspar, P.; Trčka, T.; Částková, K.; Kastyl, J.; Zvereva, I.; Wang, C.; Selimov, D.; et al. Ultrasound and water flow driven piezophototronic effect in self-polarized flexible α-Fe2O3 containing PVDF nanofibers film for enhanced catalytic oxidation. *Nano Energy* **2021**, *90*, 106586. [CrossRef]
9. Borojeni, I.A.; Gajewski, G.; Riahi, R.A. Application of Electrospun Nonwoven Fibers in Air Filters. *Fibers* **2022**, *10*, 15. [CrossRef]
10. Jankowska, K.; Zdarta, J.; Grzywaczyka, A.; Degórska, O.; Kijeńska-Gawrońska, E.; Pinelo, M.; Jesionowski, T. Horseradish peroxidase immobilised onto electrospun fibres and its application in decolourisation of dyes from model sea water. *Process Biochem.* **2020**, *102*, 10–21. [CrossRef]
11. Kijeńska-Gawrońska, E.; Bolek, T.; Bil, M.; Swieszkowski, W. Alignment and bioactive molecule enrichment of bio-composite scaffolds towards peripheral nerve tissue engineering. *J. Mater. Chem. B* **2019**, *7*, 4509–4519. [CrossRef]
12. Arslan, A.; Simşek, M.; Dalkıranoğlu Aldemir, S.; Kazaroglu, N.; Gümüşderelioğlu, M. Honey-based PET or PET/chitosan fibrous wound dressings: Effect of honey on electrospinning process. *J. Biomater. Sci. Polym. Ed.* **2014**, *25*, 999–1012. [CrossRef]
13. Kotaki, M.; Yong, T.; He, W.; Ramakrishna, S. Surface engineering of electrospun polyethylene terephthalate (PET) nanofibers towards development of a new material for blood vessel engineering. *Biomaterials* **2005**, *26*, 2527–2536.
14. Bonfim, D.P.F.; Cruz, F.G.S.; Bretas, R.E.S.; Guerra, V.G.; Aguiar, M.L. A Sustainable Recycling Alternative: Electrospun PET-Membranes for Air Nanofiltration. *Polymers* **2021**, *13*, 1166. [CrossRef] [PubMed]
15. Zander, N.E.; Gillan, M.; Sweetser, D. Recycled PET Nanofibers for Water Filtration Applications. *Polymers* **2016**, *9*, 247. [CrossRef]
16. Strain, I.N.; Wu, Q.; Pourrahimi, A.M.; Hedenqvist, M.S.; Olsson, R.T.; Andersson, R.L. Electrospinning of recycled PET to generate tough mesomorphic fibre membranes for smoke filtration. *J. Mater. Chem. A* **2014**, *3*, 1632–1640. [CrossRef]
17. Münchow, E.A.; Albuquerque, M.T.; Zero, B.; Kamocki, K.; Piva, E.; Gregory, R.L.; Bottino, M.C. Development and characterization of novel ZnO-loaded electrospun membranes for periodontal regeneration. *Dent. Mater. Off. Publ. Acad. Dent. Mater.* **2015**, *31*, 1038–1051. [CrossRef]

18. *ISO 10993-5:2009*; International Organization for Standarization. Biological Evaluation of Medical Devices—Part 5: Tests for In Vitro Cytotoxicity. ISO: Geneva, Switzerland, 2009.
19. *ISO 10993-12:2012*; International Organization for Standarization. Biological Evaluation of Medical Devices—Part 12: Sample Preparation and Reference Materials. ISO: Geneva, Switzerland, 2012.
20. EFSA. Guidelines on submission of a dossier for safety evaluation by the EFSA of a recycling process to produce recycled plastics intended to be used for manufacture of materials and articles in contact with food—Opinion of the Scientific Panel on food additives, flavourings, processing aids and materials in contact with food (AFC). *EFSA J.* **2008**, *6*, 717. [CrossRef]
21. Mejzler, D. Random degradation of chain polymers accompanied by partial dissolution of the degraded material. *J. Polym. Sci. Part A Gen. Pap.* **1964**, *2*, 1341–1354. [CrossRef]
22. Jones, F.R.; Foreman, J.P. The response of aerospace composites to temperature and humidity. In *Polymer Composites in the Aerospace Industry*, 2nd ed.; Woodhead Publishing Series in Composites Science and Engineering; Woodhead Publishing: Thorston, UK, 2020; pp. 253–287. [CrossRef]
23. Pereira, A.P.; Silva, M.; Lima, É.P., Jr.; Paula, A.; Tommasini, F. Processing and Characterization of PET Composites Reinforced with Geopolymer Concrete Waste. *Mater. Res.* **2017**, *20*, 411–420. [CrossRef]
24. Kurusu, R.S.; Demarquette, N. Surface modification to control the water wettability of electrospun mats. *Int. Mater. Rev.* **2018**, *64*, 1–39. [CrossRef]
25. Veleirinho, B.; Coelho, D.S.; Dias, P.F.; Maraschin, M.; Pinto, R.; Cargnin-Ferreira, E.; Peixoto, A.; Souza, J.A.; Ribeiro-do-Valle, R.M.; Lopes-da-Silva, J.A. Foreign body reaction associated with PET and PET/chitosan electrospun nanofibrous abdominal meshes. *PLoS ONE* **2014**, *9*, e95293. [CrossRef]
26. Additives for Mechanical Plastic Recycling. Available online: https://www.basf.com/ru/ru/media/news-releases/2013/07/p-13-307.html?fbclid=IwAR29d-CqHllifESijytqVwJsp6UXE9MXsag15QIrIuNzAO3nslsHdeg7kaE (accessed on 23 June 2022).
27. Liu, Y.; Park, M.; Ding, B.; Kim, J.; El-Newehy, M.; Al-Deyab, S.S.; Kim, H.-Y. Facile electrospun Polyacrylonitrile/poly(acrylic acid) nanofibrous membranes for high efficiency particulate air filtration. *Fibers Polym.* **2015**, *16*, 629–633. [CrossRef]
28. Wu, C.-S.; Wu, D.-Y.; Wang, S.-S. Bio-based polymer nanofiber with siliceous sponge spicules prepared by electrospinning: Preparation, characterisation, and functionalisation. *Mater. Sci. Eng. C* **2019**, *108*, 110506. [CrossRef] [PubMed]

polymers

Article

Recycled Poly(Ethylene Terephthalate) from Waste Textiles with Improved Thermal and Rheological Properties by Chain Extension

Wen-Jun Wu [1,2], Xiao-Li Sun [1,2,*], Qinghua Chen [1,2,*] and Qingrong Qian [1,2]

1 College of Environmental Science and Engineering, Fujian Normal University, Fuzhou 350007, China; 17805931286@163.com (W.-J.W.); qrqian@fjnu.edu.cn (Q.Q.)
2 Engineering Research Center of Polymer Green Recycling of Ministry of Education, Fujian Normal University, Fuzhou 350007, China
* Correspondence: sunxiaoli@fjnu.edu.cn (X.-L.S.); cqhuar@126.com (Q.C.)

Abstract: Annual production of textile fibers is continuing to rise and the substantial discharge of undegradable waste polyester fibers can cause serious environmental and even health problems. Thus, the recycling and reuse of recycled poly(ethylene terephthalate) from waste textiles (rPET-F) is highly desirable but still challenging. Here, five chain extenders with a different number of epoxy groups per molecules were used to blend with discarded PET fibers and improve its viscosity and quality loss in the recycling process. The molecule weight, thermal properties, rheological properties and macromolecular architecture of modified r-PET were investigated. It was found that all modified rPET-F samples show higher viscosities and better thermal properties. rPET-F modified by difunctional EXOP molecules show linear structure and improved rheological properties. rPET-F modified by polyfunctional commercial ADR and synthesized copolymers exhibit a long chain branched structure and better crystallization. This study reveals a deeper understanding of the chain extension and opens an avenue for the recycling of PET textiles.

Keywords: waste textile; PET fibers; chain extender; mechanical recycling; rheological property

Citation: Wu, W.-J.; Sun, X.-L.; Chen, Q.; Qian, Q. Recycled Poly(Ethylene Terephthalate) from Waste Textiles with Improved Thermal and Rheological Properties by Chain Extension. *Polymers* **2022**, *14*, 510. https://doi.org/10.3390/polym14030510

Academic Editors: Cristina Cazan and Mihaela Cosnita

Received: 9 January 2022
Accepted: 25 January 2022
Published: 27 January 2022

1. Introduction

Polyester fabrics, with poly(ethylene terephthalate) (PET) as the main ingredient, are the largest variety of chemical fibers, accounting for more than 80% of total chemical fibers [1–4]. With the development of the textile industry, the amount of waste polyester generated every year continues to increase significantly [5–8]. PET is a non-renewable petroleum resource, and it is difficult to degrade naturally in the natural environment. A large amount of waste polyester fabrics not only causes a waste of resources, but it also leads to a greater impact on the environment [9–13]. Therefore, the recycling and reuse of waste polyester fabrics is highly desired [14,15].

However, compared with PET bottles and food packaging, recycling PET fibers through mechanical recycling is still a great challenge due to its poor processability. The main problem in the recycling process of discard PET fiber is the lower inherent viscosity and viscosity loss caused by thermal and hydrolytic degradation. In addition, textile products often contain many additives, e.g., pigments, dyes, dispersants, etc., which also limit the recycling and reuse of PET fibers. Therefore, reused PET fibers are normally applied in lower value uses, e.g., reinforced fibers and fillers. In 2018, textile recovery in China was only 15% [16]. The mechanical recycling process has been well established for PET bottles. Substantial research has been carried out to upcycle PET flakes, and the most commonly used method is to add a chain extender during melt processing, which provides a reasonable solution for the recycling of discarded PET fibers [17].

The chain extender is an additive containing at least two functional groups that can react with the groups of macromolecular segments to generate new covalent bonds, which produce polymers with high molecular weight and recovery properties. As far as PET is concerned, the commonly used chain extenders with functional groups that can react with PET end groups include oxazolines [18,19], pyromellitic anhydride [20–24], organic phosphites [25–28] and epoxides [29,30]. Among them, the epoxides chain extenders are extensive, economical and efficient, offering great application prospects. Bikiaris and Karayannidis found that the excess epoxy groups would react with the hydroxyl end groups and react with the new hydroxyl groups created by the bonding of epoxides and carboxyl groups [31]. Haralabakopoulos [30] used five commercial diepoxides as chain extenders to extend PET molecules and they found the modified PET exhibited higher intrinsic viscosities than the unprocessed PET. Moreover, it is proved that the cyclic diepoxides have better chain extension effects than diglycidyl ether.

Polyfunctional chain extenders could also enhance the processability of recycled polyesters. One of the most common representative commercial chain extenders is the Joncryl® ADR series of BASF, with a number average functionality of $f_n > 4$, which are introduced in various polymers to enhance the rheological properties, melt strength and molecular weight [32]. Xiao li-ren et al. studied the effects of Joncryl® ADR 4370s on the relative molecular mass, distribution, branching and gel structure of recycled PET [33]. Supawee Makkam and Wanlop Harnnarongchai studied the effect of Joncryl® ADR 4380 on the improvement of molecular structure and elongation properties of rPET [34]. Additionally, all found that the modified PET has a long chain branched structure and increased molecular weight after modified by ADR, which can cause a significant change in its rheological properties and improving properties.

In order to achieve better recovery properties of rPET, a series of copolymers with epoxy functional groups were also synthesized [35–38]. Benvenuta Tapia et al. synthesized a series of copolymers, e.g., a triblock polymers with middle block of poly(styrene-acrylonitrile) and two end blocks of poly(styrene-glycidyl methacrylate) (SGMA-SAN-SGMA), a triblock polymer with a middle block of poly(butyl acrylate) and extreme blocks of poly(styrene-glycidyl methacrylate) (SGMA-BA-SGMA) and a random copolymer of glycidyl methacrylate and styrene (GMS-*ran*-S) as a chain extender for polyester via reversible addition-fragmentation transfer (RAFT) and nitroxide mediated polymerization (NMP), respectively [39–45]. Moreover, the application of copolymers in the processing of polyester greatly improved the molecular weight and rheological properties of polyester.

However, the chain extension and its effect on the properties of waste PET fibers was rarely investigated. Herein, 1,4-Butanediol diglycidyl ether (EPOX) with 2 epoxy groups, Joncryl® ADR-4468 with about 9–15 epoxy groups and three synthesized poly(glycidyl methacrylate-*co*-methyl methacrylate-*co*-styrene) (P(GMA-co-MMA-co-S)) with 38–132 epoxy groups per molecule, were used as a chain extender of discarded PET fibers. In this study, we present a comprehensive analysis of these five chain extenders with a different number of epoxy groups in order to obtain a deeper understanding of the influence on the molecule weight, molecular architecture and thermal properties of r-PET fibers. This study opens an avenue for the recycling of PET textiles.

2. Materials and Methods

2.1. Materials

Waste PET fibers from polyester textile wastes (referred to as rPET-F) and waste PET bottle flakes from discarded drink bottles (referred to as rPET-B) were supplied by Fujian Baichuan Resources Recycling Science and Company, Quanzhou, Fujian Province, China. rPET-F were exposed to a granulator to produce densified particles before use. Fiber-grade virgin PET particles (vPET) were purchased from DuPont Company, Wilmington, Delaware, USA. 2,2-Azobis(isobutyronitrile) (AIBN, initiator), glycidyl methacrylate (GMA, 99+%), methyl methacrylate (MMA, 99%) and styrene (St, 99%) were provided by Macklin, Shanghai, China. To remove the inhibitor and to obtain epichlorhydrin-free, GMA, MMA and St were passed through the basic alumina column. Tetrahydrofuran (THF, 99%), phenol and 1,1,2,2-tetrachloroethane were provided by Aladdin, Shanghai, China. The chain-transfer agent, S-1-odecyl-S-(α,α'-dimethyl-α''-acetic acid) trithiocarbonate (TC), was synthesized according to the literature [46]. EPOX with 2 epoxy functional groups was purchased from Aldrich and used as received. Joncryl® ADR-4468, a styrene-acrylic multi-functional epoxide oligomeric agent with about 9–15 epoxy functional groups per molecule were purchased by BASF and used as received.

2.2. Synthesis of Copolymers

The block copolymers P(GMA-*co*-MMA-*co*-S) (PGMS) were synthesized via RAFT polymerization in a schlenk tube. The copolymers were prepared by the RAFT polymerization of GMA MMA and St. AIBN as the initiator, TC as the chain transfer agent and monomers and THF as the solvent were subsequently added to the tube. The solution was degassed using freeze-pump-thaw techniques (3 cycles) and the tube was immersed in an oil bath at 80 °C. Polymerization proceeded for 24 h under magnetic stirring and then stopped by cold water. The copolymer was purified by precipitation in methanol and filtration. The light-yellow powder was obtained after drying in a vacuum oven at room temperature for 2 days. PGMS with different content of GMA (defined as PGMS1, PGMS2 and PGMS3) were prepared by adjusting the feeding molar ratio of GMA/MMA/St from 9/2/1, 7/2/1 to 5/2/1.

2.3. Blending Process of PET with Chain Extenders

EPOX, Joncryl® ADR-4468 and synthesized copolymers (PGMS1, PGMS2 and PGMS3) were used as the chain extender for rPET-F. The rPET-F was kept for 24 h in a 110 °C pre-heated oven before the blending process. Mixing of rPET-F with the chain extenders was performed in a RM-200B torque rheometer with a 50 mL internal chamber at 260 °C and 40 rpm for 7 min. 60 g of rPET-F and fixed amount of chain extender (0.5 wt%–1.5 wt% of rPET-F) were added in the intensive mixer. The melted samples after blending were cooled at room temperature.

2.4. The Melt Flow Rate

The melt flow rate (MFR) of PET samples were determined by a MFI452-A melt flow module, using an overhead weight of 2.16 kg at 257 °C. The tests of MFR were in accordance with GB/T 3682-2000.

2.5. Intrinsic Viscosity

The rPET-F, rPET-B, vPET and chain extended rPET-F were dissolved in 50/50 1,1,2,2-tetrachloroethane/phenol solution (*w*/*w*) and intrinsic viscosity (IV) values were measured by the Ubbelohde viscometer at 25 °C. The molecular weight was calculated in Equation (1) as follows [47]:

$$\mathrm{IV} = 2.1 \times 10^{-4} \times (Mn)^{0.82} \tag{1}$$

The insoluble content of the chain extended rPET-F were determined according to Xiao et al. [33].

2.6. Gel Permeation Chromatography

The molecular weights of the copolymers were monitored by gel permeation chromatography (GPC), using Agilent 1260 infinity II equipped with a G7110B isocratic pump and G7162A refractive index detector at a temperature of 50 °C at a flow rate of 1 mL/min with polystyrene standards. The samples dissolved in DMF at room temperature and filtrated by 0.2μm PTFE filter prior to GPC test.

2.7. Nuclear Magnetic Resonance Spectroscopy

Proton nuclear magnetic resonance spectroscopy, ^{1}H NMR (400.13 MHz), was performed on a Bruker 400 MHz spectrometer to measure the copolymer composition. About 10 mg of copolymer samples were dissolved in tritiated chloroform and scanned 32 times at room temperature.

2.8. Thermal Gravimetric Analysis

Thermal gravimetric analysis (TGA) was conducted in an analyzer (Q50, TA Instruments, New Castle, DE, USA) to measure the thermal stability of PLA. Approximately 8 mg of samples were heated from 30 to 600 °C with a heating ramp of 10 °C/min in an inert atmosphere (N_2).

2.9. Differential Scanning Calorimetry

The melting and crystallization behaviors of the PET samples were recorded on a TA Instruments Q20 differential scanning calorimeter (DSC). The samples (6–8 mg by weight) were sealed in aluminum hermetic pans and subjected to a heat/cool/heat cycle over the temperature range 30–280 °C with a linear heating and cooling rate of 10 °C/min. The glass transition temperature (T_g), crystallization and melting behavior of the polymer were determined from the second heating curve and analyzed using the commercially available Universal Analysis software (TA Instruments).

2.10. Dynamic Rheological Measurements

The complex viscosity and modulus were measured in a TA DHR-2 rheometer using standard 25 mm parallel plates. When the temperature stabilized at 265 °C for about 5 min, the sample was loaded between the parallel plates and melted at 265 °C for 1 min. Before each test, the parallel plate compressed the sample to a thickness of 1.00 mm. The stability of samples was checked through the dynamic time sweep test at 1 Hz and 265 °C with the stain amplitude of 1%. Frequency sweeps were performed in the range of 0.01–100 rad/s with a given strain amplitude of 1% within the linear viscoelastic region.

3. Results

3.1. Properties of the rPET-F, rPET-B and vPET

Compared with rPET-B, the recycling of rPET-F is still a challenge and rarely reported. Firstly, the rheological properties, thermal properties and chemical structure of rPET-B, rPET-F and vPET were investigated and compared. The IV value is an important index to measure the quality of polyester and it is also the most commonly used parameter to guide the actual production process. IV and MFR values of three PET samples are summarized in Table 1. The IV values of rPET-F is 0.58 dL/g, lower than rPET-B and vPET. At the same time, the MFR value obtained for rPET-F, rPET-B and vPET were 48.8 g/10 min, 18.6 g/10 min and 14.6 g/10 min. The lower IV value and higher MFR value of rPET-F implied that the decreased molecular weight and higher impurities during its use and recycling procedure [48].

Table 1. The IV, the MFR and thermal properties of PET samples.

Samples	IV (dL/g)	MFR (g/10 min)	T_g (°C)	T_m (°C)	T_c (°C)	T_{onset} (°C)
rPET-F	0.58	48.8	80.1	248.1	209.4	385.8
rPET-B	0.66	18.6	80.7	251.0	202.1	398.7
vPET	0.68	14.6	83.9	254.9	179.0	400.6

Table 1 and Figure 1a present the thermal behaviors. The thermal crystallization temperature (T_c) of rPET-F is 209.4 °C, higher than that of rPET-B and vPET, which is attributed to the additives introduced during spinning (e.g., pigments, dyes, spinning oil, etc.). These impurities serve as nucleation sites to increase the crystallization temperature. In addition, the rPET-F shows a rather broad melting peak, which corresponds to the wide distribution of lamellae thickness caused by a wider molecular weight distribution [49]. Figure 1b displays the TGA trace of the rPET-F, rPET-B and vPET. As shown in Figure 1b and Table 1, the T_{onset} of rPET-B decreases 2 °C compared with vPET. It can also be seen that the T_{onset} of rPET-F reduces 14.8 °C in relation to vPET, which means the thermal stability rPET-F is worse than rPET-B and vPET. Such a result is related to the broken molecular chain mentioned above.

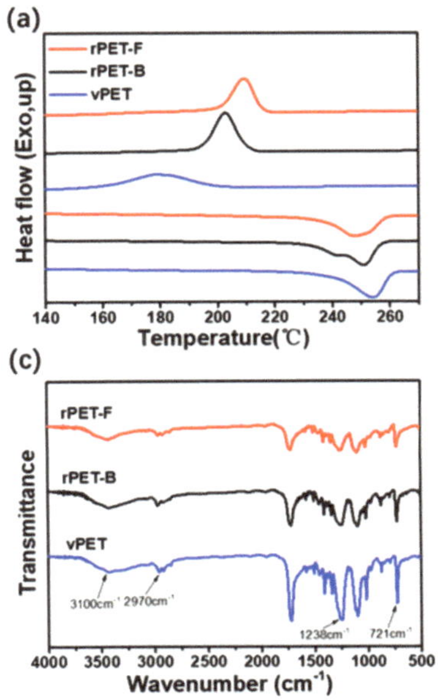

Figure 1. (**a**) DSC curves, (**b**)TGA curves, (**c**) FTIR spectra and (**d**) complex viscosity as function of frequency of rPET-F, rPET-B and vPET.

Figure 1c presents the FT-IR spectra of the PETs signals. The absorption peaks at 721 cm^{-1}, 871 cm^{-1} and 1238 cm^{-1} indicate the presence of the aromatic ring. The absorption peaks at 3100 cm^{-1} and 3500 cm^{-1} are signed to hydroxyl groups (-OH). The peak of methylene oxy (-OCH_2) appears at 1116 cm^{-1}. One can observe that no significant difference in the infrared spectra of the three kinds of PET. This indicates that the chemical structure of the three PETs have basically not changed.

Rheological behavior can well quantify the changes in the molecular weight and structure of polymers. Figure 1d displays complex viscosity as a function of angular frequency for rPET-F, rPET-B and vPET. All the samples show the Newtonian behavior throughout the whole frequency range, which indicates linear chains [50]. It also can be seen that the rheological curve for rPET-F is characterized to be the lowest complex viscosity throughout the whole frequency range. Such behavior is associated with the results observed in the intrinsic viscosity, the melt flow rate and thermal properties, where the rPET-F present a lower molecular weight and more impurities than rPET-B and vPET.

To achieve the upcycling of rPET from waste textiles, the improvement of lower viscosity and worse thermal stability of rPET-F were highly needed. Here, the chain extension of rPET-F was designed to improve their properties. Additionally, the effect of chain extenders with different structure and functional group numbers on the property improvement were investigated in detail.

3.2. Synthesis and Characterization of Chain Extender

To investigate the influence of epoxy groups on the properties and structure of the modified rPET, EPOX, ADR-4468 and PGMSx, copolymers with different GMA content were chose as chain extenders. EPOX is a small molecule, which contains 2 epoxy groups capable of reacting with the carboxyl end of the PET segment (and to a lesser extent, the hydroxyl groups). Additionally, ADR-4468 has 5–9 epoxy groups and the number average molecular weight (M_n) is ~2000. In addition, RAFT polymerization was introduced to synthesis the copolymers PGMS with controlled molecular weight and chemical composition. Figure 2 left shows the ^1H NMR spectrum for PGMS, which presents significant chemical shifts of the phenyl protons of St in the 6.8–7.4 ppm region and the methyl ester group of MMA at 3.6 ppm. Meanwhile, the chemical shifts of the methyl protons and methylene oxy (-OCH_2) protons of GMA at 0.5–1.2 and 3.7–4.5 ppm, respectively. These results suggests that the PGMS copolymers are successfully synthesized.

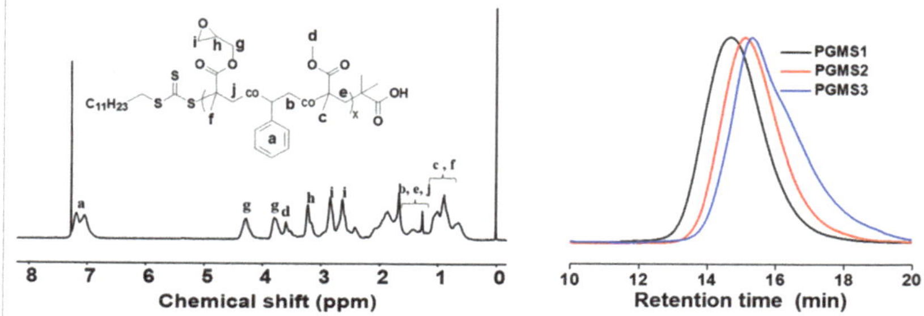

Figure 2. The ^1H NMR spectrum of PGMS in $CDCl_3$ (**left**) and the GPC curves of PGMS copolymers (**right**).

To investigate the influence of epoxy group in PGMS on the structure of the modified PET, PGMSx with different GMA content, PGMS1 with 76.0%, PGMS2 with 70.4%, and PGMS3 with 60.6%, were synthesized. The GPC traces of PGMS1, PGMS2, and PGMS3 are shown in Figure 2 right. The GPC curves of PGMSx are unimodal and symmetric. Furthermore, Table 2 presents the number-average molecular weight (M_n), weight-average molecular weight (M_w) and polymer dispersity index (PDI) of PGMSx from GPC, revealing the M_n in the range of 28,000–63,000 g/mol, the M_w in the range of 51,000–99,000 g/mol and the PDI ranged from 1.58 to 1.80. Meanwhile, the contents of each segment of the PGMSx are listed in Table 2. The average number of epoxy groups per molecule (N_{epoxy}) for PGMS1, PGMS2, and PGMS3 is 38, 70 and 152, respectively. Epoxide number of PGMS1, PGMS2, and PGMS3 is calculated as 0.24, 0.18, and 0.14 mol/100 g, which is lower than ADR (0.17–0.3 mol/100 g) and EXOP (0.99 mol/100 g). Multifunctional epoxide moieties in PGMS play important roles in the chain extension of PET.

Table 2. Chemical composition, molecular weights, and PDIs of five chain extenders.

Samples	GMA (mol%)	S (mol%)	MMA (mol%)	N_{epoxy}	*Mn* (g/mol)	*Mw* (g/mol)	Epoxide Number (mol/100 g)	PDI
PGMS1	76.0 [1]	11.0 [1]	13.0 [1]	152	63,000 [2]	99,000 [2]	0.24	1.58 [2]
PGMS2	70.4 [1]	14.1 [1]	15.5 [1]	70	38,000 [2]	61,000 [2]	0.18	1.59 [2]
PGMS3	60.6 [1]	19.4 [1]	20.0 [1]	38	28,000 [2]	51,000 [2]	0.14	1.80 [2]
ADR-4468	30–50	-	-	5–9	<2500	7250	0.17–0.3	>3
EPOX	-	-	-	2	202	-	0.99	-

[1] Determined from NMR analysis. [2] Determined from GPC analysis.

3.3. IV and MFI of the Chain Extended PET Samples

The chain extension of rPET-F samples was performed in a torque rheometer and x% EPOX/rPET-F, x% ADR/rPET-F and x% PGMSx/rPET-F present the extended rPET-F samples with x% EPOX, ADR and PGMSx as chain extender, respectively. IV and MFR of chain extended rPET-F samples are shown in Table 3. As observed in Table 1, the MFR value of rPET-F is 48.8 g/10 min. It is obvious that with the addition of the chain extender, the MFR of all modified PET samples decrease. The higher the chain extender content is, the higher the MFR values of modified PET. After the chain was extended by EPOX, the MFR of 1.5%EPOX/rPET-F decreases to 24.2 g/10 min since EPOX have a slight effect on molecular weight in the specified blending time. It also could be found that the MFR of 1.5%ADR/rPET-F obviously decreases to 13.4 g/10 min, informing the successful reaction between ADR and rPET-F. Furthermore, the best result of MFR was obtained with the addition of 1.5% PGMS1, which decreases to 10.4 g/10 min. Compared with ADR-4468, PGMS contains more epoxide moieties per chain, which increases the molecular weight and more branched chains of the modified PET chain. In addition, for chain extended rPET-F modified by copolymers, 1.5%PGMS3/rPET-F shows a lower MFR value of 13.2 g/10 min due to the only 38 epoxy groups per chain. Therefore, the epoxy numbers per chain greatly influence the MFI values of modified PETs.

Table 3. The IV and the MFR of chain extended rPET-F samples.

Samples	MFR (g/10 min)	IV (dL/g)	*Mn* (g/mol)	Insoluble Content (%)
0.5%EPOX/rPET-F	50.2	0.62	17,100	-
1.0%EPOX/rPET-F	44.6	0.66	18,400	-
1.5%EPOX/rPET-F	24.2	0.70	19,800	-
0.5%ADR/rPET-F	18.6	0.72	20,500	-
1.0%ADR/rPET-F	15.5	0.78	22,600	-
1.5%ADR/rPET-F	13.4	0.86	25,400	-
0.5%PGMS1/rPET-F	16.3	-		6.7
1.0%PGMS1/rPET-F	13.6	-		10.7
1.5%PGMS1/rPET-F	10.4	-		14.7
0.5%PGMS2/rPET-F	16.4	-		5.3
1.0%PGMS2/rPET-F	14.8	-		10.4
1.5%PGMS2/rPET-F	11.0	-		13.1
0.5%PGMS3/rPET-F	16.3	-		4.0
1.0%PGMS3/rPET-F	13.3	-		8.9
1.5%PGMS3/rPET-F	13.2	-		10.6

Due to the multifunctional groups of chain extenders, crosslinking structure are inevitably formed in the melting expansion process. It was found that rPET samples modified by PGMS are partly insoluble in the solvent during the IV measurements. Thus, an estimation of the crosslinking content of the insoluble samples was conducted and the results are shown in Table 3. It is found that both higher epoxy numbers per chain and higher content of the chain extender significantly increased the insoluble content of the sample.

3.4. Thermal Characterization of the PET Samples

The thermal behaviors of PET samples were investigated, and the results are shown in Figure 3 and Table 4. For the difunctional EPOX, the thermal crystallization temperature (T_c) of 1.5%EPOX/rPET-F reduces to 198.7 °C, 10.7 °C lower than that of rPET-F. Moreover, the T_g of EPOX/rPET-F system decrease as increasing the content of EPOX. This result could be explained by the chain growth of rPET, which hinder the mobility of macromolecules and increase intermolecular entanglement. Such increased molecular entanglement reduces the crystallization rate, the degree of crystallinity, the crystallite size of the formed crystals and the amorphous regions, which increase segmental mobility leading to the lower T_g and T_c.

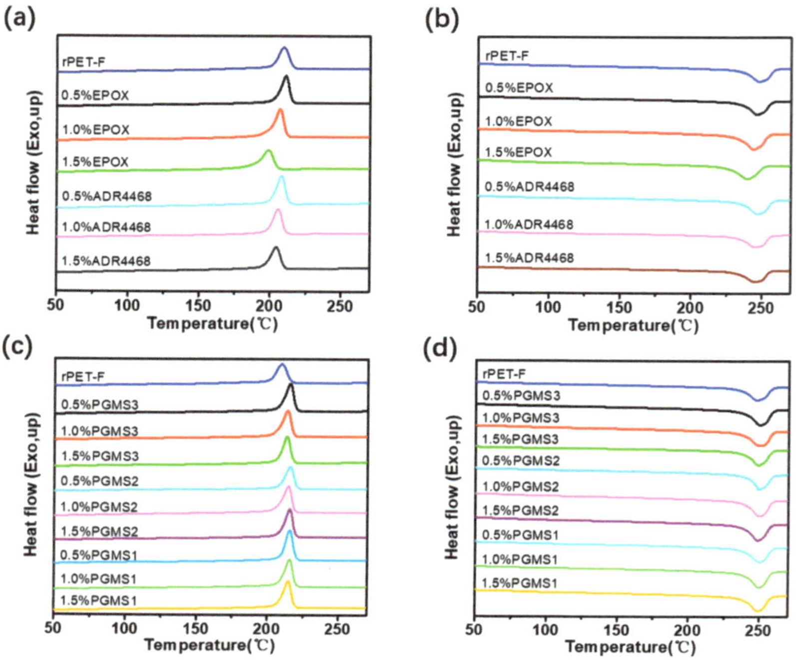

Figure 3. DSC curves for rPET-F modified by EPOX and ADR-4468, (**a**) cooling scan, (**b**) heating scan; DSC curves for rPET-F modified by different PGMS, (**c**) cooling scan, (**d**) heating scan.

Table 4. The crystallization temperature, melting point and glass transition temperature of modified rPET-F blends.

Samples	T_c (°C)	T_m (°C)	T_g (°C)
0.5%EPOX/rPET-F	210.7	246.4	75.7
1.0%EPOX/rPET-F	206.9	244.5	73.7
1.5%EPOX/rPET-F	198.7	240.2	72.1
0.5%ADR/rPET-F	208.0	247.2	77.5
1.0%ADR/rPET-F	205.8	245.5	78.1
1.5%ADR/rPET-F	204.5	245.7	77.5
0.5%PGMS1/rPET-F	215.5	250.4	78.5
1.0%PGMS1/rPET-F	215.5	249.9	77.5
1.5%PGMS1/rPET-F	214.7	249.5	77.9
0.5%PGMS2/rPET-F	215.7	248.9	77.3
1.0%PGMS2/rPET-F	214.5	250.2	79.1
1.5%PGMS2/rPET-F	214.3	249.0	77.5
0.5%PGMS3/rPET-F	215.5	250.4	76.9
1.0%PGMS3/rPET-F	213.8	249.7	78.1
1.5%PGMS3/rPET-F	213.4	249.0	77.6

The polyfunctional chain extenders, ADR-4468 and PGMSx, were also added as a chain extender to rPET-F. Interestingly, the results of PGMS/rPET-F and ADR/rPET-F exhibit completely opposite thermal behaviors and properties. As seen, T_c of PGMS/rPET-F and ADR/rPET-F do not decrease as EPOX/rPET-F. T_c of ADR/rPET-F is lower than that of rPET-F, which decreases from 208.0 °C to 204.5 °C with the increase in ADR addition. Additionally, all PGMS/rPET-F samples show higher T_c than rPET-F, ranged from 213.4 °C to 215.7 °C, which is less related with the content of PGMS. Compared with rPET-F, PGMS/rPET-F and ADR/rPET-F show a slight decrease in T_g, from 80.1 °C to 77–79 °C.

It has been demonstrated that the content of polyfunctional chain extenders within an appropriate value which did not suppress the crystallization as reported [51–53]. The addition of polyfunctional chain extender induces a branched structure, which restricts the movement of PET chains and suppresses the crystallization. Meanwhile, the branched sites could act as crystal nucleation regions which promote the crystallization of PET [51,54]. Moreover, it is interesting to note that the PS segment in the chain extender could also act as nucleating agents to promote crystallization, as previously reported [55]. In addition, the transition from linear to branched chains increases the number of free chain ends and disrupts the packing of the polymer chains, thereby reducing the impediment to segment mobility resulting in a decrease in T_g. These things work together to generate minor changes in T_c and T_g. The inhibiting effect is obvious for ADR and the positive effect is dominant for PGMS/rPET-F. Therefore, via tuning the chemical structure and N_{epoxy} of the chain extender molecule, the chain extended rPET-F samples with different thermal properties and macromolecular structures could be prepared for further applications.

3.5. Rheological Behavior of the PET Samples

The dynamic rheological frequency sweep was conducted to investigate the structures of the chain extended rPET-F samples which were modified by different chain extenders. Figure 4 shows the dynamic time sweep of modified rPET-F for a time period of 30 min at 265 °C and a frequency of 6.28 rad/s, normalized by their initial values at $t = 0$. It is clear that the viscosity of rPET-F exhibits significant reductions with increasing test time, given that a loss in molecular weight caused by the accelerated chain-scission reactions at 265 °C.

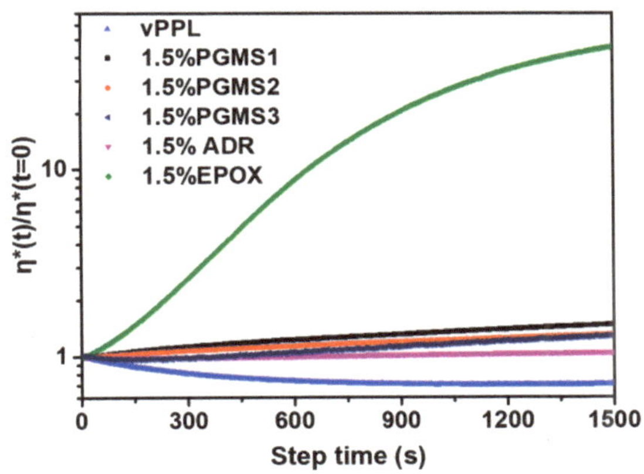

Figure 4. Dynamic time sweeps of various modified rPETs.

The chain extension of rPET-F increases its complex viscosity as shown in Figure 4. Obviously, complex viscosity of extended rPET-F by 1.5% EPOX continues increasing from 39 Pa.s to 1800 Pa.s in the dynamic rheological frequency sweep, which indicates 1.5% EPOX/rPET-F presents an insufficient reaction in the torque rheometer, and EPOX needs a long time to react with rPET-F, even under high temperature and shear force, which is caused by higher epoxide number (0.99 mol/100 g for EPOX) and more epoxy group addition. To obtain the reliable rheological date of EPOX/rPET-F, the samples were annealed at 250 °C for 30 min to accelerate the chain-extension reaction. It should be noted that in the actual processing, the extrusion in the production process of factory could not provide sufficient time for reaction of rPET-F with EPOX. The residual epoxy group continues to react and influences the properties during use.

As shown in Figure 4, the complex viscosity of 1.5% ADR/rPET-F increases only 4% over 25 min, indicating the 1.5% ADR/rPET-F maintains a stable viscoelastic response. The slight increase in complex viscosity of 1.5% ADR/rPET-F also reflected that the ADR reacted fully with rPET-F in the torque rheometer. The PGMSx/rPET-F show roughly similar results to ADR/rPET-F, which demonstrated that the rPET-F have a rapid chain extension reaction with polyfunctional chain extenders.

In order to gain a clear idea of the final polymer topology, we conducted the dynamic strain sweep tests. Figure 5 shows the linear viscoelastic functions (η^*) of chain extended rPET-F versus the frequency. It can be seen that rPET-F presents typical Newtonian behavior throughout the whole frequency range, which is characteristic of linear polymers. After 30 min annealing, EPOX/rPET-F samples show an obvious increase in the complex viscosity compared with rPET-F throughout the whole frequency region, which means the molecular chain growth. Meanwhile, for polyfunctional chain extenders, extended rPET-F shows about 5 to 300 times of complex viscosity higher than that of rPET-F in the low frequency area. Additionally, the complex viscosity of ADR/rPET-F and PGMSx/rPET-F gradually decreases with increasing frequency, which is a typical shear thinning behavior. Such behavior is due to the change in polymer chain entanglement and the enhancement of the relaxation mechanism [21,56].

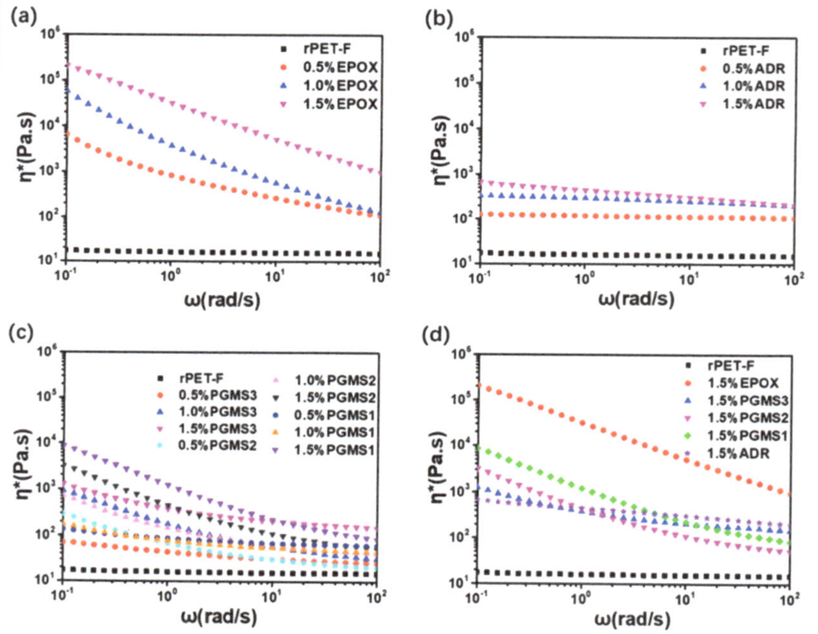

Figure 5. Complex viscosity (η^*) vs. angular frequency (ω) for various modified PETs by (**a**) EPOX, (**b**) ADR-4468, (**c**) PGMSx and (**d**) different chain extender at in the same concentration.

In particular, 1.5% PGMS1/rPET-F presents the highest complex viscosity in the low frequency area and the highest shear sensitivity and non-Newtonian behavior as shown in Figure 5c,d, which is the result of the highest level long-chain branches (LCB) caused by higher N_{epoxy} [57]. At the same time, the increase in the shear thinning make processability more efficient at high rates.

The ADR/rPET-F exhibits a different shape of the viscosity functions, which shows a smaller decrease in complex viscosity with increasing frequency and higher complex viscosity system in a high frequency region compared to PGMSx/rPET-F. It was reported that the value of η^* reflected the different branch structures of modified r-PET, especially at low frequencies [57]. Thus, the above results of ADR/rPET-F might be caused by the lower branched segments [58]. Thus, N_{epoxy} significantly influences the molecular architecture of the modified rPET-F. At the same time, insoluble chains in PGMSx/rPET-F also reveal that the highly branched structure and crosslinking structure was formed, which is consistent with the complex viscosity results.

Figure 6a shows the Han plots of rPET-F and modified rPET-F, the loss modulus (G″) as a function of storage modulus (G′). It can be seen that the curves are shifting toward to right with the increase in the amount of chain extender and number of epoxy groups on the chain extender molecule, which demonstrates that the extent of the G′ change is slightly higher than that of the G″ change with the frequency. The increase in G′ reveals that the behavior of modified PET by polyfunctional chain extenders transformed from liquid-like to solid-like with better melt elasticity. It is well known that melt elasticity is directly related to melt strength, which means that the chain extender improves the melt strength of the r-PET [39]. Figure 6b presents the damping factor (tan δ) of all samples, which referred to the angle that the strain lagged behind the stress. It is obvious that tan δ of rPET-F is much higher than that of modified rPET-F by polyfunctional chain extenders, which demonstrated that rPET-F exhibit more viscous than elastic. On the other hand, the tan δ of modified rPET-F decrease with the increase in the amount of chain extender and number of epoxy groups present on the chain extender molecule, which ascribes to the formation of a more branched structure [58–60].

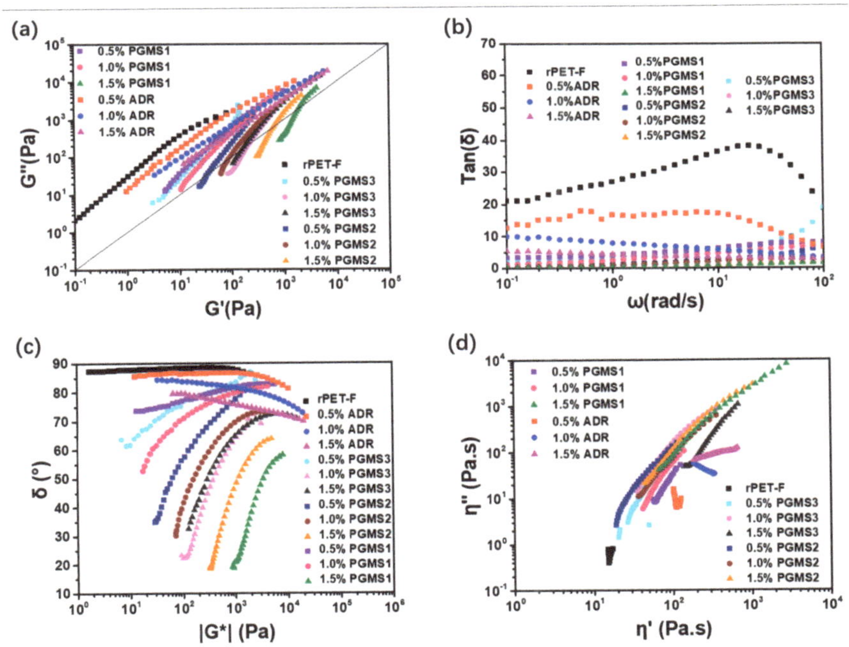

Figure 6. Han plots for various modified PETs (**a**), dynamic loss tangent (tan δ) as a function of ω for various modified PETs (**b**), vGP plots for various modified PETs (**c**) and Cole–Cole plots for various modified PETs (**d**).

To further distinguish the differences in the molecular structure of rPET-F modified by different chain extenders, the van Gurp-Palmen (vGP) plot and the Cole-Cole plot were drawn. By plotting phase angle verse complex modulus, the van Gurp-Palmen (vGP) plot is frequently used to characterize the topology of polymers [61–63]. It was known from the literature that the lower phase angle indicates the higher level of LCB [64]. As shown in Figure 6c, the rPET-F exhibits the typical behavior of a linear polymer, with a plateau of 90° in the whole |G*| range. For the 0.5% ADR/rPET-F, the vGP plots shows the same trend as rPET-F, suggesting that ADR/rPET-F system with the 0.5% ADR-4468 is a linear polymer. Moreover, the 1.0% ADR/rPET-F and 1.5% ADR/rPET-F present a lower δ in the higher value interval, which indicates that the existence of LCB with high content of ADR-4468. The vGP plots of the PGMS/rPET-F system deviate from the curve of the rPET-F and moves to a lower phase angle position. Therefore, 1.5% PGMS/rPET-F presents the highest LCB degree.

Previous research [58,60,65] has shown that the Cole-Cole plots of a linear chain is semi-circular and the diameter of the semicircle become larger as the molecular weight increases. The Cole-Cole plots of r-PET samples are shown in Figure 6d, and it can be seen that the Cole-Cole plots deviate from the semi-circular shape and gradually rise in the low-frequency region, which is due to the rheological changes in the longer relaxation time in the low frequency region, which is caused by the appearance of long-branched structures [66–68]. The irregular shape of the semicircle is caused by the low viscosity of rPET-F. The radius of the PGMS/rPET-F curve gradually increases, and the upward warping trend becomes more obvious with the increase in the amount of chain extender and N_{epoxy}, which demonstrates an increase in the long-branched structure. However, the radius of the Cole-Cole curve of the ADR/rPET-F increased without the upward warping trend, indicating a lower degree of branching. Based on the results above, it can be seen that more functional groups lead to a more branched chain of the modified rPET-F.

According to the rheological behavior results of modified rPET-F, the average number of epoxy groups, as the paramount reaction site in chain extension, plays a pivotal role in the branch structures formation in modified rPET-F samples. Hence, branching is more likely to happen when ADR and copolymers are used with a high number of epoxy groups, as shown in Scheme 1. The above results show that the addition of chain extenders increases the complex viscosity of rPET-F. Moreover, the van Gurp-Palmen plot and Cole-Cole plot of modified rPET-F indicate higher levels of branching in the chain extender with more epoxy groups.

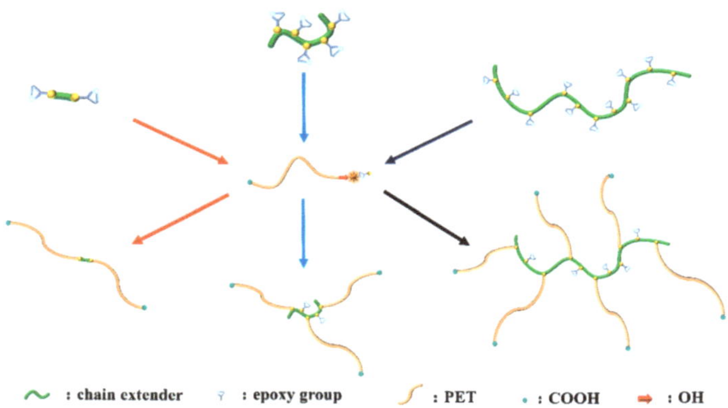

Scheme 1. Schematic representation of chain extended rPET-F in this study.

4. Conclusions

In this study, in order to achieve the upcycling of rPET-F, we synthesized a series of reactive copolymers PGMS with a high content of epoxy groups via RAFT polymerization as the chain extender of rPET-F, and compared this with EPOX and ADR-4468, which contain 2 and 5–9 epoxy groups, respectively. The effect of the number of epoxy groups on the thermal properties, rheological properties, and molecular structure changes of modified rPET-F were explored. Chain extension obviously improves MFR and IV values, and the crystallization temperature of EPOX/rPET-F and ADR-4468/rPET-F shifts to lower temperature, but PGMS addition leads to a higher crystallization rate compared to rPET-F. The rheological analysis shows that ADR-4468 and PGMS present faster reactive rates than

EPOX, implying that ADR-4468 and PGMS possess better prospects in the actual production process in a factory setting. ADR-4468/rPET-F exhibit slight shear thinning behavior and a low degree of branching rheological results. On the other hand, the PGMS/rPET-F present more pronounced shear thinning behavior with higher LCB levels, resulting from more epoxy groups per chain. The study provides an effective strategy for the high-value utilization of rPET from waste textiles, suggesting that rPET-F with different qualities can be modified by chain extenders with a different number of epoxy groups, and can be used in a wide range of applications, such as spinning, film and foam materials, etc.

Author Contributions: Methodology, validation and formal analysis, W.-J.W. and X.-L.S.; writing—original draft preparation, W.-J.W. and X.-L.S.; writing—review and editing, X.-L.S., Q.Q. and Q.C. All authors have read and agreed to the published version of the manuscript.

Funding: This research was funded by the National Key Research and Development Program of China (grant number 2019YFC1904500), National Natural Science Foundation of China (grant number 21801251), Natural Science Foundation of Fujian Province (grant number 2021J01199) and Key Project of Science and Technology Innovation of Fujian Province (grant number 2021G02022).

Institutional Review Board Statement: Not applicable.

Informed Consent Statement: Not applicable.

Data Availability Statement: The data presented in this study are available on request from the corresponding authors.

Conflicts of Interest: The authors declare no conflict of interest.

References

1. Damayanti, D.; Wulandari, L.A.; Bagaskoro, A.; Rianjanu, A.; Wu, H.S. Possibility routes for textile recycling technology. *Polymers* **2021**, *13*, 3834. [CrossRef] [PubMed]
2. Al-Salem, S.M.; Lettieri, P.; Baeyens, J. Recycling and recovery routes of plastic solid waste (PSW): A review. *Waste Manag.* **2009**, *29*, 2625–2643. [CrossRef] [PubMed]
3. Aizenshtein, E.M. Polyester fibres in the post-crisis period. *Fibre Chem.* **2011**, *42*, 341–349. [CrossRef]
4. Zou, Y.; Reddy, N.; Yang, Y. Reusing polyester/cotton blend fabrics for composites. *Compos. Part B Eng.* **2011**, *42*, 763–770. [CrossRef]
5. Mäkelä, M.; Rissanen, M.; Sixta, H. Machine vision estimates the polyester content in recyclable waste textiles. *Resour. Conserv. Recycl.* **2020**, *161*, 105007. [CrossRef]
6. Sandin, G.; Peters, G.M. Environmental impact of textile reuse and recycling—A review. *J. Clean. Prod.* **2018**, *184*, 353–365. [CrossRef]
7. Harmsen, P.; Scheffer, M.; Bos, H. Textiles for circular fashion: The logic behind recycling options. *Sustainability* **2021**, *13*, 9714. [CrossRef]
8. Keßler, L.; Matlin, S.A.; Kümmerer, K. The contribution of material circularity to sustainability—Recycling and reuse of textiles. *Curr. Opin. Green Sustain. Chem.* **2021**, *32*, 100535. [CrossRef]
9. Taniguchi, I.; Yoshida, S.; Hiraga, K.; Miyamoto, K.; Kimura, Y.; Oda, K. Biodegradation of PET: Current status and application aspects. *ACS Catal.* **2019**, *9*, 4089–4105. [CrossRef]
10. Fortuna, L.M.; Diyamandoglu, V. Optimization of greenhouse gas emissions in second-hand consumer product recovery through reuse platforms. *Waste Manag.* **2017**, *66*, 178–189. [CrossRef]
11. Thiounn, T.; Smith, R.C. Advances and approaches for chemical recycling of plastic waste. *J. Polym. Sci.* **2020**, *58*, 1347–1364. [CrossRef]
12. Shamsi, R.; Abdouss, M.; Sadeghi, G.M.M.; Taromi, F.A. Synthesis and characterization of novel polyurethanes based on aminolysis of poly(ethylene terephthalate) wastes, and evaluation of their thermal and mechanical properties. *Polym. Int.* **2009**, *58*, 22–30. [CrossRef]
13. Burgess, S.K.; Leisen, J.E.; Kraftschik, B.E.; Mubarak, C.R.; Kriegel, R.M.; Koros, W.J. Chain mobility, thermal, and mechanical properties of poly(ethylene furanoate) ccompared to poly(ethylene terephthalate). *Macromolecules* **2014**, *47*, 1383–1391. [CrossRef]
14. Ghisellini, P.; Cialani, C.; Ulgiati, S. A review on circular economy: The expected transition to a balanced interplay of environmental and economic systems. *J. Clean. Prod.* **2016**, *114*, 11–32. [CrossRef]
15. Mao, Y.; Li, Q.; Wu, C. Surface modification of PET fiber with hybrid coating and its effect on the properties of PP composites. *Polymers* **2019**, *11*, 1726. [CrossRef] [PubMed]
16. Zhu, X.O. The circular economy opportunity for urban & industrial innovation in China. *Circ. Econ. Perspect. Ser.* 2018. Available online: https://ellenmacarthurfoundation.org/urban-and-industrial-innovation-in-china (accessed on 18 September 2018).

17. Standau, T.; Nofar, M.; Dörr, D.; Ruckdäschel, H.; Altstädt, V. A review on multifunctional epoxy-based Joncryl® ADR chain extended thermoplastics. *Polym. Rev.* **2021**, 1–55. [CrossRef]
18. Cardi, N.; Po, R.; Giannotta, G.; Occhiello, E.; Garbassi, F.; Messina, G. Chain extension of recycled poly(ethylene terephthalate) with 2,2′-Bis (2-oxazoline). *J. Appl. Polym. Sci.* **1993**, *50*, 1501–1509. [CrossRef]
19. Karayannidis, G.P.; Psalida, E.A. Chain extension of recycled poly(ethylene terephthalate) with 2,2′-(1,4-phenylene) Bis (2-oxazoline). *J. Appl. Polym. Sci.* **2000**, *77*, 2206–2211. [CrossRef]
20. Kruse, M.; Wagner, M.H. Rheological and molecular characterization of long-chain branched poly(ethylene terephthalate). *Rheol. Acta* **2017**, *56*, 887–904. [CrossRef]
21. Härth, M.; Kaschta, J.; Schubert, D.W. Shear and elongational flow properties of long-chain branched poly(ethylene terephthalates) and correlations to their molecular structure. *Macromolecules* **2014**, *47*, 4471–4478. [CrossRef]
22. Kruse, M.; Wang, P.; Shah, R.S.; Wagner, M.H. Analysis of high melt-strength poly(ethylene terephthalate) produced by reactive processing by shear and elongational rheology. *Polym. Eng. Sci.* **2019**, *59*, 396–410. [CrossRef]
23. Forsythe, J.S.; Cheah, K.; Nisbet, D.R.; Gupta, R.K.; Lau, A.; Donovan, A.R.; O'Shea, M.S.; Moad, G. Rheological properties of high melt strength poly(ethylene terephthalate) formed by reactive extrusion. *J. Appl. Polym. Sci.* **2006**, *100*, 3646–3652. [CrossRef]
24. Daver, F.; Gupta, R.; Kosior, E. Rheological characterisation of recycled poly(ethylene terephthalate) modified by reactive extrusion. *J. Mater. Process. Technol.* **2008**, *204*, 397–402. [CrossRef]
25. Jacques, B.; Devaux, J.; Legras, R.; Nield, E. Investigation on model molecules of the reactions induced by triphenyl phosphite addition during polyester processing. *Macromolecules* **1996**, *29*, 3129–3138. [CrossRef]
26. Bimestre, B.H.; Saron, C. Chain extension of poly(ethylene terephthalate) by reactive extrusion with secondary stabilizer. *Mater. Res.* **2012**, *15*, 467–472. [CrossRef]
27. Cicero, J.A.; Dorgan, J.R.; Dec, S.F.; Knauss, D.M. Phosphite stabilization effects on two-step melt-spun fibers of polylactide. *Polym. Degrad. Stab.* **2002**, *78*, 95–105. [CrossRef]
28. Jacques, B.; Devaux, J.; Legras, R.; Nield, E. Reactions induced by triphenyl phosphite addition during melt mixing of PET/PBT blends: Chromatographic evidence of a molecular weight increase due to the creation of bonds of two different natures. *Polymer* **1997**, *38*, 5367–5377. [CrossRef]
29. Guo, B.; Chan, C. Chain Extension of Poly(butylene terephthalate) by reactive extrusion. *J. Appl. Polym. Sci.* **1999**, *71*, 1827–1834. [CrossRef]
30. Haralabakopoulos, A.A.; Tsiourvas, D.; Paleos, C.M. Chain extension of poly(ethylene terephthalate) by reactive blending using diepoxides. *J. Appl. Polym. Sci.* **1999**, *71*, 2121–2127. [CrossRef]
31. Achilias, D.S.; Bikiaris, D.N.; Karavelidis, V.; Karayannidis, G.P. Effect of silica nanoparticles on solid state polymerization of poly(ethylene terephthalate). *Eur. Polym. J.* **2008**, *44*, 3096–3107. [CrossRef]
32. Villalobos, M.; Awojulu, A.; Greeley, T.; Turco, G.; Deeter, G. Oligomeric chain extenders for economic reprocessing and recycling of condensation plastics. *Energy* **2006**, *31*, 3227–3234. [CrossRef]
33. Xiao, L.; Wang, H.; Qian, Q.; Jiang, X.; Liu, X.; Huang, B.; Chen, Q. Molecular and structural analysis of epoxide-modified recycled poly(ethylene terephthalate) from rheological data. *Polym. Eng. Sci.* **2012**, *52*, 2127–2133. [CrossRef]
34. Makkam, S.; Harnnarongchai, W. Rheological and mechanical properties of recycled PET modified BY reactive extrusion. *Energy Procedia* **2014**, *56*, 547–553. [CrossRef]
35. Awaja, F.; Pavel, D. Recycling of PET. *Eur. Polym. J.* **2005**, *41*, 1453–1477. [CrossRef]
36. Chen, C.W.; Liu, P.H.; Lin, F.J.; Cho, C.J.; Wang, L.Y.; Mao, H.I.; Chiu, Y.C.; Chang, S.H.; Rwei, S.P.; Kuo, C.C. Influence of different molecular weights and concentrations of poly(glycidyl methacrylate) on recycled poly(ethylene terephthalate): A thermal, mechanical, and rheological study. *J. Polym. Environ.* **2020**, *28*, 2880–2892. [CrossRef]
37. Tan, Z.; Liu, S.; Cui, X.; Sun, S.; Zhang, H. Application of macromolecular chain extender and contribution to the toughening of poly(ethylene terephthalate). *J. Thermoplast. Compos. Mater.* **2016**, *29*, 833–849. [CrossRef]
38. Montava-Jorda, S.; Lascano, D.; Quiles-Carrillo, L.; Montanes, N.; Boronat, T.; Martinez-Sanz, A.V.; Ferrandiz-Bou, S.; Torres-Giner, S. Mechanical recycling of partially bio-based and recycled polyethylene terephthalate blends by reactive extrusion with poly(styrene-co-glycidyl methacrylate). *Polymers* **2020**, *12*, 174. [CrossRef]
39. Benvenuta Tapia, J.J.; Tenorio-López, J.A.; Martínez-Estrada, A.; Guerrero-Sánchez, C. Application of RAFT-synthesized reactive tri-block copolymers for the recycling of post-consumer R-PET by melt processing. *Mater. Chem. Phys.* **2019**, *229*, 474–481. [CrossRef]
40. Benvenuta-Tapia, J.J.; González-Coronel, V.J.; Soriano-Moro, G.; Martínez-De la Luz, I.; Vivaldo-Lima, E. Recycling of poly(ethylene terephthalate) by chain extension during reactive extrusion using functionalized block copolymers synthesized by RAFT polymerization. *J. Appl. Polym. Sci.* **2018**, *135*, 46771. [CrossRef]
41. Benvenuta-Tapia, J.J.; Vivaldo-Lima, E.; Guerrero-Santos, R. Effect of copolymers synthesized by nitroxide-mediated polymerization as chain extenders of postconsumer poly(ethylene terephthalate) waste. *Polym. Eng. Sci.* **2019**, *59*, 2255–2264. [CrossRef]
42. Benvenuta-Tapia, J.J.; Vivaldo-Lima, E. Reduction of molar mass loss and enhancement of thermal and rheological properties of recycled poly(lactic acid) by using chain extenders obtained from RAFT chemistry. *React. Funct. Polym.* **2020**, *153*, 104628. [CrossRef]

43. Benvenuta Tapia, J.J.; Hernández Valdez, M.; Cerna Cortez, J.; Díaz García, V.M.; Landeros Barrios, H. Improving the rheological and mechanical properties of recycled PET modified by macromolecular chain extenders synthesized by controlled radical polymerization. *J. Polym. Environ.* **2018**, *26*, 4221–4232. [CrossRef]
44. Vargas, M.A.; Benvenuta Tapia, J.J.; Sánchez, A.; Manero, O. Asphalt modified with reactive tri-block polymers obtained by via reversible addition-fragmentation chain transfer polymerization. *J. Appl. Polym. Sci.* **2019**, *136*, 47201. [CrossRef]
45. Benvenuta-Tapia, J.J.; Champagne, P.; Tenorio-López, J.A.; Vivaldo-Lima, E.; Guerrero-Santos, R. Improving recycled poly(Lactic acid) biopolymer properties by chain extension using block copolymers synthesized by nitroxide-mediated polymerization (nmp). *Polymers* **2021**, *13*, 2791. [CrossRef] [PubMed]
46. Lai, J.T.; Filla, D.; Shea, R. Functional polymers from novel carboxyl-terminated trithiocarbonates as highly efficient RAFT agents. *Macromolecules* **2002**, *35*, 6754–6756. [CrossRef]
47. Allen, N.S.; Edge, M.; Mohammadian, M.; Jones, K. Physicochemical aspects of the environmental degradation of poly(ethylene terephthalate). *Polym. Degrad. Stab.* **1994**, *43*, 229–237. [CrossRef]
48. Bremner, T.; Rudin, A.; Cook, D.G. Melt flow index values and molecular weight distributions of commercial thermoplastics. *J. Appl. Polym. Sci.* **1990**, *41*, 1617–1627. [CrossRef]
49. Włochowicz, A.; Eder, M. Distribution of lamella thicknesses in isothermally crystallized polypropylene and polyethylene by differential scanning calorimetry. *Polymer* **1984**, *25*, 1268–1270. [CrossRef]
50. Wood-Adams, P.M.; Dealy, J.M.; DeGroot, A.W.; Redwine, O.D. Effect of molecular structure on the linear viscoelastic behavior of polyethylene. *Macromolecules* **2000**, *33*, 7489–7499. [CrossRef]
51. Nofar, M.; Zhu, W.; Park, C.B.; Randall, J. Crystallization kinetics of linear and long-chain-branched polylactide. *Ind. Eng. Chem. Res.* **2011**, *50*, 13789–13798. [CrossRef]
52. Nofar, M.; Zhu, W.; Park, C.B. Effect of dissolved CO_2 on the crystallization behavior of linear and branched PLA. *Polymer* **2012**, *53*, 3341–3353. [CrossRef]
53. Nofar, M. Synergistic effects of chain extender and nanoclay on the crystallization behaviour of polylactide. *Int. J. Mater. Sci. Res.* **2018**, *1*, 1–8. [CrossRef]
54. Nofar, M.; Oğuz, H. Development of PBT/recycled-PET blends and the influence of using chain extender. *J. Polym. Environ.* **2019**, *27*, 1404–1417. [CrossRef]
55. Hung, C.Y.; Wang, C.C.; Chen, C.Y. Enhanced the thermal stability and crystallinity of polylactic acid (PLA) by incorporated reactive PS-b-PMMA-b-PGMA and PS-b-PGMA block copolymers as chain extenders. *Polymer* **2013**, *54*, 1860–1866. [CrossRef]
56. Härth, M.; Dörnhöfer, A.; Kaschta, J.; Münstedt, H.; Schubert, D.W. Molecular structure and rheological properties of a poly(ethylene terephthalate) modified by two different chain extenders. *J. Appl. Polym. Sci.* **2021**, *138*, 1–14. [CrossRef]
57. Arayesh, H.; Golshan Ebrahimi, N.; Khaledi, B.; Khabazian Esfahani, M. Introducing four different branch structures in PET by reactive processing—A rheological investigation. *J. Appl. Polym. Sci.* **2020**, *137*, 49243. [CrossRef]
58. Chen, J.; Wei, W.; Qian, Q.; Xiao, L.; Liu, X.; Xu, J.; Huang, B.; Chen, Q. The structure and properties of long-chain branching poly(trimethylene terephthalate). *Rheol. Acta* **2014**, *53*, 67–74. [CrossRef]
59. Graebling, D. Synthesis of branched polypropylene by a reactive extrusion process. *Macromolecules* **2002**, *35*, 4602–4610. [CrossRef]
60. Tian, J.; Yu, W.; Zhou, C. The preparation and rheology characterization of long chain branching polypropylene. *Polymer* **2006**, *47*, 7962–7969. [CrossRef]
61. Trinkle, S.; Friedrich, C. Van Gurp-Palmen-plot: A way to characterize polydispersity of linear polymers. *Rheol. Acta* **2001**, *40*, 322–328. [CrossRef]
62. Trinkle, S.; Walter, P.; Friedrich, C. Van Gurp-Palmen plot II—Classification of long chain branched polymers by their topology. *Rheol. Acta* **2002**, *41*, 103–113. [CrossRef]
63. Yang, Z.; Xin, C.; Mughal, W.; Li, X.; He, Y. High-melt-elasticity poly(ethylene terephthalate) produced by reactive extrusion with a multi-functional epoxide for foaming. *J. Appl. Polym. Sci.* **2018**, *135*, 45805. [CrossRef]
64. Liu, J.; Lou, L.; Yu, W.; Liao, R.; Li, R.; Zhou, C. Long chain branching polylactide: Structures and properties. *Polymer* **2010**, *51*, 5186–5197. [CrossRef]
65. García-Franco, C.A.; Srinivas, S.; Lohse, D.J.; Brant, P. Similarities between gelation and long chain branching viscoelastic behavior. *Macromolecules* **2001**, *34*, 3115–3117. [CrossRef]
66. Hao, Y.; Yang, H.; Pan, H.; Ran, X.; Zhang, H. The effect of MBS on the heat resistant, mechanical properties, thermal behavior and rheological properties of PLA/EVOH blend. *J. Polym. Res.* **2018**, *25*, 1–9. [CrossRef]
67. Vargas, M.A.; Herrera, R.; Manero, O. Modeling of the linear viscoelastic behavior of partially hydrogenated polymer-modified asphalts. *Rubber Chem. Technol.* **2007**, *80*, 340–364. [CrossRef]
68. Li, Y.; Jia, S.; Du, S.; Wang, Y.; Lv, L.; Zhang, J. Improved properties of recycled polypropylene by introducing the long chain branched structure through reactive extrusion. *Waste Manag.* **2018**, *76*, 172–179. [CrossRef]

MDPI AG
Grosspeteranlage 5
4052 Basel
Switzerland
Tel.: +41 61 683 77 34

Polymers Editorial Office
E-mail: polymers@mdpi.com
www.mdpi.com/journal/polymers

MDPI